Classical and Celestial Mechanics

Classical and Celestial Mechanics

The Recife Lectures

Edited by Hildeberto Cabral and Florin Diacu

Princeton University Press

Princeton and Oxford

Published by Princeton University Press, 41 William Street, Princeton, New Jersey 08540
In the United Kingdom: Princeton University Press, 3 Market Place, Woodstock, Oxfordshire OX20 1SY

Library of Congress Cataloging-in-Publication Data

Classical and celestial mechanics: the Recife lectures / edited by Hildeberto Cabral and Florin Diacu.
p. cm.
Includes bibliographical references and index.
ISBN 0-691-05022-8 (alk. paper)
1. Many-body problem. 2. Celestial mechanics. 3. Mechanics. I. Cabral, Hildeberto, 1940- II. Diacu, Florin, 1959-

QB362.M3 C52 2002
521-dc21   2002072263

British Library Cataloging-in-Publication Data is available

The publisher would like to acknowledge the editors of this volume for providing the camera-ready copy from which this book was printed.

Printed on acid-free paper. ∞

www.pupress.princeton.edu

Printed in the United States of America

10 9 8 7 6 5 4 3 2 1

Contents

• The Motion of the Moon
Dieter Schmidt

Foreword

Just south of the equator, near where the easternmost tip of South America protrudes into the Atlantic, is the colonial city of Recife, Brazil. The rich history of this region traces back to when the country was discovered 500 years ago by the Portuguese explorer Pedro Alvares Cabral.

What a marvelous, beautiful area! The beaches of Recife, in particular the Boa Viagem (often called the "Copacabana of Pernambuco"), easily rank among the very best that I have ever experienced; indeed, I prefer them to Copacabana.

These wide stretches of clean beaches bordered by palm trees are accompanied by a consistent warm and gentle weather, bright sun, warm ocean waves tempered by the reefs, friendly relaxed residents, frequent street dancing, some of the best folk art in Brazil, an unbelievable assortment of exotic, delightful fruits and flowers, and soft living. This area comes close to resembling paradise. The only missing factor for an academic is an exchange of new research ideas. But now even that is available; Hildeberto Cabral, a Professor of Mathematics at the Federal University of Pernambuco in Recife and one of the editors of this volume, has been successfully creating a mathematical center for the study of classical and celestial mechanics.

The theme of classical and celestial mechanics is both a personal one for Cabral and a natural choice for a mathematical emphasis. It is a personal choice because Hildeberto's 1972 Berkeley Ph.D. thesis in mathematics, written under the direction of his advisor Steve Smale, emphasized aspects of the dynamics of the Newtonian N-body problem. In this thesis, Cabral made nice advances in our understanding of the topology of the energy and angular momentum surfaces for the three-dimensional N-body problem. As he proved, changing the emphasis from the planar two-dimensional problem to the full three-dimensional problem introduces some fascinating new twists into the description.

The theme is a natural one for a mathematics center if only because of the historic importance these research topics have played in the development of mathematics. Indeed, a portion of this history describing the role of the three-body problem in the discovery of "chaotic motion" and how this motion plays a central role in the contemporary study of the N-body problem is nicely captured in the Princeton University Press book *Celestial Encounters*, authored by Florin Diacu, the other editor of the current volume, and Phil Holmes.

When I first met Hildeberto in December 1981 at a conference on celestial mechanics held in São Paulo, Brazil, he told me about his dream to develop a research interest in celestial mechanics at the mathematics department at his home university in Recife. He has done much more; he is on his way toward creating an international center for celestial and classical mechanics. A surprisingly large number of experts in these areas have fond memories and new research ideas coming from their "Recife lectures." This book captures some of that spirit by printing a selection of the presentations.

Anyone who knows the history of South America and the struggles of Brazil to conquer runaway inflation can appreciate the difficulties of securing funding to create anything as ambitious as a center back in the 1980s. It was a time of such volatile exchange rates that only small amounts of money were exchanged at any time; after all, a dollar at noon could bring in more Brazilian currency than it had that same morning! Yet, in spite of these discouraging difficulties, Cabral made continual efforts. For instance, somehow he found support to use me as a "guinea pig" to start off his visitor program.

I arrived in Recife in January 1983 to give a five-week series of lectures on my work about the "velocity decomposition" and its consequences for the Newtonian N-body problem. The excellent group of faculty, visitors, and advanced students provided an intellectual setting that encouraged a nice exchange of research ideas and ensured a delightful, productive visit. A bit later in the 1980s, Huseyin Koçak, from Miami University, also spent time working in Recife.

As part of his project to build this research emphasis, Hildeberto and the full Cabral family spent the 1987/1988 academic year at Northwestern University, where Hildeberto played an active role in our Celestial Mechanics Discussion group. But, what a shock this visit must have been to the Cabral family's systems! Being fully accustomed to a steady Recife temperature which stays around the low 80s all year long, they ran into a climate of extremes: an Evanston winter with wind chills advertised at less than negative 80 degrees, and a summer when the temperatures would climb to over 100! Years later, Hildeberto would still shake his head and exclaim, "A hundred degree Celsius differential!"

The real beginning of the Recife program was probably in 1991. This was a time of some stabilization of Brazilian economics. So, at an international conference, "HAMSYS", on Hamiltonian Systems and Celestial Mechanics which was held in Guanajuato, Mexico, Hildeberto floated the idea of visitors coming to Recife for an extended period in order to give a series of lectures on some theme while fully participating in the research program. He found strong support. A rough outline of the foreign researchers who spent at least a month giving lectures follows.

- Zhihong "Jeff" Xia, then at Georgia Tech, December 1992: Lectures on Arnold diffusion in the three-body problem.
- Donald Saari, Northwestern University, January 1993: Lectures on the n-body problem.
- Dieter Schmidt, University of Cincinnati, in February 1993: Lectures on relative equilibria of the n-body problem and on the motion of the moon.
- Alain Albouy, CNRS, France, January–December 1993 and February–May 1999: Lectures on geometry and mechanics.
- Francesco Fasso, then at University of Trento, November–December 1993: Lectures on integrable and nonintegrable Hamiltonian systems, Nekhoroshev theory.
- Clark Robinson, Northwestern University, January 1995: Lectures on symbolic dynamics.
- Ernesto Perez-Chavela, UAM-Iztapalapa, Mexico, February 1995: Lectures on Poincaré compactification and application to celestial mechanics.
- Kenneth Meyer, University of Cincinnati, February 1995: Lectures on periodic solutions of the n-body problem.
- Mark Levi, then at Rensselaer Polytechnic Institute, February 1996: Lectures on geometrical mechanics.

- Ernesto Lacomba, UAM-Iztapalapa, Mexico, February 1996: Lectures on ejection-collision orbits and invariant tori in Hill's problem.
- Joaquin Delgado, UAM-Iztapalapa, Mexico, January–February 1997: Course on Riemannian geometry at the doctorate level for the graduate program in Recife.
- Jose Guadalupe Reyes Victoria, UAM-Iztapalapa, Mexico, January–February 1998: Course on differential geometry for students in the master's program.
- Sergey Zybin, High Energy Institute, Moscow, February 1998: Lectures on parallel numerical algorithms.
- Florin Diacu, University of Victoria, Canada, February 1999: Lectures on singularities in the n-body problem.
- Christopher McCord, University of Cincinnati, February 2000: Lectures on the topology of the n-body problem.
- Stefanella Boatto, Bureau des Longitudes, Paris, November–December 2000: Seminars on vortex dynamics.

To get a better appreciation of the growth of research activity in Recife, note that the list of shorter term visitors includes Jaume Llibre (UAB-Barcelona), Alain Chenciner (Université de Paris VII), Atsuro Sannami (University of Hokkaido), Konstantin Mischaikow (Georgia Institute of Technology), and Kurt Ehlers (University of California at Santa Cruz).

This is just a list of foreign visitors; when one adds the large number of Brazilians who have passed through the department and the students receiving their graduate degrees from this program, one gains a better appreciation of the intellectual activity that now resides in this beautiful corner of the world.

Let me offer my congratulations to Hildeberto Cabral for successfully overcoming all of the obstacles in order to create this mathematical center and my gratitude to Cabral and Florin Diacu for making these Recife Lectures available to a wider international audience!

Donald G. Saari
February 2001

Preface

The material we present here comes from a special lecture series offered at the Federal University of Pernambuco (UFPE), in Recife, Brazil, between 1993 and 1999. The Brazilian Government sponsored this initiative through CNPq and CAPES grants (the latter via UFPE) obtained by the event's host, Hildeberto Cabral, a mathematics professor in Recife. With the financial help of these governmental agencies, Cabral could attract several internationally established researchers, who lectured on classical or celestial mechanics. The audience consisted of professors, postdoctoral fellows, researchers, and graduate students from Brazil and other countries.

After giving his invited Recife lectures in February 1999, Florin Diacu suggested this book. What motivated the project and simplified the process of assembling the material was that, at Cabral's initiative, some of the earlier lectures had been gathered in reprints of small circulation; others were still in progress. But because of the high quality of the results and the general interest in these topics, Diacu thought that they should be made available to interested readers around the world.

Cabral reacted with enthusiasm to this suggestion. He contacted all those who had written lectures and asked if they would be willing to participate. Most of them were. Then Diacu wrote an extended proposal and submitted it together with two sample lectures to Trevor Lipscombe, the mathematics editor at Princeton University Press, who became interested in the project and helped it reach conclusion. Cabral and Diacu edited the material, including each other's lectures. They are indebted to Scott Dumas from the University of Cincinnati, who translated the original French lectures of Alain Albouy.

Not all Recife lectures are published here. The course on periodic solutions of the n-body problem offered by Kenneth Meyer in February 1995 already appeared in the Springer-Verlag Lecture Notes series (no. 1719), whereas the contributions of Christopher McCord, Francesco Fasso, Ernesto Lacomba, Clark Robinson, Donald Saari, Zhihong Xia, and Sergey Zybin are not in final form yet.

The final step of putting everything together in book format was the work of Oscar Pereira da Silva Neto, the system administrator of the Mathematics Department at the Federal University of Pernambuco. It was his idea to include pictures of Recife as a way to convey the atmosphere of this old colonial city with its warm and candid people. Some pictures were taken during the traditional February carnival, a rare collective expression of happiness and joy.

The material is ordered by subject matter, with the celestial mechanics lectures first. The order in which these lectures were given is as follows:

> *Dieter Schmidt*, February 1993
> *Ernesto Pérez-Chavela*, February 1995
> *Mark Levi*, February 1996
> *Jack Hale and Plácido Táboas*, January–February 1996
> *Jair Koiller*, January–February 1998
> *Hildeberto Cabral*, September–October 1998
> *Florin Diacu*, February 1999
> *Alain Albouy*, February–May 1999

The editors hope that this book will inform readers about some modern trends in classical and celestial mechanics. Many people have helped us in this project. We would like to thank the colleagues and students of the Celestial Mechanics Group in Recife for the stimulating environment and all the visitors for helping to consolidate this research group. Thanks also to Oscar Neto for preparing the book format.

We would like to thank our editor, Vickie Kearn and her colleagues at Princeton University Press, especially Beth Gallagher and Anne Reifsnyder, who painstakingly copy edited the manuscript. Our thanks also to Trevor Lipscomb, now at John Hopkins University Press, for demonstrating confidence in the project and for his stimulating words of encouragement.

Also, we would like to thank CNPq, CAPES, and UFPE for the financial support that made this project possible, as well as the Secretary of Social Communication of the city of Recife for the permission to use the photos that illustrate the book.

Hildeberto Cabral is especially indebted to Donald Saari for his enthusiastic support, beginning with a 1984 visit to UFPE, when he offered a remarkable course on the n-body problem. Both editors are also thankful for his independent introduction to this book.

> *Hildeberto Cabral and Florin Diacu*
> Recife and Victoria, 20 December 2000

Photograph: Ivan Feitosa

Central Configurations and

Relative Equilibria

for the N-Body Problem

Dieter Schmidt

DEPARTMENT OF COMPUTER SCIENCE
UNIVERSITY OF CINCINNATI
CINCINNATI, OH 45221-0008, USA
e-mail: dieter.schmidt@uc.edu

1   Introduction

Given are N point particles of mass $m_j \in \mathbf{R}^+$ at the positions $q_j \in \mathbf{R}^3$, with the velocities $\dot{q}_j \in \mathbf{R}^3$ for $j = 1, \ldots, N$. The bodies attract each other in accordance with Newton's law. The problem consists in describing their future motion. With appropriate units for time, distance, and mass, the equations of motion have the form

$$m_j \ddot{q}_j = \frac{\partial U}{\partial q_j}, \qquad j = 1, \ldots, N. \tag{1}$$

The function

$$U(q) = \sum_{i<j} \frac{m_i m_j}{|q_i - q_j|}$$

will be called the Newtonian potential, although actually it is the negative of the potential. The equations of motion are then

$$\ddot{q}_j = \sum_{i \neq j} \frac{m_i(q_i - q_j)}{|q_i - q_j|^3}.$$

In matrix form the equations are given by

$$\ddot{q} = M^{-1} \frac{\partial U}{\partial q}$$

with $q = (q_1, \ldots, q_N) \in \mathbf{R}^{3N}$ and

$$M = \mathrm{diag}(m_1, m_1, m_1, \ldots, m_N, m_N, m_N).$$

The differential equations allow 10 integrals. Six of them come from the fact that the center of mass of the system moves with constant velocity along a straight line. One can account for these integrals by using a coordinate system where the center of mass resides at the origin at all times, that is, $m_1 q_1 + \cdots + m_N q_N = 0$. Three integrals are due to the preservation of the angular momentum for the entire system, that is, $\sum m_j q_j \times \dot{q}_j = \text{const}$, and an additional integral corresponds to the preservation of the total energy, that is,

$$\frac{1}{2} \sum_{i=1}^{N} m_i |\dot{q}_i|^2 + U(q) = \text{const.}$$

It is possible to reduce the order of the system of differential equations by utilizing these last 4 integrals, but the resulting set of equations is too complicated, so this reduction is seldom done.

Relative equilibria and central configurations seem to be the only solutions for the general N-body problem that can be given in closed form. In both cases one assumes that at all times the acceleration vector for each body points toward the origin. For central configurations one assumes that the motion takes place on straight lines going through the origin. In other word, one looks for solutions to (1) of the form

$$q_j = r(t)Q_j$$

with $r(t)$ a scalar function and Q_j constant vectors. Such a solution can be found if $r(t)$ satisfies Kepler's equation,

$$\ddot{r} + \lambda/r^2 = 0, \tag{2}$$

for some constant λ and if the vectors Q_j, $j = 1, \ldots, N$, are a solution to the algebraic set of equations

$$\sum_{i<j} \frac{Q_i - Q_j}{|Q_i - Q_j|^3} = \lambda Q_j, \qquad j = 1, \ldots, N. \tag{3}$$

In matrix form these equations are

$$M^{-1}\frac{\partial U}{\partial Q} = \lambda Q,$$

and since the potential function $U(q)$ is a homogeneous function of degree -1, it follows from Euler's theorem that

$$\lambda = -\frac{U(q)}{q^t M q}.$$

The constant λ can also be viewed as a Lagrange multiplier so that the solutions of (3) can be viewed as the extrema of the potential $U(q)$ under the condition that the moment of inertia,

$$I(q) = \frac{1}{2}\sum_{i=1}^{N} m_i |q_i|^2,$$

remains constant. This point of view will be explored later on. It will allow us to use bifurcation theory to find new central configurations near very special solutions of (3). Only for $N \leq 3$ can (3) be analyzed completely.

One type of solution for (2) is

$$r(t) = (-9\lambda/2)^{1/3} t^{2/3},$$

indicating that all bodies will collide at time $t = 0$ at the origin. Although this solution is very special, it is interesting, because it has been shown that any other collision orbits approach it asymptotically.

2   Dziobek's Coordinates

Our approach uses bifurcation theory to search for new relative equilibria or central configurations near known ones. In order to do this, we have to find degenerate central configurations. In this context a central configuration is called degenerate if it is not isolated. But in fact, no configuration is isolated, because near each configuration is another one obtained from the first by a simple rotation. This is a manifestation of the fact that we have not yet accounted for the integrals which give the preservation of the angular momentum. For this reason one introduces equivalence classes of central configurations. Two configurations are called equivalent if they can be transformed into each other by a rotation around the center of mass. Degeneracy then means degeneracy among equivalence classes, that is, mod$SO(2)$ or mod$SO(3)$.

Difficulties of this nature can often be avoided with the choice of an appropriate coordinate system. The mutual distances between the different bodies is such a choice. The distances determine a configuration, but they do not change with a rotation. They have been used by Dziobek [1] already and were exploited further in [3] and [6]. Let r_{ij} be the distance between the ith and the jth body. The potential function is then

$$U = \sum_{i<j} \frac{m_i m_j}{r_{ij}},$$

and with the center of mass at the origin, the moment of inertia is

$$I = \frac{1}{2M} \sum_{i<j} m_i m_j r_{ij}^2,$$

with $M = \sum m_i$ the total mass of the system.

In the following sections we will use the coordinates of Dziobek to derive some of the classical results for the three-, four-, and five-body problems, and we include our results, which were obtained with the help of bifurcation analysis.

3 Configurations for the Three-Body Problem

For three bodies, the distances r_{12}, r_{23}, and r_{13} determine the configuration up to a reflection, provided that the obvious geometric constraint in the form of the triangle inequality is met.

It was mentioned in section 1 that central configurations are extrema of U under the condition that the moment of inertia is constant. It means that they are the extrema of the function $U + \lambda(I - I_0)$ or the solution to the equations

$$-\frac{m_i m_j}{r_{ij}^2} + \frac{\lambda}{M} m_i m_j r_{ij} = 0, \qquad 1 \leq i < j \leq 3,$$

$$\sum_{1 \leq i < j \leq 3} m_i m_j r_{ij}^2 - 1 = 0,$$

where we have selected $I_0 = 1/2M$. The solutions to these equations give at once the equilateral solutions of Lagrange:

$$r_{12} = r_{23} = r_{13} = (m_1 m_2 + m_2 m_3 + m_3 m_1)^{-1/2}.$$

On the other hand, the collinear solutions of Euler are not even indicated due to the fact that the given potential function does not have an extrema at the collinear solutions. The collinear solutions can be found if we search extrema under the constraint that all three bodies lie on a line.

Although the collinear solutions can be found directly from the differential equations, we present their derivation with the help of the coordinates of Dziobek for completeness sake and to illustrate the method we will use for the problem with four and five bodies. The three bodies will lie on a straight line if we insist that the area of the triangle they form is zero. The determinant

$$F(r_{12}^2, r_{13}^2, r_{23}^2) = \begin{vmatrix} 0 & 1 & 1 & 1 \\ 1 & 0 & r_{12}^2 & r_{13}^2 \\ 1 & r_{12}^2 & 0 & r_{23}^2 \\ 1 & r_{13}^2 & r_{23}^2 & 0 \end{vmatrix}$$

is related to the area A of the triangle via $F = 32A^2$. When $F = 0$ it has the property that

$$\frac{\partial F}{\partial r_{ij}^2} = 4\delta_i \delta_j,$$

where δ_i and δ_j are the lengths of the sides opposite to the points P_i and P_j, respectively. These lengths have to be taken with the appropriate orientation, that is, relative to an outward normal of the triangle. Selecting that P_2 lies between P_1 and P_3, we require that

$$r_{13} = r_{12} + r_{23},$$

from which it follows that

$$\frac{\partial F}{\partial r_{12}^2} = -4r_{13}r_{23}, \qquad \frac{\partial F}{\partial r_{23}^2} = -4r_{12}r_{13}, \qquad \frac{\partial F}{\partial r_{13}^2} = 4r_{12}r_{23}.$$

In order to find the extrema of U under the additional constraint $F = 0$, let σ be another Lagrangian multiplier, so that we have to find the extrema of

$$\sum_{1 \le i < j \le 3} m_i m_j / r_{ij} + \lambda(I - I_0) + \sigma F.$$

The condition for collinear extrema is then

$$m_i m_j (\lambda r_{ij} - r_{ij}^{-2}) + 2\sigma \frac{\partial F}{\partial r_{ij}} = 0, \qquad 1 \le i < j \le 3,$$

$$I - I_0 = 0, \qquad F = 0.$$

The first three equations, slightly modified, are

$$m_1 m_2 m_3 (\lambda - r_{12}^{-3}) - 8\sigma m_3 r_{13} r_{23} = 0,$$

$$m_1 m_2 m_3 (\lambda - r_{23}^{-3}) - 8\sigma m_1 r_{12} r_{13} = 0,$$

$$m_1 m_2 m_3 (\lambda - r_{13}^{-3}) + 8\sigma m_2 r_{12} r_{23} = 0.$$

First eliminate σ to obtain

$$\frac{\lambda - r_{12}^{-3}}{m_3 r_{13} r_{23}} = \frac{\lambda - r_{23}^{-3}}{m_1 r_{12} r_{13}} = \frac{-\lambda + r_{13}^{-3}}{m_2 r_{12} r_{23}}$$

and then solve for λ by using the first and the third expressions and then again the second and the third, so that

$$\lambda = \frac{m_2 r_{12}^{-2} + m_3 r_{13}^{-2}}{m_2 r_{12} + m_3 r_{13}} = \frac{m_2 r_{23}^{-2} + m_1 r_{13}^{-2}}{m_2 r_{23} + m_1 r_{13}}.$$

Let ρ be a factor between 0 and 1 so that

$$r_{12} = \rho r_{13} \qquad \text{and} \qquad r_{23} = (1 - \rho) r_{13}.$$

Then

$$\lambda r_{13}^3 = \frac{m_2 \rho^{-2} + m_3}{m_2 \rho + m_3} = \frac{m_2 (1 - \rho)^{-2} + m_1}{m_2 (1 - \rho) + m_1}$$

and in the interval $0 < \rho < 1$, the difference

$$f(\rho) = \frac{m_2 \rho^{-2} + m_3}{m_2 \rho + m_3} - \frac{m_2 (1 - \rho)^{-2} + m_1}{m_2 (1 - \rho) + m_1}$$

is a monotonically decreasing function. It tends to $+\infty$ for $\rho \to 0$ and to $-\infty$ for $\rho \to 1$. Therefore, there exists a unique solution for ρ in $0 < \rho < 1$. The solution is the only real solution of the following fifth-order polynomial

$$(m_1 + m_3)\rho^5 - (2m_1 + 3m_3)\rho^4 + (m_1 + 2m_2 + 3m_3)\rho^3$$
$$-(m_1 + 3m_2)\rho^2 + (2m_1 + 3m_2)\rho - m_1 - m_2 = 0.$$

The value of r_{13} is found from the condition that the moment of inertia is preserved. The other collinear solutions are found by interchanging the points P_1, P_2, and P_3.

4   Configurations in the Four-Body Problem

Between four bodies there are six mutual distances. Assume that these distances are given and that they meet the necessary geometric constraint so that they define a tetrahedron in $\mathbf{R}^3$. We are looking for extrema of

$$\sum_{1 \leq i < j \leq 4} m_i m_j \left(1/r_{ij} + \frac{\lambda}{2}(r_{ij}^2 - 1) \right).$$

They follow from the six equations

$$m_i m_j (\lambda - r_{ij}^{-3}) = 0, \qquad 1 \leq i < j \leq 4,$$

and from the constraint

$$\sum m_i m_j (r_{ij}^2 - 1) = 0.$$

It is at once apparent that the only solution to these equations determines a regular tetrahedron in $\mathbf{R}^3$. If we are looking for planar solutions we have to impose the additional constraint that the volume of the tetrahedron is zero, that is, that

$$F = \begin{vmatrix} 0 & 1 & 1 & 1 & 1 \\ 1 & 0 & r_{12}^2 & r_{13}^2 & r_{14}^2 \\ 1 & r_{12}^2 & 0 & r_{23}^2 & r_{24}^2 \\ 1 & r_{13}^2 & r_{23}^2 & 0 & r_{34}^2 \\ 1 & r_{14}^2 & r_{24}^4 & r_{34}^4 & 0 \end{vmatrix} = 0. \tag{4}$$

Similar to the collinear for the three-body problem we have to look for the extrema of

$$U + \lambda(I - I_0) + \frac{\sigma}{32} F.$$

It leads to the six equations

$$m_i m_j (\lambda - r_{ij}^{-3}) + \sigma A_i A_j = 0 \tag{5}$$

together with

$$I - I_0 = 0 \tag{6}$$

and

$$F = 0. \tag{7}$$

As explained in Appendix A, A_i is the oriented area of the triangle, which does not include the point P_i. When the masses are given, the above eight equations can be used to determine the eight quantities λ, σ, $r_{12}, \ldots, r_{34}$. Unfortunately these equations are nonlinear, so little is known about the solution manifolds in the general case. Instead special cases are considered or questions of a different nature are asked. One of these questions is, Given a configuration in the plane, is it possible to find four positive masses with which this configuration can be realized?

To answer the above question, we display the first six equations in the following form:

$$m_1 m_2 (r_{12}^{-3} - \lambda) = \sigma A_1 A_2, \qquad m_3 m_4 (r_{34}^{-3} - \lambda) = \sigma A_3 A_4,$$

$$m_1 m_3 (r_{13}^{-3} - \lambda) = \sigma A_1 A_3, \qquad m_2 m_4 (r_{24}^{-3} - \lambda) = \sigma A_2 A_4,$$

$$m_1 m_4 (r_{14}^{-3} - \lambda) = \sigma A_1 A_4, \qquad m_2 m_3 (r_{23}^{-3} - \lambda) = \sigma A_2 A_3.$$

The equations have been grouped so that when they are multiplied together pairwise, their right-hand sides are identical. This leads to the well-known condition

$$(r_{12}^{-3} - \lambda)(r_{34}^{-3} - \lambda) = (r_{13}^{-3} - \lambda)(r_{24}^{-3} - \lambda) = (r_{14}^{-3} - \lambda)(r_{23}^{-3} - \lambda). \tag{8}$$

This is a necessary and sufficient condition for solving equations (5)–(7) for the four masses, when the distances are given. It is sufficient because (8) shows that one can solve for λ, from which the ratio of the masses can be found. The condition (8) is necessary, as it guarantees that the three expressions for λ will be equal. They follow from (8) by solving different pairs of equations, and they are with the abbreviation $s_{ij} = r_{ij}^{-3}$:

$$\lambda = \frac{s_{12} s_{34} - s_{13} s_{24}}{s_{12} + s_{34} - s_{13} - s_{24}} = \frac{s_{13} s_{24} - s_{14} s_{23}}{s_{13} + s_{24} - s_{14} - s_{23}} = \frac{s_{14} s_{23} - s_{12} s_{34}}{s_{14} + s_{23} - s_{12} - s_{34}}.$$

From the different ratios of two masses that can be found from (5), we obtain the following list:

$$\frac{m_1 A_2}{m_2 A_1} = \frac{s_{23} - \lambda}{s_{13} - \lambda} = \frac{s_{24} - \lambda}{s_{14} - \lambda} = \frac{s_{23} - s_{24}}{s_{13} - s_{14}}, \tag{9}$$

$$\frac{m_1 A_3}{m_3 A_1} = \frac{s_{23} - \lambda}{s_{12} - \lambda} = \frac{s_{34} - \lambda}{s_{14} - \lambda} = \frac{s_{23} - s_{34}}{s_{12} - s_{14}}, \tag{10}$$

$$\frac{m_1 A_4}{m_4 A_1} = \frac{s_{24} - \lambda}{s_{12} - \lambda} = \frac{s_{34} - \lambda}{s_{13} - \lambda} = \frac{s_{24} - s_{34}}{s_{12} - s_{13}}, \tag{11}$$

$$\frac{m_2 A_3}{m_3 A_2} = \frac{s_{13} - \lambda}{s_{12} - \lambda} = \frac{s_{34} - \lambda}{s_{24} - \lambda} = \frac{s_{13} - s_{34}}{s_{12} - s_{24}}, \tag{12}$$

$$\frac{m_2 A_4}{m_4 A_2} = \frac{s_{14} - \lambda}{s_{12} - \lambda} = \frac{s_{34} - \lambda}{s_{23} - \lambda} = \frac{s_{14} - s_{34}}{s_{12} - s_{23}}, \tag{13}$$

$$\frac{m_3 A_4}{m_4 A_3} = \frac{s_{14} - \lambda}{s_{13} - \lambda} = \frac{s_{24} - \lambda}{s_{23} - \lambda} = \frac{s_{14} - s_{24}}{s_{13} - s_{23}}. \tag{14}$$

Condition (8) guarantees only that the mass ratios can be computed. It does not say that the masses will be positive. With the help of equations (9)–(14) this additional information can be obtained.

Two cases have to be considered separately. These are the convex and the concave configurations. In the concave case one body lies inside the triangle formed by the other three bodies. In the convex case this does not occur. The degenerate case where three bodies lie on a line leads to one mass being zero as long as not all four bodies lie on a line. This follows from the following argument. Assume that the three bodies with nonzero masses m_1, m_2, and m_3 lie on a line. From the assumptions it follows that for the triangle that does not contain the fourth body we must have $A_4 = 0$. Since none of the other triangles are degenerate, it follows from (11), (13), and also (14) that $s_{14} = s_{24} = s_{34}$. This would require that at least two bodies coincide. If this is not the case, then at least one of the four masses has to be zero.

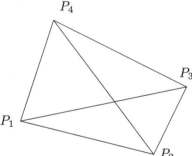

Figure 1. Convex configuration.

We start with the convex case and assume that the bodies are arranged as in Figure 1. The signs of the four areas are then

$$A_1 = \Delta\, P_2 P_3 P_4 > 0,$$

$$A_2 = \Delta\ P_1 P_4 P_3 < 0,$$
$$A_3 = \Delta\ P_1 P_2 P_4 > 0,$$
$$A_4 = \Delta\ P_1 P_3 P_2 < 0.$$

If we assume that $s_{12} - \lambda > 0$ and ask for positive masses, we find from equations (9)–(14) that

$$s_{12}, s_{23}, s_{34}, s_{14} > \lambda > s_{13}, s_{24}.$$

With the notation $\lambda = \rho^{-3}$, the above inequality can be written in terms of distances

$$r_{12}, r_{23}, r_{34}, r_{14} < \rho < r_{13}, r_{24}. \tag{15}$$

It means that all exterior sides have to be smaller than the diagonals. If we had assumed that $s_{12} - \lambda < 0$, then the inequalities in (15) would be reversed and the configuration could not be realized geometrically. If we make the assumption that r_{12} is the smallest side, then the last expressions in (10) and in (13) give the conditions that $s_{12} > s_{23} > s_{34}$ and $s_{12} > s_{14} > s_{34}$, or equivalently that $r_{12} < r_{23} < r_{34}$ and $r_{12} < r_{14} < r_{34}$. It means that the longest and the shortest sides have to lie opposite each other. This result can already be found in [1].

Limits on the ratios of the diagonals of a feasible central configuration can be obtained by starting the construction of the quadrilateral with the diagonal of length $r_{13} > \rho$. Draw circles of radius ρ around P_1 and P_3, respectively. The other two masses have to lie in the common area of intersection. Since also $r_{24} > \rho$, these masses have to be outside the two circles around Q_1 and Q_2. Since

$$\left(\frac{r_{13}}{2}\right)^2 + \left(\frac{\overline{Q_1 Q_2}}{2}\right)^2 = \rho^2$$

but $\rho < r_{24} < \overline{Q_1 Q_2}$, we find that

$$r_{13} < \sqrt{3}\rho < \sqrt{3}r_{24}.$$

By an analogous argument

$$r_{24} < \sqrt{3}\rho < \sqrt{3}r_{13}.$$

Together, then, we have the following theorem, also found in [2].

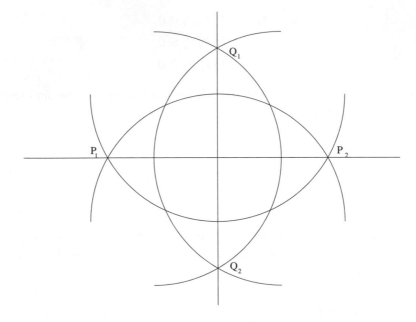

Figure 2. Limits on the ratio of the diagonals.

Theorem 4.1 *For a convex central configuration with positive masses the ratio of the diagonals is limited by*

$$1/\sqrt{3} < r_{13}/r_{24} < \sqrt{3}.$$

Furthermore, the size of the interior angles is restricted to the interval from 30 to 120 degrees.

The last statement of the theorem can be verified easily by looking at the extreme cases which with Figure 2 could be drawn and still meeting the condition (15).

For concave configurations, let m_4 be the interior mass. With $A_4 < 0$ and $A_1, A_2, A_3 > 0$ we find from (9)–(14) that

$$s_{12}, s_{23}, s_{13} < \lambda < s_{14}, s_{24}, s_{34}$$

or

$$r_{12}, r_{23}, r_{13} > \rho > r_{14}, r_{24}, r_{34}.$$

It means that all interior edges are longer than the exterior edges. If we assume the ordering $r_{12} > r_{23} > r_{13}$ for the lengths of the exterior edges, then we find that $r_{34} > r_{14} > r_{24}$ and we can state the following theorem.

Theorem 4.2 *In a concave configuration with four positive masses, the longest exterior side lies opposite the longest interior edge. The same statement also holds for the shortest exterior and interior edges.*

By considering each exterior edge one at a time, we can even state in which of the half-planes defined by the perpendicular bisector the body m_4 has to lie. If its mass is to be positive, it has to lie in the half-plane, which does not include the third exterior vertex. The interior body is therefore restricted to the intersection of three half-planes. Since the three perpendicular bisectors of a triangle meet at one point, this intersection is not empty, but it may lie outside of the given triangle. This will not happen if the center of the circumscribing circle lies inside the given triangle. This restriction can be expressed algebraically by the inequalities $a^2 < b^2 + c^2$, where a, b, and c are any permutation of r_{12}, r_{23}, and r_{13}.

Equations (9)–(14) allow one also to look for degenerate configurations. Degeneracy is indicated by indeterminate expressions on the right-hand sides. Assuming that m_4 cannot be determined, we obtain at once that $s_{12} = s_{23} = s_{13}$ and $s_{14} = s_{24} = s_{34}$. It represents an equilateral triangle with three equal masses $m_1 = m_2 = m_3$ at the vertices and a variable mass m_4 at its center. Let the exterior masses be of unit value and $r_{12} = r_{23} = r_{13} = \sqrt{3}$, so that $r_{14} = r_{24} = r_{34} = 1$. We will vary the interior mass m_4 and look for a value m_c so that the configuration becomes degenerate and bifurcation analysis can be applied.

In order to explain the details of our calculations, we introduce the eight-dimensional vector

$$z = (\lambda, \sigma, r_{12}, r_{13}, r_{23}, r_{14}, r_{24}, r_{34}).$$

Set $m_1 = m_2 = m_3 = 1$ and let $m_4 = m$ be indeterminate so far. The eight equations of (5), (6), and (7) are

$$F = 0,$$

$$I - I_0 = 0,$$

$$(\lambda - r_{12}^{-3}) + \sigma A_1 A_2 = 0,$$
$$(\lambda - r_{13}^{-3}) + \sigma A_1 A_3 = 0,$$
$$(\lambda - r_{23}^{-3}) + \sigma A_2 A_3 = 0, \tag{16}$$
$$m(\lambda - r_{14}^{-3}) + \sigma A_1 A_4 = 0,$$
$$m(\lambda - r_{24}^{-3}) + \sigma A_2 A_4 = 0,$$
$$m(\lambda - r_{34}^{-3}) + \sigma A_3 A_4 = 0.$$

Denote these equations by $W(z, m) = 0$. For $I_0 = (9 + 3m)/2$, the equilateral triangular family is given by $z = a$, where

$$a = \left(\frac{3m + \sqrt{3}}{6(3m + 9)}, \frac{4(9 - \sqrt{3})m}{27(3 + m)}, \sqrt{3}, \sqrt{3}, \sqrt{3}, 1, 1, 1 \right).$$

Thus $W(a, m) = 0$ for all m. With the aid of a computer algebra program we find that the Jacobian

$$\left| \frac{\partial W}{\partial z} \right|_{z=a} = (133 - 60\sqrt{3})m_c^2(249m_c - 64\sqrt{3} - 81)^2/20667$$

is positive definite except for

$$m_4 = m_c = \frac{81 + 64\sqrt{3}}{249},$$

a value already found by Palmore [5]. For $m_4 = m_c$ the rank of the Hessian reduces by 2. A nonzero subdeterminant of the Hessian is obtained by deleting the last two rows and columns. By the implicit function theorem, we can thus solve the first six equations for $\lambda, \sigma, r_{12}, r_{13}, r_{23}, r_{14}$ in terms of r_{24}, r_{34} for m near m_c.

Set $m = m_c + \varepsilon$ and $z = a + \varepsilon b + \varepsilon^2 c + \cdots$ and solve the set of equations $W(z, m) = 0$ order by order in ε. Let $\beta = (b_1, b_2, b_3, b_4, b_5, b_6)$; then, from first-order terms, we get $\beta = \beta(b_7, b_8)$, with the equations W_7, W_8 at order 1 satisfied automatically. Similarly, for second-order terms let $\gamma = (c_1, \ldots, c_6)$, and they can be found as functions of b_7, b_8, c_7, and c_8. The last two equations now deliver the bifurcation equations. They are in factored form:

$$(2b_7 + b_8)(b_8 + p) = 0,$$
$$(b_7 + 2b_8)(b_7 + p) = 0, \tag{17}$$

with

$$p = \frac{4531167 - 3089347\sqrt{3}}{18889832}.$$

There are three sets of nontrivial solutions:

$$b_7 = -p, \ b_8 = -p,$$
$$b_7 = -p, \ b_8 = 2p,$$
$$b_7 = 2p, \ b_8 = -p.$$

They are a consequence of the threefold symmetry of the problem, and we can concentrate on the first solution. The Jacobian of (17) at that solution is $-p^2 \neq 0$ and higher order terms can be found. It means that the last two distances r_{24} and r_{34} can be expanded as series in ε and are $r_{24} = r_{34} = 1 - \varepsilon p + \mathcal{O}(\varepsilon^2)$. Previously we had found the first six variables, so that $z = a + \varepsilon b + \mathcal{O}(\varepsilon^2)$ could be computed order by order.

For the distances within the triangle we have through order 1

$$r_{12} = r_{13} = \sqrt{3} - \frac{81 + 64\sqrt{3}}{83} p\varepsilon + \cdots,$$
$$r_{23} = \sqrt{3} + \frac{81 + 64\sqrt{3}}{83} p\varepsilon + \cdots,$$
$$r_{14} = 1 + 2p\varepsilon + \cdots,$$
$$r_{24} = r_{34} = 1 - p\varepsilon + \cdots.$$

5 Configurations in the Five-Body Problem

With five bodies there are ten mutual distances r_{ij}, $1 \leq i < j \leq 5$. Three constraints are needed if the distances define a planar configuration. Only one constraint is needed if the configuration is to be realized in $\mathbf{R}^3$. The two cases will be discussed separately, and we will start with the planar case.

Let $F(i, j, k, l)$ be the determinant given in (4), with (i, j, k, l) replacing $(1, 2, 3, 4)$ in that order. We choose the following constraints to insure that the configuration is planar:

$$F_1 = F(1, 2, 3, 4), \qquad F_2 = F(1, 2, 3, 5), \qquad F_3 = F(2, 3, 4, 5).$$

Consider
$$V = U + \lambda(I - I_0) + \sigma_1 F_1 + \sigma_2 F_2 + \sigma_3 F_3$$
so that V is a function of the 14 variables given in the vector
$$z = (\lambda, \sigma_1, \sigma_2, \sigma_3, r_{12}, r_{13}, r_{23}, r_{14}, r_{2,4}, r_{34}, r_{15}, r_{25}, r_{35}, r_{45}).$$

The equations to be solved are $\partial V / \partial z = 0$. They are similar to those in (16) and again denoted by $W(z, m) = 0$. With four equal masses $m_1 = m_2 = m_3 = m_4 = 1$ at the vertices of a square, we have a solution to these equations with

$$z = a = \left(\frac{1 + 2\sqrt{2} + 4m}{4m + 16}, \frac{-1 + 2\sqrt{2} + m(2\sqrt{2} - 4)}{256 + 64m}, \frac{(15 - 2\sqrt{2})m}{128 + 32m}, \right.$$

$$\left. \frac{(15 - 2\sqrt{2})m}{128 + 32m}, \sqrt{2}, 2, \sqrt{2}, \sqrt{2}, 2, \sqrt{2}, 1, 1, 1, 1 \right)$$

with any arbitrary mass $m_5 = m$ at the center of the square. The Jacobian of the system of equations is then

$$73728(67\sqrt{2} - 84)(4m + 1)(12m - 13 - 11\sqrt{2})^2,$$

which is nonzero for all positive masses except for one critical value

$$m = m_c = \frac{13 + 11\sqrt{2}}{12}.$$

A nonzero subdeterminant can be found by deleting the last two rows and columns. By the implicit function theorem we can thus solve the first 12 equations of $W(z, m) = 0$ for m near m_c in terms of r_{35} and r_{45}. Set

$$m = m_c + \delta\varepsilon^2 \quad \text{with} \quad \delta = \pm 1 \quad \text{and let} \quad z = a + \varepsilon b + \varepsilon^2 c + \varepsilon^3 d + \cdots.$$

Set $Z = (W_{13}, W_{14})$ to be the last two equations of $W(z, m)$ and $u = (r_{35}, r_{45})$ so that the problem is reduced to the study of the bifurcation equations

$$Z(u, \varepsilon) = 0.$$

The first nonzero terms occur at order ε^3. These are, except for some constants,

$$(\delta q_1 - q_2 b_{35}^2 + q_3 b_{45}^2)b_{35} = 0, \tag{18}$$

$$(\delta q_1 + q_3 b_{35}^2 - q_2 b_{45}^2)b_{45} = 0. \tag{19}$$

where

$$q_1 = 122691244 + 411830784\sqrt{2},$$
$$q_2 = 291093160040 + 211659410269\sqrt{2},$$
$$q_3 = 1204699032744 + 860002887155\sqrt{2}.$$

For $\delta = +1$, the solutions are

$$b_{35} = 0, \qquad b_{45} = \pm\sqrt{q_1/q_2}, \qquad \text{or} \qquad b_{35} = \pm\sqrt{q_1/q_2}, \qquad b_{45} = 0.$$

For $\delta = -1$, the solutions are

$$b_{35} = b_{45} = \pm\sqrt{q_3 - q_2} \qquad \text{and} \qquad b_{35} = -b_{45} = \pm\sqrt{q_3 - q_2}.$$

At first order in ε, we had found

$$b_{12} = -b_{34} = \frac{82 + 75\sqrt{2}}{62}(b_{35} + b_{45}),$$
$$b_{14} = -b_{23} = \frac{82 + 75\sqrt{2}}{62}(b_{35} - b_{45}),$$
$$b_{13} = \quad b_{24} = 0, \quad b_{15} = -b_{35}, \quad b_{25} = -b_{45},$$

so that the solution for $\delta = +1$ gives central configurations that look like a kite, and $\delta = -1$ corresponds to central solutions in the form of a trapezoid; see Figure 3. The multiple choice in the signs accounts for the symmetries of the square configuration. A configuration in the shape of a rhombus would only occur for a negative mass, $m_5 = -0.25$.

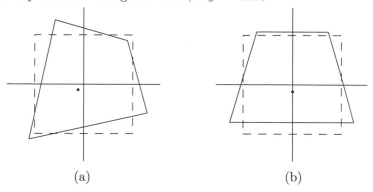

(a) (b)

Figure 3. Bifurcations for the five-body problem: (a) $\delta = -1$; (b) $\delta = +1$.

6   Palmore's Coordinates

The number of constraints which insures that a configuration can be realized geometrically in $\mathbf{R}^2$ or $\mathbf{R}^3$ increases quadratically with the number of the bodies. Therefore, mutual distances as coordinates become less attractive. Instead, Cartesian coordinates are used. The complexity of the problem is reduced by starting with regular configurations and looking for configurations that might bifurcate from them.

Place N bodies of unit mass at the vertices of a regular N-gon and a mass of size m at the origin. We will vary m to search for a new configuration nearby. Let $w = e^{2\pi i/N}$; then $q_j = w^j$, $j = 0, \ldots, N-1$, are the positions of the N unit masses and $q_N = 0$ is the position of the body at the origin in this $(N+1)$-body problem. Configurations nearby are described most conveniently by a coordinate system used by Palmore [5] and possibly by others before him. Let $\Omega = (w^0, w^1, \ldots, w^{N-1}, 0)^T$ be the position vector for the $N+1$ bodies in the regular polygon configuration and $q = \Omega + Az$ be the position vector for the configuration in the vicinity. Here $z = (z_0, \ldots, z_N)^T$ and

$$
A = \begin{pmatrix}
1 & 1 & 1 & \cdots & 1 & 1 \\
1 & w^1 & w^2 & \cdots & w^{N-1} & 1 \\
1 & w^2 & w^4 & \cdots w^{2(N-1)} & \\
\vdots & \vdots & \vdots & & \vdots & \vdots \\
1 & w^{(N-1)} & w^{2(N-1)} & \cdots & w^{(N-1)^2} & 1 \\
0 & 0 & 0 & \cdots & 0 & -N/m
\end{pmatrix}.
$$

The individual positions of the bodies are given in complex form by

$$
q_j = w^j + \sum_{i=0}^{N-1} w^{ij} z_i + z_N, \qquad j = 0, \ldots, N-1,
$$

and for the body in the center by

$$
q_N = -\frac{N}{m} z_N.
$$

The center of mass is found to be

$$\sum_{j=0}^{N} m_j q_j - \sum_{j=0}^{N-1} \left(\omega^j + \sum_{i=0}^{N-1} (\omega^{ij} z_i + z_N) \right) - m\frac{N}{m} z_N = N z_0.$$

Setting $z_0 = 0$ fixes the center of mass at the origin.

The first step is to compute the Hessian of the potential function. The calculations are somewhat tedious and error prone, and for this reason we give some of the details here. We write the potential function as

$$U = \sum_{0 \le i < j \le N} \frac{m_i m_j}{|q_i - q_j|}$$

$$= \frac{1}{2} \sum_{j=1}^{N-1} \sum_{i=0}^{N-1} \frac{1}{|q_i - q_{i+j}|} + \sum_{j=0}^{N-1} \frac{m}{|q_j - q_N|}$$

$$= U_1 + m U_2.$$

With the abbreviation $d_j = 1 - \omega^j$ and

$$|d_j| = |\omega^{j/2} - \omega^{-j/2}| = 2 \sin j\pi/N,$$

the distances between two bodies of unit mass is

$$|q_i - q_{i+j}| = \left| \omega^i - \omega^{i+j} + \sum_{k=0}^{N-1} (\omega^{ik} - \omega^{(i+j)k}) z_k \right|$$

$$= \left[d_j d_{-j} + \sum_{k=0}^{N-1} (\omega^{i(k-1)} d_{jk} d_{-j} z_k + \omega^{i(1-k)} d_{-jk} d_j \bar{z}_k) \right.$$

$$\left. + \sum_{k,l} \omega^{i(k-l)} d_{jk} d_{-jl} z_k \bar{z}_l \right]^{\frac{1}{2}}.$$

When $z = 0$, we thus have $|q_i - q_{i+j}| = |d_j|$ for $i, j = 0, \ldots, N-1$. We found it easiest to set $z_j = x_j + \sqrt{-1} y_j$ and to calculate first the derivatives of U_1 with respect to x_j and y_j, $j = 1, \ldots, N-1$. In this way the results can also be used for the N-body problem, when the body in the center does not exist.

For $1 \le r \le N-1$, we find

$$\frac{\partial U_1}{\partial x_r} = -\frac{1}{4} \sum_{j=1}^{N-1} \sum_{i=0}^{N-1} \left\{ \omega^{i(r-1)} d_{jr} d_{-j} + \omega^{i(1-r)} d_{-jr} d_j \right.$$

$$+ \sum (\omega^{i(k-r)} d_{jk} d_{-jr} z_k + \omega^{i(r-k)} d_{jr} d_{-jk} \bar{z}_k) \bigg\} / |q_i - q_{i+j}|^3,$$

$$\frac{\partial U_1}{\partial y_r} = -\frac{\sqrt{-1}}{4} \sum_{j=1}^{N-1} \sum_{i=0}^{N-1} \bigg\{ \omega^{i(r-1)} d_{jr} d_{-j} - \omega^{i(1-r)} d_{-jr} d_j$$

$$+ \sum (\omega^{i(r-k)} d_{jr} d_{-jk} \bar{z}_k - \omega^{i(k-r)} d_{jk} d_{-jr} z_k) \bigg\} / |q_i - q_{i+j}|^3.$$

We then compute for $1 \le s \le N - 1$, and setting $z = 0$,

$$\frac{\partial^2 U_1}{\partial x_r \partial x_s} = \frac{3}{8} \sum_{j=1}^{N-1} \sum_{i=0}^{N-1} \frac{\omega^{i(r+s-2)} d_{jr} d_{js} d_{-j}^2 + \omega^{-i(r+s-2)} d_{-jr} d_{-js} d_j^2}{|d_j|^5}$$

$$+ \frac{1}{8} \sum_{j=1}^{N-1} \sum_{i=0}^{N-1} \frac{\omega^{i(s-r)} d_{js} d_{-jr} + \omega^{i(r-s)} d_{jr} d_{-js}}{|d_j|^3},$$

$$\frac{\partial^2 U_1}{\partial x_r \partial y_s} = \frac{3\sqrt{-1}}{8} \sum_{j=1}^{N-1} \sum_{i=0}^{N-1} \frac{\omega^{i(r+s-2)} d_{jr} d_{js} d_{-j}^2 - \omega^{i(2-r-s)} d_{-jr} d_{-js} d_j^2}{|d_j|^5}$$

$$+ \frac{\sqrt{-1}}{8} \sum_{j=1}^{N-1} \sum_{i=0}^{N-1} \frac{\omega^{i(s-r)} d_{js} d_{-jr} - \omega^{i(r-s)} d_{jr} d_{-js}}{|d_j|^3},$$

$$\frac{\partial^2 U_1}{\partial y_r \partial y_s} = -\frac{3}{8} \sum_{j=1}^{N-1} \sum_{i=0}^{N-1} \frac{\omega^{i(r+s-2)} d_{jr} d_{js} d_{-j}^2 + \omega^{-i(r+s-2)} d_{-jr} d_{-js} d_j^2}{|d_j|^5}$$

$$+ \frac{1}{8} \sum_{j=1}^{N-1} \sum_{i=0}^{N-1} \frac{\omega^{i(s-r)} d_{js} d_{-jr} + \omega^{i(r-s)} d_{jr} d_{-js}}{|d_j|^3}.$$

It turns out that most second-order derivatives are zero. Only for a few indices do we find nonzero values. They are, for $s = r$,

$$\frac{\partial^2 U_1}{\partial x_r^2} = \frac{N}{4} \sum_{j=1}^{N-1} \frac{|d_{jr}|^2}{|d_j|^3} = \frac{N}{8} \sum_{j=1}^{N-1} \frac{\sin^2 \pi j r / N}{\sin^3 \pi j / N},$$

$$\frac{\partial^2 U_1}{\partial y_r^2} = \frac{N}{4} \sum_{j=1}^{N-1} \frac{|d_{jr}|^2}{|d_j|^3} = \frac{N}{8} \sum_{j=1}^{N-1} \frac{\sin^2 \pi j r / N}{\sin^3 \pi j / N},$$

and for $s = N + 2 - r$,

$$\frac{\partial^2 U_1}{\partial x_r \partial x_s} = \frac{3N}{8} \sum_{j=1}^{N-1} \frac{d_{jr} d_{js} d_{-j}^2 + d_{-jr} d_{-js} d_j^2}{|d_j|^5}$$

$$= -\frac{3N}{8} \sum_{j=1}^{N-1} \frac{\sin\left(\pi jr/N\right)\sin \pi j(r-2)/N}{\sin^3 \pi j/N},$$

$$\frac{\partial^2 U_1}{\partial y_r \partial y_s} = -\frac{3N}{8} \sum_{j=1}^{N-1} \frac{d_{jr} d_{js} d_{-j}^2 + d_{-jr} d_{-js} d_j^2}{|d_j|^5}$$

$$= \frac{3N}{8} \sum_{j=1}^{N-1} \frac{\sin\left(\pi jr/N\right)\sin \pi j(r-2)/N}{\sin^3 \pi j/N}.$$

Since U_1 does not depend on z_N, the nonzero partial derivatives involving x_N and y_N come only from U_2, which is

$$U_2 = \sum_{j=0}^{N-1} \frac{1}{|q_j - q_N|}.$$

The individual distances to the mass at the center are

$$|q_j - q_N| = \left| \omega^j + \sum_{k=0}^{N-1} \omega^{jk} z_k + \left(1 + \frac{N}{m}\right) z_N \right|$$

$$= \left[1 + \sum_k (\omega^{j(k-1)} z_k + \omega^{j(1-k)} \overline{z}_k) + \sum_{k,l} \omega^{j(k-l)} z_k \overline{z}_l \right.$$

$$+ \left(1 + \frac{N}{m}\right)\left(\omega^{-j} z_N + \sum_k \omega^{-jk} \overline{z}_k z_N + \omega^j \overline{z}_N + \sum_k \omega^{jk} z_k \overline{z}_N \right)$$

$$\left. + \left(1 + \frac{N}{m}\right)^2 z_N \overline{z}_N \right]^{1/2}.$$

The computations of the second-order derivatives are slightly simpler than those for U_1, since for $z = 0$ we have $|q_j - q_N| = 1$, and with it we find

$$\frac{\partial^2 U_2}{\partial x_r^2} = \frac{N}{2}, \; \frac{\partial^2 U_2}{\partial x_r \partial x_s} = \frac{3N}{2}, \; \frac{\partial^2 U_2}{\partial x_2 x_N} = \frac{3N}{2}\left(1 + \frac{N}{m}\right),$$

$$\frac{\partial^2 U_2}{\partial x_N^2} = \frac{N}{2}\left(1 + \frac{N}{m}\right)^2, \; \frac{\partial^2 U_2}{\partial y_r^2} = \frac{N}{2}, \; \frac{\partial^2 U_2}{\partial y_r \partial y_s} = -\frac{3N}{2},$$

$$\frac{\partial^2 U_2}{\partial y_2 y_N} = -\frac{3N}{2}\left(1+\frac{N}{m}\right), \quad \frac{\partial^2 U_2}{\partial y_N^2} = \frac{N}{2}\left(1+\frac{N}{m}\right)^2.$$

The Hessian of U with respect to $(x_1,\ldots,x_N,y_1,\ldots,y_N)$ therefore has the following relatively simple form

$$\Delta = \begin{pmatrix} B+C & 0 \\ 0 & B-C \end{pmatrix}, \tag{20}$$

with the diagonal matrix

$$B = \begin{pmatrix} b_1 & & & & & \\ & b_2 & & & & \\ & & b_3 & & & \\ & & & \cdot & & \\ & & & & b_{N-1} & \\ & & & & & b_N \end{pmatrix}$$

and the cross-diagonal matrix

$$C = C^T = \begin{pmatrix} 0 & 0 & 0 & \cdot & 0 & 0 \\ 0 & 0 & 0 & \cdot & 0 & c_2 \\ 0 & 0 & 0 & \cdot & c_3 & 0 \\ \cdot & \cdot & & \cdot & \cdot & \cdot \\ 0 & 0 & c_{N-1} & \cdot & 0 & 0 \\ 0 & c_N & 0 & \cdot & 0 & 0 \end{pmatrix}.$$

The nonzero entries were found above and are

$$b_r = \frac{N}{8}\sum_{j=1}^{N-1}\frac{\sin^2 \pi jr/N}{\sin^3 \pi j/N} + \frac{mN}{2}, \qquad 1 \le r \le N-1,$$

$$b_N = \frac{mN}{2}\left(1+\frac{N}{m}\right)^2,$$

$$c_r = -\frac{3N}{8}\sum_{j=1}^{N-1}\frac{\sin \pi jr/N \sin \pi (r-2)j/N}{\sin^3 \pi j/N} + \frac{3mN}{2},$$

$$c_2 = c_N = \frac{3mN}{2}\left(1+\frac{N}{m}\right), \qquad 3 \le r \le N-1.$$

The moment of inertia is given by

$$I = \frac{1}{2} \sum_{j=0}^{N-1} |q_j|^2 + \frac{m}{2} |q_N|^2$$

$$= \frac{N}{2} \left(1 + z_1 + \bar{z}_1 + \sum_{j=0}^{N} |z_j|^2 + \frac{N}{m} |z_N|^2 \right). \tag{21}$$

The manifold $I = I_0 = N/2$ is therefore in its first approximation determined by

$$\frac{1}{2}(z_1 + \bar{z}_1) = \text{Re } z_1 = 0.$$

By imposing z_1 to be real, we select a representative from the rotational equivalence classes. Setting $z_0 = 0$, Im $z_1 = 0$, and Re $z_1 = x_1$, we find that $\partial I/\partial x_1|_{z=0} = N$, so that we can solve $I - I_0 = 0$ for x_1 as a function of the remaining variables, that is,

$$x_1 = \phi(z_2, \ldots, z_N).$$

Changing variables by

$$x_1' = x_1 - \phi(z_2, \ldots, z_N),$$
$$z_2' = z_2,$$
$$\vdots$$
$$z_N' = z_N$$

brings the manifold $I = I_0$ locally into the hyperplane $x_1' = 0$. Thus $z_2', \ldots, z_N'$ are valid local coordinates for a representative of an equivalence class. The Hessian for it will give us the desired information. Thus we need the derivatives with respect to $x_2', \ldots, x_N', y_2', \ldots, y_N'$, and they are

$$\frac{\partial U}{\partial x_r'} = \frac{\partial U}{\partial x_r} + \frac{\partial U}{\partial x_1} \frac{\partial \Phi}{\partial x_r}, \qquad 2 \leq r \leq N,$$

$$\frac{\partial^2 U}{\partial x_r' \partial x_s'} = \frac{\partial}{\partial x_s'} \left(\frac{\partial U}{\partial x_r} + \frac{\partial U}{\partial x_1} \frac{\partial \Phi}{\partial x_r} \right) + \frac{\partial}{\partial x_1} \left(\frac{\partial U}{\partial x_r} + \frac{\partial U}{\partial x_1} \frac{\partial \Phi}{\partial x_r} \right) \frac{\partial \Phi}{\partial x_s}.$$

Fortunately all first-order and most second-order derivatives of Φ are zero for $z_2' = \cdots = z_n' = 0$. The only nonzero ones, as found from (21), are

$$\frac{\partial^2 \Phi}{\partial x_r'^2} = \frac{\partial^2 \Phi}{\partial y_r'^2} = -1, \qquad 2 \leq r \leq N,$$

$$\frac{\partial^2 \Phi}{\partial x_N'^2} = \frac{\partial^2 \Phi}{\partial y_N'^2} = -\left(1 + \frac{N}{m}\right).$$

It shows that only the terms on the diagonal of the Hessian in (20) get modified. For this we also need

$$\left.\frac{\partial U}{\partial x_1}\right|_{z'=0} = -\frac{N}{4} \sum_{j=1}^{N-1} \frac{1}{\sin \pi j/N} - mN.$$

Combining all of this we see that the new Hessian Δ has the same form as in (20), except that the matrices B and C are now of size $(N-1) \times (N-1)$ and of the form

$$B = \begin{pmatrix} b_2 & & & & \\ & b_3 & & & \\ & & \cdot & & \\ & & & b_{N-1} & \\ & & & & b_N \end{pmatrix}, \quad C = \begin{pmatrix} & & & & c_2 \\ & & & c_3 & \\ & & \cdot & & \\ & c_{N-1} & & & \\ c_N & & & & \end{pmatrix}.$$

In order to simplify the notation we introduce the following abbreviations for sums that have appeared already:

$$S = \frac{1}{2} \sum_{j=1}^{N-1} \frac{1}{\sin \pi j/N},$$

$$R_r = \frac{1}{4} \sum_{j=1}^{N-1} \frac{\sin^2 \pi r j/N}{\sin^3 \pi j/N} + S,$$

$$T_r = \frac{3}{4} \sum_{j=1}^{N-1} \frac{\sin(\pi r j/N) \sin \pi j(r-2)/N}{\sin^3 \pi j/N}.$$

With it we can write the entries for our new Hessian as

$$b_r = \frac{N}{2}[R_r + 3m], \qquad 2 \le r \le N-1,$$

$$b_N = \frac{N}{2}\left(1 + \frac{N}{m}\right)[3m + N + S],$$

$$c_r = \frac{N}{2}[-T_r + 3m], \qquad 3 \le r \le N-1,$$

$$c_2 = c_3 = \frac{3mN}{2}\left(1 + \frac{N}{m}\right).$$

The Hessian Δ becomes singular when $B \pm C$ is singular. Due to the structures of B and C, this can be learned from 2×2 submatrices or from a single term. The former is seen from the submatrices

$$D_\pm(N, r) = \begin{pmatrix} b_r & \pm c_r \\ \pm c_r & b_s \end{pmatrix}$$

with $2 \le r < N/2+1$ and $s = N+2-r$. When N is even, the two diagonals cross in one position at an element with index $N/2 + 1$. Also denote these 1×1 submatrices by $D_\pm(N, r)$, where $2r = N + 2$ now.

In most cases the determinants of $D_\pm(N, r)$ are singular for a unique value of m, which we will denote by $m(N, r)$. One exception is when N is even and the two diagonals cross. In this case only $D_+(N, N/2+1)$ contains m and we obtain

$$m(N, r) = \frac{T_r - R_r}{6}, \qquad 2r = N + 2. \tag{22}$$

For $3 \le r < N/2 + 1$ with $s = N + 2 - r$, the corresponding value is

$$m(N, r) = \frac{T_s^2 - R_r R_s}{3(R_r + R_s + 2T_r)}.$$

It is easily seen that this formula is also valid for (22). For $r = 2$ the formula is different due to the occurence of the factor $1 + N/m$. Nevertheless, here too we find a unique value for m, where the two submatrices are singular. It is given by

$$m(N, 2) = \frac{R_2(N + S)}{3(2N - R_2 - S)}.$$

For $N = 3$ and 4 these values have to agree with those obtained previously in sections 4 and 5. We leave this verification to the reader.

For larger N it becomes impractical to evaluate these formulas by hand. For this reason we give the values generated by machine. Figure 4 also shows which masses are positive and thus meaningful for the N-body problem.

For a central mass near a critical value, that is, $m = m(N, k) + \varepsilon$, we can then look for new relative equilibria, which bifurcate from the regular N-gon for ε near zero. The Hopf bifurcation theorem and the symmetry

N	k = 2	3	4	5	6	7	8	9	10
2	0.138								
3	0.770								
4	2.379	−0.250							
5	6.478	−0.244							
6	20.906	−0.220	0.006						
7	−643.284	−0.181	0.324						
8	−37.929	−0.130	0.698	0.996					
9	−25.443	−0.069	1.119	1.773					
10	−21.723	0.001	1.581	2.641	3.012				
11	−20.270	0.079	2.079	3.588	4.391				
12	−19.738	0.165	2.610	4.605	5.893	6.337			
13	−19.671	0.258	3.172	5.686	7.506	8.457			
14	−19.872	0.357	3.760	6.825	9.220	10.734	11.251		
15	−20.242	0.462	4.374	8.018	11.025	13.152	14.251		
16	−20.726	0.572	5.011	9.261	12.916	15.701	17.440	18.030	
17	−21.290	0.687	5.671	10.551	14.885	18.371	20.802	22.048	
18	−21.913	0.807	6.351	11.883	16.929	21.153	24.325	26.287	26.950
19	−22.581	0.931	7.050	13.257	19.041	24.041	27.997	30.730	32.124

Figure 4. Critical values $m(N, k)$ of the central mass in the regular N-gon.

of U can be used to show the existence of these new relative equilibria. Unfortunately, this result is rather weak, since the bifurcation could be degenerate. For this reason it is necessary to normalize the higher order terms in U in order to see that the bifurcation really occurs. This work was carried out in [4], and we refer the reader there for the details.

APPENDIX A: The Area of a Triangle

Consider two triangles with the vertices

$$(x_0, y_0), \quad (x_1, y_1), \quad (x_2, y_2)$$

and

$$(x_0', y_0'), \quad (x_1', y_1'), \quad (x_2', y_2').$$

Their areas A and A' with appropriate signs are given by

$$2A = \begin{vmatrix} x_1 & y_1 & 1 \\ x_2 & y_2 & 1 \\ x_3 & y_3 & 1 \end{vmatrix} \quad \text{and} \quad 2A' = \begin{vmatrix} x_1' & y_1' & 1 \\ x_2' & y_2' & 1 \\ x_3' & y_3' & 1 \end{vmatrix}.$$

The determinants can be rewritten as

$$-2A = \begin{vmatrix} 0 & 0 & 0 & 1 \\ x_1 & y_1 & 1 & 0 \\ x_2 & y_2 & 1 & 0 \\ x_3 & y_3 & 1 & 0 \end{vmatrix} \quad \text{and} \quad 2A' = \begin{vmatrix} 0 & 0 & 1 & 0 \\ x'_1 & y'_1 & 0 & 1 \\ x'_2 & y'_2 & 0 & 1 \\ x'_3 & y'_3 & 0 & 1 \end{vmatrix}.$$

By multiplying these two determinants row by row, one obtains

$$-4AA' = \begin{vmatrix} 0 & 1 & 1 & 1 \\ 1 & x_1x'_1 + y_1y'_1 & x_1x'_2 + y_1y'_2 & x_1x'_3 + y_1y'_3 \\ 1 & x_2x'_1 + y_2y'_1 & x_2x'_2 + y_2y'_2 & x_2x'_3 + y_2y'_3 \\ 1 & x_3x'_1 + y_3y'_1 & x_3x'_2 + y_3y'_2 & x_3x'_3 + y_3y'_3 \end{vmatrix}.$$

If one sets

$$r_{ij}^2 = (x_i - x'_j)^2 + (y_i - y'_j)^2$$

for the distance between points in the different triangles, one can write

$$x_ix'_j + y_iy'_j = -\frac{1}{2}r_{ij}^2 + \frac{1}{2}(x_i^2 + y_i^2) + \frac{1}{2}(x_j'^2 + y_j'^2).$$

In the determinant for $-4AA'$, subtract from the ith row $(i - 2, 3, 4)$ the first row after multiplying it with $\frac{1}{2}(x_i^2 + y_i^2)$, and subtract from the jth column $(j = 2, 3, 4)$ the first column after multiplying it with $\frac{1}{2}(x_i'^2 + y_i'^2)$. One obtains

$$-4AA' = \begin{vmatrix} 0 & 1 & 1 & 1 \\ 1 & -r_{11}^2/2 & -r_{12}^2/2 & -r_{13}^2/2 \\ 1 & -r_{21}^2/2 & -r_{22}^2/2 & -r_{23}^2/2 \\ 1 & -r_{31}^2/2 & -r_{32}^2/2 & -r_{33}^2/2 \end{vmatrix}.$$

Finally, multiply the last three rows with -2 and the first column with $-1/2$ so that

$$-16AA' = \begin{vmatrix} 0 & 1 & 1 & 1 \\ 1 & r_{11}^2 & r_{12}^2 & r_{13}^2 \\ 1 & r_{21}^2 & r_{22}^2 & r_{23}^2 \\ 1 & r_{31}^2 & r_{32}^2 & r_{33}^2 \end{vmatrix}. \tag{23}$$

When the two triangles coincide, that is, when

$$(x_i, y_i) = (x_i', y_i'), \qquad i = 1, 2, 3,$$

we obtain the desired formula for the area of a triangle:

$$-16A^2 = \begin{vmatrix} 0 & 1 & 1 & 1 \\ 1 & 0 & r_{12}^2 & r_{13}^2 \\ 1 & r_{12}^2 & 0 & r_{23}^2 \\ 1 & r_{13}^2 & r_{23}^2 & 0 \end{vmatrix}.$$

By expanding this determinant the standard formula

$$16A^2 = (r_{12} + r_{23} + r_{13})(-r_{12} + r_{23} + r_{13})(r_{12} - r_{23} + r_{13})(r_{12} + r_{23} - r_{13})$$

for the area of a triangle is found, which is typically given in the form

$$A^2 = s(s - r_{12})(s - r_{23})(s - r_{13})$$

with

$$s = (r_{12} + r_{23} + r_{13})/2.$$

Of importance in this derivation is that the ideas can be extended to higher dimensional objects as long as the factors are adjusted accordingly. For example, the volume V of a tetrahedron is given by $288V^2 = F$, where

$$F = \begin{vmatrix} 0 & 1 & 1 & 1 & 1 \\ 1 & 0 & r_{12}^2 & r_{13}^2 & r_{14}^2 \\ 1 & r_{12}^2 & 0 & r_{23}^2 & r_{24}^2 \\ 1 & r_{13}^2 & r_{23}^2 & 0 & r_{34}^2 \\ 1 & r_{14}^2 & r_{24}^2 & r_{34}^2 & 0 \end{vmatrix}.$$

Differentiating this determinant, for example, with respect to r_{12}^2, leads to

$$\frac{\partial F}{\partial r_{12}^2} = -2 \begin{vmatrix} 0 & 1 & 1 & 1 \\ 1 & r_{12}^2 & r_{23}^2 & r_{24}^2 \\ 1 & r_{13}^2 & 0 & r_{34}^2 \\ 1 & r_{14}^2 & r_{34}^2 & 0 \end{vmatrix}.$$

In accordance with (23) this determinant can be interpreted as the product of the area of two triangles containing the points $(2, 3, 4)$ and $(1, 3, 4)$, respectively. Denoting by A_i the triangle where the ith point is omitted, we find in general that

$$\frac{\partial F}{\partial r_{ij}^2} = 32 A_i A_j.$$

One point of caution has to be made. The way in which formula (23) was used requires that the two triangles lie in the same plane. In other words, the above formula is only valid when the volume of the tetrahedron is zero.

APPENDIX B: *Mathematica* Code for the Four-Body Problem

```
m   = { 1, 1, 1, mc + Epsilon };
z = {Lambda, Sigma, r[1, 2], r[1, 3], r[1, 4],
         r[2, 3], r[2, 4], r[3, 4]};

rplz0 = {Lambda -> Lambda0, Sigma -> Sigma0,
     r[1, 2] -> Sqrt[3], r[1, 3] -> Sqrt[3], r[1, 4] -> 1,
     r[2, 3] -> Sqrt[3], r[2, 4] -> 1,
     r[3, 4] -> 1 };

F[r12_, r13_, r14_, r23_, r24_, r34_] :=
     Det[{{0, 1,    1,    1,    1},
          {1, 0,    r12^2, r13^2, r14^2},
          {1, r12^2, 0,    r23^2, r24^2},
          {1, r13^2, r23^2, 0,    r34^2},
          {1, r14^2, r24^2, r34^2, 0}}] ;
Inrt = 1/2*Sum[Sum[ m[[i]]*m[[j]]*r[i, j]^2,{i,1,j-1}],{j,2,4}] ;
Inrt0 = Inrt/. rplz0 ;
U = Sum[Sum[m[[i]]*m[[j]]/r[i, j], {i,1,j-1}], {j,2,4} ] ;
F =   F[r[1, 2], r[1, 3], r[1, 4], r[2, 3], r[2, 4], r[3, 4]] ;
V =   U + Lambda * (Inrt - Inrt0) + Sigma*F/2;
Do[eq[i] = D[ V, z[[i]] ], {i, 1, 8}];
```

Obtain λ and σ at order 0, and determine critical mass value:

```
Do [eq0[i] = Simplify[eq[i] /. rplz0], {i, 1, 8}];
rplLam0Sig0 =
  Part[Solve[{eq0[5] == 0, eq0[6] == 0}, {Lambda0, Sigma0}], 1]
```

```
Wronskian =
  Simplify[Det[
      Table[D[V, z[[i]], z[[j]]] /. rplz0 /.
            rplLam0Sig0 /. Epsilon -> 0, {i, 1, 8}, {j, 1, 8}]]]
rplmc = Part[Solve[Wronskian == 0, mc], 4]
rplz = Table[
    z[[i]] -> Simplify[z[[i]] /. rplz0 /. rplLam0Sig0 /. rplmc]
          + Epsilon*b[i] + Epsilon^2 *c[i], {i, 1, 8}]
Do[equ[i] = Series[ eq[i] /. rplz /. rplmc, {Epsilon,0,2}],{i,1,8}]
```

First-order terms in ϵ:

```
Do[eq1[i] = FullSimplify[Coefficient[equ[i], Epsilon, 1]],{i,1,8}]
eqs = Table[eq1[i] == 0, {i, 1, 6}];
ord1 = FullSimplify[Part[Solve[eqs, Array[b, 6]], 1]]
Do[Print["eq1[", i, "]=", FullSimplify[eq1[i] /. ord1]],{i,7,8}]
```

Second-order terms with bifurcation equations:

```
Do[eq2[i] = FullSimplify[Coefficient[equ[i], Epsilon, 2] /. ord1],
    {i, 1, 8}];
eqs = Table[eq2[i] == 0, {i, 1, 6}];
ord2 = Simplify[Part[Solve[eqs, Array[c, 6]], 1]]
bifeq1 = Factor[FullSimplify[eq2[7] /. ord2]]
bifeq2 = Factor[FullSimplify[eq2[8] /. ord2]]
bif = FullSimplify[Solve[{bifeq1 == 0, bifeq2 == 0}, {b[7], b[8]}]]
```

Determine distances for one of the new configurations:

```
rplrij = Drop[rplz, 2];
Do[rplrij = rplrij /. c[j] -> 0, {j, 3, 8}]
rplrij = FullSimplify[rplrij /. ord1 /. Part[bif, 3]]
```

APPENDIX C: *Mathematica* **Code for the Planar Five-Body Problem**

The code is very similar to the one for the four-body problem. The major difference is the additional constraints, which insure that the configuration can be realized in the plane. The *Mathematica* code for central configurations in $\mathbf{R}^3$ is not listed here, but it is very similar to the one shown. The difference is that only one constraint is needed, which insures that the configuration can be realized in $\mathbf{R}^3$. This constraint is given by a 6×6 determinant, which is the obvious extension of the one listed in the program.

```
m={ 1, 1, 1, 1, mc + Delta*Epsilon^2 };
rplz0={Lambda -> Lambda0,
        Sigma1 -> Sigma10, Sigma2 -> Sigma20, Sigma3 -> Sigma30,
      r[1, 2] -> Sqrt[2],r[1,3] -> 2,r[1,4] -> Sqrt[2],r[1,5] -> 1,
      r[2, 3] -> Sqrt[2],r[2,4] -> 2,r[2,5] -> 1,
      r[3, 4] -> Sqrt[2],r[3,5] -> 1,
      r[4, 5] -> 1};

z={Lambda, Sigma1, Sigma2, Sigma3, r[1,2], r[1,3], r[1,4], r[1,5],
        r[2,3], r[2,4], r[2,5], r[3,4], r[3,5], r[4,5]};
F[r12_, r13_, r14_, r23_, r24_, r34_] :=
      Det[{{0, 1,  1,  1,  1},
           {1, 0, r12^2, r13^2, r14^2},
           {1, r12^2, 0, r23^2, r24^2},
           {1, r13^2, r23^2, 0, r34^2},
           {1, r14^2, r24^2, r34^2, 0}}] ;
Inrt = 1/2*Sum[Sum[ m[[i]]*m[[j]]*r[i, j]^2,{i,1,j-1}],{j,2,5}] ;
Inrt0 = Inrt/. rplz0 ;
U = Sum[Sum[m[[i]]*m[[j]]/r[i, j], {i,1,j-1}], {j,2,5}] ;
F1 =  F[r[1, 2], r[1, 3], r[1, 4], r[2, 3], r[2, 4], r[3, 4]] ;
F2 =  F[r[1, 2], r[1, 3], r[1, 5], r[2, 3], r[2, 5], r[3, 5]] ;
F3 =  F[r[2, 3], r[2, 4], r[2, 5], r[3, 4], r[3, 5], r[4, 5]] ;
V  =  U+Lambda*(Inrt-Inrt0)+(Sigma1*F1+Sigma2*F2+Sigma3*F3)/2;
Do[eq[i] = D[ V, z[[i]] ], {i, 1, 14}];
```

Obtain Lambdas and Sigmas at order 0, and determine critical mass value:

```
Do[ eq0[i] = Simplify[eq[i] /. rplz0 ], {i, 1, 14}]
rplLam0Sig0 =
  FullSimplify[ Part[ Solve[
  Table[eq0[i] = 0, {i,1,14}], {Lambda0,Sigma10,Sigma20,Sigma30}],1]]
Wronskian =
  Simplify[Det[Table[Simplify[
      D[V, z[[i]]], z[[j]]] /. rplz0 /. rplLam0Sig0 /. Epsilon -> 0],
          {i, 1, 14}, {j, 1, 14}]]]
rplmc = Last[Solve[Wronskian == 0, mc]]
rplz = Table[
    z[[i]] -> Simplify[z[[i]] /. rplz0 /. rplLam0Sig0 /. rplmc]+
          Epsilon*b[i]+Epsilon^2*c[i]+Epsilon^3*d[i],{i,1,14}]
Do[equ[i] = Series[ eq[i] /. rplz /. rplmc, {Epsilon,0,3}], {i,1,14}]
```

Solve equations at first order. One needs to solve linear systems of equations in two stages, as *Mathematica* is not very good at solving larger systems. It does not simplify intermediate results on its own.

```
Do[eq1[i] = FullSimplify[Coefficient[equ[i],Epsilon,1]],{i,1,14}]
rplb8to12 =
   FullSimplify[Part[Solve[Array[eq1, 5]=0,Table[b[i],{i,8,12}]],1]]
eq1a6to14 = FullSimplify[Table[eq1[i] /. rplb8to12, {i, 6, 14}]]
rplb1to7 =
   FullSimplify[
     Part[Solve[Table[eq1a6to14[[i]] = 0, {i,1,9}], Array[b,7]],1]]
ord1 = Join[rplb1to7, FullSimplify[rplb8to12 /. rplb1to7]]
```

Check that the remaining two equations are satisfied:

```
Do[Print["eq1[", i, "]=", FullSimplify[eq1[i] /. ord1]], {i,13,14}]
```

Solve equations at second order:

```
Do[eq2[i] =
   FullSimplify[Coefficient[equ[i], Epsilon, 2] /. ord1],{i,1,14}];
rplb8to12 =
   FullSimplify[Part[Solve[Array[eq2,5]=0,Table[c[i],{i,8,12}]],1]];
eq1a6to14 = FullSimplify[Table[eq2[i] /. rplb8to12, {i,6,14}]];
rplb1to7 =
   FullSimplify[
     Part[Solve[Table[eq1a6to14[[i]] = 0,{i,1,9}], Array[c,7]],1]];
ord2 = Join[rplb1to7, FullSimplify[rplb8to12 /. rplb1to7]];
Do[Print["eq2[", i, "]=", FullSimplify[eq2[i] /. ord2]],{i,13,14}]
```

Solve equations at third order:

```
Do[eq3[i] =
   FullSimplify[Coefficient[equ[i],Epsilon,3]/. ord2/. ord1],{i,1,14}];
rplb8to12 =
   FullSimplify[Part[Solve[Array[eq3,5]=0,Table[d[i],{i,8,12}]],1]];
eq1a6to14 = FullSimplify[Table[eq3[i] /. rplb8to12, {i, 6, 14}]];
rplb1to7 =
   FullSimplify[
     Part[Solve[Table[eq1a6to14[[i]] = 0,{i,1,7}], Array[d,7]],1]];
ord3 = Join[rplb1to7, FullSimplify[rplb8to12 /. rplb1to7]]

(* Solve bifurcation equations  *)

bifeq1 =  FullSimplify[eq3[13] /. ord3]
bifeq2 = FullSimplify[eq3[14] /. ord3]
rplbif = Solve[{bifeq1 == 0, bifeq2 == 0}, {b[13], b[14]}]
```

Choose Delta = +1 or -1 and, depending on that, one or the other of the solutions to the bifurcation equations.

References

[1] O. Dziobek. Über einen merkwürdigen Fall des Vielkörperproblems. *Astro. Nach.*, 152:32–46, 1900.

[2] W. D. MacMillan and W. Bartky. Permanent configurations in the problem of four bodies. *Trans. Amer. Math. Soc.*, 34:838–875, 1932.

[3] K. R. Meyer and D. S. Schmidt. Bifurcations of relative equilibria in the 4 and 5 body problem. *Ergodic Theory and Dynamical Systems*, 8:215–225, 1988.

[4] K. R. Meyer and D. S. Schmidt. Bifurcations of relative equilibria in the N-body and Kirchhoff problems. *SIAM J. Math. Anal.*, 19:1295–1313, 1988.

[5] J. I. Palmore. Measure of degenerate relative equilibria I. *Ann. of Math.*, 104:421–429, 1976.

[6] D. S. Schmidt. Central configurations in $\mathbf{R}^2$ and $\mathbf{R}^3$. In K. R. Meyer and D. G. Saari, editors, *Hamitonian Dynamical Systems*, pages 59–76, American Mathematical Society, Providence, RI, 1988.

Singularities of the N-Body Problem

Florin Diacu

PACIFIC INSTITUTE FOR THE MATHEMATICAL SCIENCES
AND
DEPARTMENT OF MATHEMATICS AND STATISTICS
UNIVERSITY OF VICTORIA
VICTORIA, BC, CANADA
e-mail: diacu@math.uvic.ca

1   Introduction

In these lectures we present some fundamental results on the singularities of the classical N-body problem. The notion of *singularity* has two connotations in celestial mechanics. One refers to the values of the dependent variable for which the system of differential equations describing the motion loses its meaning, for example, when some denominator becomes 0. These are called *singularities of the equations*. The other one regards the finite value of the independent variable for which a certain solution blows up. This is a singularity that corresponds to a solution, so we will call it *solution singularity*. In section 3 we will define these notions and see what connections exist between them.

The main object of our lectures is the dynamics of N bodies (point particles) of masses $m_1, m_2, \ldots, m_N$, which move in the three-dimensional Euclidean space $\mathbb{R}^3$ under the influence of a mutually attracting inverse-square force law. With respect to an absolute coordinate system, the position of the body m_i is assigned by the three-dimensional vector $\mathbf{q}_i = (q_i^1, q_i^2, q_i^3)$, $i = 1, 2, \ldots, N$. The *configuration* of the particle system is given by the $3N$-dimensional vector $\mathbf{q} = (\mathbf{q}_1, \mathbf{q}_2, \ldots, \mathbf{q}_N)$. When the points move in space, $\mathbf{q}$ is a function of the time t. The *momentum* of the body m_i is $\mathbf{p}_i = m_i \dot{\mathbf{q}}_i$, $i = 1, 2, \ldots, N$, where the overdot denotes differentiation with respect to t. The momentum of the entire particle system is then $\mathbf{p} = (\mathbf{p}_1, \mathbf{p}_2, \ldots, \mathbf{p}_N)$. We define the *potential function* (which is also the negative *potential energy*) of the particle system as

$$U : \mathbb{R}^{3N} \setminus \Delta \to (0, \infty), \quad U(\mathbf{q}) = \sum_{1 \le i < j \le N} \frac{m_i m_j}{|\mathbf{q_i} - \mathbf{q_j}|},$$

where $|\,.\,|$ is the Euclidean norm and Δ represents the *collision set*,

$$\Delta = \bigcup_{1 \le i < j \le N} \Delta_{ij}, \quad \Delta_{ij} = \{\mathbf{q} \in \mathbb{R}^{3N} \,|\, \mathbf{q}_i = \mathbf{q}_j\}.$$

If we choose the units such that the gravitational constant is 1, the equations describing the motion of the N bodies under their mutual gravitational attraction is given by the $6N$-dimensional system

$$\begin{cases} \dot{\mathbf{q}} = \mathbf{M}^{-1}\mathbf{p}, \\ \dot{\mathbf{p}} = \nabla U(\mathbf{q}), \end{cases} \tag{1}$$

where $\mathbf{M}$ is the $(3N \times 3N)$-matrix having the elements $m_1, m_1, m_1, m_2, m_2,$ $m_2, \ldots, m_N, m_N, m_N$ on the main diagonal and 0 everywhere else and $\mathbf{M}^{-1}$ represents its inverse. The $6N$-dimensional set of the variables $(\mathbf{q}, \mathbf{p})$ is called *phase space*. The $3N$-dimensional set of the coordinates is called *configuration space*.

The one who derived this system for $N = 2$ and $N = 3$ was Isaac Newton, in his masterpiece *Principia*, first published in 1687. But Newton used a geometric language, so his equations had a different formal appearance than (1). Almost a century later, Leonhard Euler wrote the equations of motion in a similar form as above. Many notable mathematicians have since studied this system. Johann Bernoulli, Joseph Louis Lagrange, Pierre Simon Laplace, Carl Gustav Jacobi, Siméon Denis Poisson, Spiru Haretu, George Hill, George Darwin, Ernest Heinrich Bruns, Karl Weierstrass, Henri Poincaré, Paul Painlevé, Forest Ray Moulton, George David Birkhoff, and Jean Chazy are only some of those belonging to past generations that have made remarkable contributions towards understanding the N-body problem. Though thousands of papers have been published on this subject, we still know very little about the solutions of (1). Poincaré was the first to approach the system geometrically, and his ideas opened up a new direction of work on the qualitative methods used today in the study of chaotic phenomena and of other dynamical aspects encountered in the theory of nonlinear systems (see [DH,1996]).

2   First Integrals

An important result that signaled the end of the quantitative era in celestial mechanics and the beginning of the qualitative one was obtained by the German mathematician Ernst Heinrich Bruns. (Painlevé and Poincaré proved related results soon thereafter.) In a long paper that contains some gaps in the proofs [B,1887], Bruns showed that system (1) has no more than 10 linearly independent integrals that are algebraic with respect to $\mathbf{q}$ and $\mathbf{p}$. At that time a common method of approaching systems of differential equations was obtaining *first integrals*, which are real-valued functions that are constant along solutions. From the geometric point of view a first integral foliates an n-dimensional space in $(n-1)$-dimensional slices. If n

algebraic integrals are found, the system of n first-order differential equations reduces to finding the solution of an n-dimensional algebraic system. But if at least one of the integrals is transcendent, any hope of solving the algebraic system fades.

Previous work on system (1) had revealed 10 algebraic first integrals: the *energy integral*,

$$T(\mathbf{p}) - U(\mathbf{q}) = h, \tag{2}$$

where $T : \mathbb{R}^{3N} \to [0, \infty)$, $T(\mathbf{p}) = \frac{1}{2} \sum_{i=1}^{N} m_i^{-1} |\mathbf{p}_i|^2$ is the *kinetic energy*, and h is the energy constant, the three integrals of the *angular momentum*,

$$\sum_{i=1}^{N} \mathbf{q}_i \times \mathbf{p}_i = \mathbf{c}, \tag{3}$$

where $\times$ denotes the vector product and $\mathbf{c}$ is a three-dimensional constant vector, the three integrals of the *momentum*,

$$\sum_{i=1}^{N} \mathbf{p}_i = \mathbf{a}, \tag{4}$$

where $\mathbf{a}$ is a three-dimensional constant vector, and the three integrals of the *center of mass*,

$$\sum_{i=1}^{N} m_i \mathbf{q}_i - \mathbf{a}t = \mathbf{b}, \tag{5}$$

where $\mathbf{a}$ and $\mathbf{b}$ are three-dimensional constant vectors. Since there are no further linearly independent algebraic integrals, system (1) reduces only to a system of dimension $6N - 10$. In fact, using some symmetries, the reduction can be improved to $6N - 12$. Obviously, this is not enough for solving any of the N-body problems with $N \geq 3$, but the integrals are still useful for reducing the dimension of the system and for understanding certain qualitative aspects of the solutions. The complete reduction can be found in [W,1988].

The integrals (4) and (5) imply that, without loss of generality, the study of system (1) can be restricted to the *invariant set* $\mathbf{Q} \times \mathbf{P}$, where

$$\mathbf{Q} = \left\{ \mathbf{q} \,\middle|\, \sum_{i=1}^{N} m_i \mathbf{q}_i = \mathbf{0} \right\} \quad \text{and} \quad \mathbf{P} = \left\{ \mathbf{p} \,\middle|\, \sum_{i=1}^{N} \mathbf{p}_i = \mathbf{0} \right\}.$$

From the physical point of view this means that we fix the origin of the frame in the center of mass of the particle system.

3 Singularities

The study of singularities arises naturally from the most basic problem of the theory of differential equations: existence and uniqueness of global solutions for initial value problems. Indeed, notice that for given initial conditions $(\mathbf{q}, \mathbf{p})(0) \in (\mathbb{R}^{3N} \setminus \Delta) \times \mathbb{R}^{3N}$, standard results applied to system (1) ensure the local existence of a unique analytic solution $(\mathbf{q}, \mathbf{p})$ defined on some interval $[0, t^+)$. This solution can be analytically extended to an interval $[0, t^*)$, with $0 < t^+ \le t^* \le \infty$. If $t^* = \infty$, the solution is globally defined. If $t^* < \infty$, the solution is *singular* and t^* is said to be a *singularity* of the solution. On the other hand, system (1) has *singularities* whenever $\mathbf{q} \in \Delta$. Indeed, since at least one of the denominators in $\nabla U(\mathbf{q})$ vanishes, equations (1) lose their meaning.

Obviously, there is a close connection between singular solutions and singularities of the system. In what follows in this section, we will show that every singular solution $(\mathbf{q}, \mathbf{p})$ is such that $\mathbf{q}(t) \to \Delta$ when $t \to t^*$, for if not, the solution would be globally defined. Since Δ represents the collision set, we might rush to the conclusion that every singularity is a collision. But this is true only if $\mathbf{q}$ tends to an element of Δ. It may happen that $\mathbf{q}$ tends to Δ without asymptotic phase, that is, by oscillating between various elements but failing to reach a definite position. This means that there might exist collisionless singular solutions. Let us now proceed with the details.

The beginnings of the study of singularities are historically unclear. In his fundamental treatise [Wi,1941], Aurel Wintner noted that, in the 19th century, Proposition 3.1 was basic knowledge in the research community.

Let us denote
$$\rho(\mathbf{q}) = \min_{1 \le i < j \le N} q_{ij},$$
where $q_{ij} = |\mathbf{q}_i - \mathbf{q}_j|$ represents the distance between the particles of masses m_i and m_j. We can now prove the following result.

Proposition 3.1 *If* $(\mathbf{q}, \mathbf{p})$ *is an analytic solution of (1) defined on* $[0, t^*)$, *then* t^* *is a singularity if and only if*

$$\liminf_{t \to t^*} \rho(\mathbf{q}(t)) = 0.$$

Proof. Let us first prove the direct statement. For this we assume that there exists an $a > 0$ such that $\liminf_{t \to t^*} \rho(\mathbf{q}(t)) \geq a$. This implies that there is a t_0 in $[0, t^*)$ and a β in $[0, a)$ such that $q_{ij}(t) \geq \beta$ for all t in $[t_0, t^*)$. Since from (1),

$$\ddot{\mathbf{q}}_i = \sum_{j=1, j \neq i}^{N} m_j(\mathbf{q}_j - \mathbf{q}_i) q_{ij}^{-3}, \quad i = 1, 2, \ldots, N,$$

it follows that $|\ddot{\mathbf{q}}_i| \leq \beta^{-2}(\sum_{j=1}^{N} m_j)$, $i = 1, 2, \ldots, N$, so $\ddot{\mathbf{q}}$ is bounded. Writing the configuration $\mathbf{q}$ as a Taylor series about t_0 with integral remainder,

$$\mathbf{q}(t) = \mathbf{q}(t_0) + (t - t_0)\dot{\mathbf{q}}(t_0) + \int_{t_0}^{t} (t - \tau)\ddot{\mathbf{q}}(\tau)d\tau,$$

and using the fact that $\ddot{\mathbf{q}}$ is bounded, we can draw the conclusion that there exist vectors $\mathbf{q}^*$ and $\mathbf{p}^*$ such that $\lim_{t \to t^*} \mathbf{q}(t) = \mathbf{q}^*$ and $\lim_{t \to t^*} \mathbf{p}(t) = \mathbf{p}^*$, that is, that $\mathbf{q}$ and $\mathbf{p}$ tend to limiting positions. If we denote $q_{ij}^* = |\mathbf{q}_i^* - \mathbf{q}_j^*|$, it follows that $q_{ij}^* \geq \beta$ for all $i, j = 1, 2, \ldots, N$, with $i \neq j$. Physically this means that the mutual distances between particles are strictly positive at the time t^*, so the bodies can still move. This already suggests that we are in contradiction with the hypothesis by showing that the solution is analytic at t^*. Mathematically speaking, the domain in which the solution $(\mathbf{q}, \mathbf{p})$ is defined depends on β, and therefore on a, but is independent of the initial conditions. So if we choose the data close enough to the initial conditions $t^*, \mathbf{q}^*$, and $\mathbf{p}^*$, which produce the solution $(\mathbf{q}, \mathbf{p})$, the corresponding solution is still analytic at t^*. This contradicts the hypothesis and proves the direct statement.

Let us now prove the converse statement. For this we will assume that $\liminf_{t \to t^*} \rho(\mathbf{q}(t)) = 0$. Obviously, if $\ddot{\mathbf{q}}$ becomes unbounded as $t \to t^*$, then t^* is a singularity. Let us assume therefore that $\ddot{\mathbf{q}}(t)$ is bounded for all t in $[0, t^*)$. Using this, it follows from system (1) that $\nabla U(\mathbf{q})$ is also bounded.

Since

$$\frac{d}{dt}U(\mathbf{q}(t)) = [\nabla U(\mathbf{q}(t))]^T \dot{\mathbf{q}}(t),$$

$\frac{d}{dt}U(\mathbf{q}(t))$ is also bounded, therefore $U(\mathbf{q}(t))$ is bounded. But the assumption we made at the beginning, that $\liminf_{t \to t^*} \rho(\mathbf{q}(t)) = 0$, implies that $\limsup_{t \to t^*} U(\mathbf{q}(t)) = \infty$, which contradicts the previous conclusion. This completes the proof of the converse statement.

In 1895, at the invitation of King Oscar II of Sweden and Norway, Paul Painlevé gave a series of lectures at the University of Stockholm, the first of which was attended by the king himself. In these lectures Painlevé dealt with several topics in differential equations, including the subject of singularities of the N-body problem. In 1897 he published his lecture notes [P,1897], which contain the first systematic study of these singularities. A first result he came up with improved Proposition 3.1. Painlevé managed to prove that t^* is a singularity if and only if $\mathbf{q}(t) \to \Delta$ as $t \to t^*$. This is equivalent to the following statement.

Proposition 3.2 *(Painlevé, 1895). If $(\mathbf{q}, \mathbf{p})$ is an analytic solution of equations (1) defined on $[0, t^*)$, then t^* is a singularity if and only if*

$$\lim_{t \to t^*} \rho(\mathbf{q}(t)) = 0.$$

Proof. The converse statement follows from Proposition 3.1. So we are left to prove only the direct statement. For this let us assume that there is an $a > 0$ such that $\limsup_{t \to t^*} \rho(\mathbf{q}(t)) \geq a$. This implies the existence of a sequence of times, $(t_n)_n$, with $t_n \to t^*$ such that $q_{ij}(t_n) \geq a > 0$ for all i, j with $1 \leq i < j \leq n$. Therefore, the potential function is bounded at these points; that is, there is a $b > 0$ such that $U(\mathbf{q}(t_n)) \leq b$ for all positive integer numbers n. From the energy integral (2) it follows that $T(\mathbf{p}(t_n)) \leq b + h$ for all n. Consequently, there is a constant $\gamma > 0$ such that $|\mathbf{p}(t_n)| \leq \gamma$ for all n. As we have shown in the proof of Proposition 3.1, the domain of the solution is independent of the choice of the initial conditions. So for some initial data $t_n, \mathbf{q}(t_n)$, and $\mathbf{p}(t_n)$ with t_n close enough to t^*, we can show that the solution is analytic at t^*. This contradicts the fact that t^* is a singularity and proves the direct statement.

4   Collisions

Proposition 3.2 allows us to define t^* as a *collision singularity* if $\mathbf{q}(t)$ has a definite limit when $t \to t^*$, that is, if $\mathbf{q}(t)$ tends to Δ by ultimately reaching an element of it. A singularity t^* which is not of this type will be called a *pseudocollision* or a *noncollision singularity*. A pseudocollision can appear only when $\mathbf{q}(t)$ tends to Δ without reaching any element of it.

   The existence of collision singularities is easy to show. For example, when all the particles move on a line, at least a binary collision must occur in the future or in the past. We will prove this result rigorously in section 6. However, the existence of pseudocollisions is far from clear. It is not difficult to show that in the two-body problem there are no pseudocollisions. In 1895 Painlevé proved that this is also true for the three-body problem. Let us show this in detail.

Theorem 4.1 *(Painlevé, 1895). In the three-body problem all singularities are due to collisions.*

Proof. We will show that if $\lim_{t \to t^*} \rho(\mathbf{q}(t)) = 0$, then at least one of the mutual distances $q_{12}(t), q_{23}(t)$, or $q_{31}(t)$ tends to 0 when $t \to t^*$. Let us assume that this is not the case, that is, that $\limsup_{t \to t^*} q_{ij}(t) > 0$ for all $i, j = 1, 2, 3$ with $i \neq j$. Then $\rho(\mathbf{q}(t)) \to 0$ only if at least two of the mutual distances, say q_{13} and q_{23}, interchange the role of being the minimum distance. Let $(t_n)_n$ be the sequence of times when this role is changed. This means that $q_{13}(t_n) = q_{23}(t_n)$ for all positive integers n. Since $t_n \to t^*$ when $n \to \infty$, it follows that $q_{13}(t_n) \to 0$, $q_{23}(t_n) \to 0$, and from the triangle inequality also $q_{12}(t_n) \to 0$. Therefore, $J(\mathbf{q}(t_n)) \to 0$, where

$$J(\mathbf{q}) = \frac{1}{2} \sum_{i=1}^{N} m_i |\mathbf{q}_i|^2 \tag{6}$$

is the *moment of inertia*. Indeed, using the fact that the center of mass of the particle system is at the origin of the coordinate system, it is easy to show that the moment of inertia can be written in terms of the mutual distances as

$$J(\mathbf{q}) = \frac{1}{2}(m_1 + m_2 + m_3)^{-1} \sum_{1 \leq i < j \leq 3} m_i m_j q_{ij}^2,$$

a relation that proves that $J(\mathbf{q}(t_n)) \to 0$ when $n \to \infty$. With the help of the Lagrange–Jacobi relation,

$$\ddot{J}(\mathbf{q}(t)) = U(\mathbf{q}(t)) + 2h,$$

where h is the energy constant, and it can be shown that J always has a limit J^* (finite or infinite) when $t \to t^*$. Therefore $J(\mathbf{q}(t)) \to 0$ when $t \to t^*$, so $q_{12}(t), q_{13}(t), q_{23}(t) \to 0$ when $t \to t^*$, which contradicts the initial assumption and proves the theorem.

All of Painlevé's attempts to prove a similar result for more than three bodies failed. He then thought that there may exist pseudocollisions, but had no idea how to prove the existence of such solutions. Therefore, he concluded his analysis of singularities with the following conjecture.

Painlevé's Conjecture. *For $N \geq 4$ there exist solutions $(\mathbf{q}, \mathbf{p})$ of system (1) defined on $[0, t^*)$, with t^* a noncollision singularity.*

It would take almost a century until Painlevé's Conjecture was proved to be true. We will discuss this achievement in section 8.

5 Pseudocollisions

The next step toward understanding whether pseudocollisions can exist was taken by the Swedish mathematician Edvard Hugo von Zeipel. After receiving his doctoral degree from Uppsala University, he spent some time in Paris on a fellowship. He probably led discussions with Painlevé on singularities, and upon his return back home published a short paper [vZ,1908] containing a surprising result: to encounter a pseudocollision, a solution cannot tend to a definite limit and must necessarily become unbounded in finite time. He expressed this property using the moment of inertia (6), which is a measure of the particle's distribution in space.

Theorem 5.1 *(von Zeipel, 1908). If $(\mathbf{q}, \mathbf{p})$ is a solution of system (1) defined on $[0, t^*)$, with t^* singularity and such that $\lim_{t \to t^*} J(\mathbf{q}(t))$ is finite, then t^* is due to a collision.*

Having appeared in an obscure journal, von Zeipel's result went practically unnoticed for more than three decades. In 1920, Jean Chazy also announced it [C,1920] without proof and without any reference to von Zeipel. The first to mention von Zeipel's contribution was Aurel Wintner [Wi,1941] who claimed that the proof had gaps, and therefore it could not be validated. A first detailed argument came in 1970, signed by Hans-Jürgen Sperling [S,1970]. Finally in 1986, Richard McGehee [Mc,1986] filled in the details of von Zeipel's original proof, showing that the initial argument had been entirely correct.

Von Zeipel's proof is based on a cluster decomposition of the particles that leads to some technical and cumbersome notation. Therefore, instead of presenting its details here, we will outline the arguments of a more general result due to Donald Saari [S,1973], which sheds some more light on the motion's behavior for solutions leading to pseudocollisions.

Theorem 5.2 *(Saari, 1973). If* $(\mathbf{q}, \mathbf{p})$ *is a solution of system (1) defined on* $[0, t^*)$, *with* t^* *singularity and such that* $J(\mathbf{q})$ *is a slowly varying function of* t *as* $t \to t^*$, *then* t^* *is due to a collision.*

Outline of proof. Recall first that a function $f : [0, t^*) \to \mathbb{R}$ is called *slowly varying* as $t \to t^*$ if, for all $a > 0$,

$$\lim_{t \to t^*} \frac{f(a(t - t^*))}{f(t - t^*)} = 1.$$

The idea of the proof is to show that a pseudocollision may occur only if some kind of oscillatory motion takes place, more precisely if there exist distinct indices i and j such that

$$\liminf_{t \to t^*} q_{ij}(t) J^{-\frac{1}{2}}(\mathbf{q}(t)) = 0 \quad \text{and} \quad \limsup_{t \to t^*} q_{ij}(t) J^{-\frac{1}{2}}(\mathbf{q}(t)) > 0.$$

This is achieved by proving that there is a sequence $(t_n)_n$, $t_n \to t^*$, such that for all elements of the sequence, $q_{ij}(t_n) < \epsilon J^{\frac{1}{2}}(\mathbf{q}(t_n))$ for some indices i and j, where $\epsilon > 0$ is arbitrarily small, whereas $q_{ij}(t_n) > bJ^{\frac{1}{2}}(\mathbf{q}(t_n))$ for the other indices, where $b > 0$ is a fixed constant. This means that some particles form clusters while their distances to the other particles are bounded from below. The details of this proof can be found in [S,1973].

6   Particular Cases

In the same paper [S,1973], Donald Saari proved that if the gravitational motion of n particles is confined to a line, then every solution encounters a collision but cannot encounter pseudocollisions. In his proof he used the following result.

Convexity Lemma. *There are no functions $f : \mathbb{R} \to \mathbb{R}$, at least twice differentiable, such that $f > 0$ and $\ddot{f} < 0$ (or $f < 0$ and $\ddot{f} > 0$) on $\mathbb{R}$.*

The proof of the lemma follows without difficulty. Let us now state and prove Saari's result.

Theorem 6.1 *(Saari, 1973). In the rectilinear N-body problem, every solution has a singularity in the past or in the future, which is due to a collision. No solution encounters pseudocollisions.*

Proof. The class of rectilinear solutions is an invariant set for system (1). Every particle has only one coordinate, q_i, $i = 1, 2, \ldots, N$. Let us assume that $q_1 \leq q_2 \leq \cdots \leq q_N$. Using the integral (5) and the fact that the origin of the coordinate system is in the center of mass, we can draw the conclusion that $\sum_{i=1}^{N} m_i q_i = 0$, which implies that $q_1 \leq 0 \leq q_N$. According to system (1),

$$\ddot{q}_N = \sum_{i=1}^{N-1} m_i(q_i - q_N) q_{iN}^{-3},$$

and since $q_i - q_N < 0$, it follows that $\ddot{q}_N < 0$. If m_N and m_{N-1} do not collide, then $q_N > 0$. But the Convexity Lemma implies that the function q_N cannot be defined on $\mathbb{R}$, only on some subset of it. Therefore, there is a singularity t^* at which the motion stops. Also, q_N is bounded by 0 from below and by a positive constant from above (because $\ddot{q}_N < 0$).

By Theorem 5.1, $J(\mathbf{q}(t)) \to \infty$ as $t \to t^*$. It is also easy to see that there exist positive constants A and B such that if $R(t) = \max_{1 \leq i < j \leq N} q_{ij}(t)$, then

$$AR^2(t) \leq J(\mathbf{q}(t)) \leq BR^2(t)$$

for all t for which the solution is defined. Therefore, $\lim_{t \to t^*} R(t) = \infty$. But $R(t) = q_{1N}(t)$, so $\lim_{t \to t^*} q_{1N}(t) = \infty$ and therefore $\limsup_{t \to t^*} q_N(t) = \infty$.

This contradicts the fact that q_N is bounded and thus proves the result.

The singular trapezoidal solutions of the four-body problem with equal masses are of the same type as the rectilinear ones described above. The proof of the following result can be found in [D,1992].

Theorem 6.2 *(Diacu, 1992). In the trapezoidal four-body problem with equal masses, every solution has a singularity (in the future or in the past), which is always a collision and never a pseudocollision.*

Part of Theorem 6.2 can be generalized to all two-degrees-of-freedom systems. The proof of the following result can be found in [SD,1994].

Theorem 6.3 *(Saari and Diacu, 1994). There is no two-degrees-of-freedom invariant set of system (1) that contains solutions leading to pseudocollisions.*

7   Clustered Configurations

Another class of solutions free of pseudocollisions are the *clustered configurations*. We say that a solution of system (1) is a clustered configuration if there exist positive constants c_1 and c_2 such that whenever

$$\rho(t) < c_1 < c_2 < R(t),$$

where $R(t) = \max_{1 \le i < j \le N} q_{ij}(t)$ and $\rho(t) = \min_{1 \le i < j \le N} q_{ij}(t)$, the particles can be partitioned into two or more subsets, such that at least one subset has more than one particle and the distance between particles in the same subset is less than $\frac{1}{3}R(t)$. For a solution, different partitions can be defined along different intervals of time.

In the three-body problem, for example, whenever

$$\rho(t) < 1 < 10 < R(t),$$

the triangle inequality implies that $q_{ij}(t) < 1 < \frac{R(t)}{3}$ and $q_{ik}(t), q_{jk}(t) > 9$ for the partition $\{\{m_i, m_j\}, \{m_k\}\}$. Of course, different partitions, say

$\{\{m_2, m_3\}, \{m_1\}\}$ and $\{\{m_1, m_2\}, \{m_3\}\}$, may correspond to different time intervals.

We can now prove the following result, published in [SD,1994]. It mainly shows that if the particles move in clusters, they cannot reach a pseudocollision. This implies that noncollision singularities can be attained only if at least one particle travels back and forth between clusters to transfer the necessary kinetic energy that makes the system become unbounded in finite time.

Theorem 7.1 *(Saari and Diacu, 1994). A clustered-configuration solution of system (1) cannot lead to pseudocollisions.*

Proof. By Theorem 5.1, if t^* is a noncollision singularity, then $R(t) \to \infty$ and $\rho(t) \to 0$ as $t \to t^*$. So if for a solution leading to a pseudocollision, $\rho(t_0) < c_1 < c_2 < R(t_0)$ for some t_0, then $\rho(t) < c_1 < c_2 < R(t)$ for all t in the interval $[t_0, t^*)$. Therefore, once a certain partition of the particles occurs, it is maintained all along the solution.

Let us assume that this partition defines the clusters $S_1, S_2, \ldots, S_n$, with $n < N$. The center of mass of the cluster S_j is $C_j = (\sum_{i \in S_j} m_i)^{-1} \sum_{i \in S_j} m_i \mathbf{q}_i$. In the expression of $\ddot{C}_j$, the terms regarding S_j cancel each other, so $\ddot{C}_j$ contains only terms involving the gravitational attraction between particles from S_j and other clusters. This implies that $|\ddot{C}_j|$ is bounded from above, therefore $\dot{C}_j$ and C_j have limits as $t \to t^*$. Let $\lim_{t \to t^*} C_j(t) = C_j^*$, $j = 1, 2, \ldots, n$. As $t \to t^*$, the distance q_{is} between a particle m_i from S_j and a particle m_s from S_k satisfies the inequality

$$q_{is}(t) \leq |\mathbf{q}_i(t) - C_j^*| + |\mathbf{q}_s(t) - C_k^*| + |C_j^* - C_k^*| \leq \frac{2R(t)}{3} + |C_j^* - C_k^*|.$$

If $C = \max_{1 \leq i < j \leq n} |C_j^* - C_k^*|$, it follows that for any m_i and m_s as above, $q_{is}(t) \leq \frac{2}{3}R(t) + C$, therefore $R(t) \leq \frac{2}{3}R(t) + C$. Dividing this relation by $R(t)$ and then letting t tend to t^*, it follows that

$$1 \leq \frac{2}{3} + \frac{C}{R(t)} \to \frac{2}{3},$$

which is a contradiction. This completes the proof.

8   Examples of Pseudocollisions

In 1975, John Mather and Richard McGehee came up with an example of a pseudocollision in the rectilinear four-body problem [MM,1975]. Two particles at one extreme formed a cluster, the one at the other extreme a different (one-particle) cluster, while the remaining particle moved back and forth between the two clusters. The authors proved that with properly chosen initial conditions, the two clusters escape to infinity in finite time. However, the particle oscillating between clusters encounters a collision every time it changes direction. The collisions are "regularized"; that is, the motion is analytically continued after collision on the same total-energy level. (We will discuss this kind of *regularization* in section 10.) Therefore, this example does not prove Painvevé's Conjecture, which requires that the pseudocollision be the first encountered singularity.

In 1984, Joe Gerver brought some arguments that a pseudocollision may appear in the planar five-body problem [G,1984]. But he never presented a complete proof in this sense.

In 1987, Zhihong Xia, a Ph.D. student under the supervision of Donald Saari at Northwestern University, announced the complete confirmation of Painlevé's Conjecture [X,1992]. Xia's example involved a spatial five-body problem in which two pairs of bodies formed clusters and the third oscillated between them. In a long and complicated proof, Xia managed to show that there exist initial conditions that lead to a pseudocollision.

Just before the publication of Xia's paper, Gerver produced the first example in the plane for a $3N$-body problem with N large [G,1991]. Gerver's arguments involve many technical computations. None of the above proofs is easily accessible to the beginner in the field. For more descriptive details, we refer the reader to [D,1992] or directly to the original papers cited earlier.

9   Relationships between Singularities

It is natural to ask whether there is some relationship between the set of initial conditions leading to collisions and the one leading to pseudocollisions. To answer this question, let $\mathbf{x} \in \mathbb{R}^{6N}$ be a point of the phase space and consider the solution $(\mathbf{q}, \mathbf{p})$ of system (1) that goes through $\mathbf{x}$, more precisely such that $(\mathbf{q}, \mathbf{p})(0) = \mathbf{x}$. Consider the sets

$$C_\tau = \{\mathbf{x} \,|\, (\mathbf{q}, \mathbf{p}) \text{ leads to a collision at time } t^* \le \tau\},$$

$$N_\tau = \{\mathbf{x} \,|\, (\mathbf{q}, \mathbf{p}) \text{ leads to a pseudocollision at time } t^* \le \tau\},$$

which represent all the points in phase space that lie along solutions leading to collisions and pseudocollisions, respectively, and

$$E_\tau = \{(\mathbf{x}, t^*) \,|\, \mathbf{x} \in C_\tau\},$$

which pairs every point of C_τ with the corresponding singularity t^*. Also let us denote by π the projection from E_τ (or its topological closure, $\bar{E}_\tau$) to C_τ,

$$\pi : E_\tau (\text{or } \bar{E}_\tau) \to C_\tau, \quad \pi(\mathbf{x}, s) = \mathbf{x}.$$

We can now prove the following result due to Saari and Xia [SX,1996].

Theorem 9.1 *(Saari and Xia, 1996). All limit points of E_τ projected to the phase space lie along solutions leading to pseudocollisions (i.e., $\pi(\bar{E}_\tau \setminus E_\tau)$ is included in N_τ).*

Proof. If $\mathbf{x}$ is in $\pi(\bar{E}_\tau \setminus E_\tau)$, then there is an s such that $(\mathbf{x}, s)$ belongs to $\bar{E}_\tau \setminus E_\tau$. Recall that $(\mathbf{q}, \mathbf{p})$ is the solution with $(\mathbf{q}, \mathbf{p})(0) = \mathbf{x}$. Then there exists a time $\sigma \le s$ such that $[0, \sigma)$ is the minimal interval in which $(\mathbf{q}, \mathbf{p})$ is defined. Indeed, if $\sigma > s$, then the continuity of solutions with respect to initial data ensures that for every $\epsilon > 0$ there is a neighborhood U of $\mathbf{x}$ such that all solutions starting in U exist on $[0, \sigma - \epsilon)$. But this contradicts the fact that $(\mathbf{x}, s)$ is a limit point of E_τ. Since $\mathbf{x}$ does not define a collision orbit, $(\mathbf{q}, \mathbf{p})$ must lead to a pseudocollision. This completes the proof.

The following result, also proved by Saari and Xia, shows that solutions leading to pseudocollisions pass arbitrarily close to collisions. Its proof can be found in [SX,1996].

Theorem 9.2 *(Saari and Xia, 1996). Let $\mathbf{x}$ be a point of N_τ that leads to a pseudocollision at time τ. Then for any $\epsilon > 0$, there is a t_0 in $[0, \tau)$ and* *a* **y** *in* C_τ *such that the distance between the solutions corresponding to* **x** *and* **y** *at time* t_0 *is smaller than or equal to* ϵ. *Moreover,* $t^* \leq \epsilon$, *where* t^* *is the collision time corresponding to* **y**.

Theorems 9.1 and 9.2 are also true if we replace C_τ (and correspondingly E_τ and N_τ) with the set T_τ in which, instead of the word *collision*, we write *at least a triple collision*. By this we mean that the collision is not binary or multiple binary, so at least three bodies collide at some point of the space. This provides an even clearer relationship between collisions and pseudocollisions.

We can summarize the results of this section as follows. In the set of singularities that occur at some finite time that does not exceed a fixed value τ, the set of pseudocollisions is formed by limit points of the set of collision singularities that involve at least three bodies. Moreover, pseudocollision solutions pass arbitrarily close to collisions involving at least three bodies.

10   Extensions beyond Collision

When two or more point masses collide, system (1) loses its meaning. But we would like to understand what happens after an elastic bounce which takes place without loss or gain of total energy. Thus, one of the important issues arising in connection with the singularities of the N-body problem is that of *regularizing* collisions, that is, continuing the motion after collision. From the mathematical point of view, this implies extending the solution beyond the singularity t^* "in a meaningful way." But what is meaningful in this case?

One point of view regarding regularization was that of obtaining global analytic solutions of system (1) defined all over $\mathbb{R}$. This was Weierstrass's formulation of the main question for the Swedish-Norwegian prize of King Oscar II, awarded in 1889 to Poincaré (see [DH,1996]). But in spite of the important contributions made in the prized paper, Poincaré did not specifically solve the regularization problem. The first to obtain an analytic continuation of binary collisons in the three-body problem was a Finnish astronomer of Swedish origin, Karl Sundman [Su,1912], who provided a power-series solution[1] in $t^{1/3}$, convergent for all t. The only case he could not cover was that of the zero-angular momentum, $\mathbf{c} = \mathbf{0}$ (which is a necessary condition for the occurrence of triple collisions [SM,1971]). Later on, the German mathematician Carl Ludwig Siegel proved that except for a discrete set of masses, triple collision solutions cannot be analytically extended [Si,1941]. However, the length and the technicalities of these proofs do not allow us to present further details here.

A different point of view was that of the Italian mathematician Tullio Levi-Civita [LC,1920], who considered that a meaningful regularization must be related to nearby orbits. He thus requested that continuity of solutions with respect to initial data be satisfied through the collision. For example, binary collisions in the Newtonian two-body problem can occur only if the motion takes place on a line [Wi,1941]. But the rectilinear motion is a limit case of a very elongated ellipse, so continuity with respect to initial conditions is satisfied. In the Manev two-body problem, however, which adds a relativistic correction term to the Newtonian potential, a binary collision can take place on spiraling orbits, which do not approximate an ellipse [DD,1996], so the motion is not regularizable with respect to initial data. Levi-Civita's idea of regularization is of more practical importance since it allows one to understand orbits passing close to collisions. A modern point of view regarding this kind of regularization was developed by the American mathematicians Charles Conley, Robert Easton, and Richard McGehee (see, e.g., [E,1971]) and is known in the literature as *block regularization*. Let us discuss it in more detail.

11 Block Regularization

To present the notion of block regularization, we first need to introduce several notations and definitions. Let $\mathbf{M}$ be a C^∞ manifold of dimension d, $\mathbf{\Delta}$ a closed set contained in $\mathbf{M}$, and $\mathbf{V} : \mathbf{M} \setminus \mathbf{\Delta} \to \mathbf{T}(\mathbf{M} \setminus \mathbf{\Delta})$ a C^1 vector

---

[1]Unfortunately, Sundman's solution of the three-body problem proves important only from the point of view of the regularization question. Otherwise it is of little importance from both the theoretical and the practical point of view. This is because the rate of convergence is so slow that it would be necessary to take at least $10^{8,000,000}$ terms in the power series to make it useful in practical astronomical matters (see also [D,1996] and [Po, 1993, p. I25]).

field, which defines the system of differential equations

$$\dot{\mathbf{x}} = \mathbf{V}(\mathbf{x}).$$

If $\mathbf{x}$ is a solution of the initial value problem given by the condition $\mathbf{x}(0) = \gamma$, let us also denote $\mathbf{x}(t)$ by γt. Also consider the subsets $\mathbf{A}$ of $\mathbf{M} \setminus \mathbf{\Delta}$ and Λ of $\mathbb{R}$. If γt is defined for each $\mathbf{x}$ in $\mathbf{A}$ and t in Λ, let us denote

$$\mathbf{A}\Lambda = \{\gamma t \,|\, \mathbf{x} \in \mathbf{A}, t \in \Lambda\}.$$

We will call γt the *action* induced by the vector field $\mathbf{V}$ on the manifold $\mathbf{M} \setminus \mathbf{\Delta}$. Assume further that $\mathbf{N}$ is a compact C^∞ submanifold of $\mathbf{M}$ of the same dimension d. Let $\mathbf{n}$ denote the boundary of $\mathbf{N}$ and suppose that $\mathbf{n}$ and $\mathbf{\Delta}$ are disjoint. Let us further denote by

$$\mathbf{n}^+ = \{\gamma \in \mathbf{n} \,|\, \text{for some } t > 0, \gamma(-t, 0) \cap \mathbf{N} = \emptyset\}$$

the set of *ingress* points of $\mathbf{n}$ and by

$$\mathbf{n}^- = \{\gamma \in \mathbf{n} \,|\, \text{for some } t > 0, \gamma(0, t) \cap \mathbf{N} = \emptyset\}$$

the set of *egress* points of $\mathbf{n}$. In other words, $\mathbf{n}^+$ and $\mathbf{n}^-$ are the entrance and exit points into and out of $\mathbf{N}$, respectively, for the solutions of the system $\dot{\mathbf{x}} = \mathbf{V}(\mathbf{x})$. Similarly we denote by

$$\tau = \{\gamma \in \mathbf{n} \,|\, \mathbf{V} \text{ is tangent to } \mathbf{n} \text{ at } \gamma\}$$

the set of *tangency* points of the solutions to $\mathbf{n}$. Notice that $\mathbf{n}^+ \cup \mathbf{n}^- \cup \tau = \mathbf{n}$. We will say that $\mathbf{N}$ is an *isolating block* for $\mathbf{V}$ if $\tau = \mathbf{n}^+ \cap \mathbf{n}^-$ and τ is a codimension-1 C^∞ submanifold of $\mathbf{n}$.

Remark 1. In general, $\mathbf{n}^+ \cap \mathbf{n}^-$ is not a subset of τ and τ is not a subset of $\mathbf{n}^+ \cap \mathbf{n}^-$.

Remark 2. If $\mathbf{N}$ is an isolating block for $\mathbf{V}$, then $\mathbf{n}^+$ and $\mathbf{n}^-$ are submanifolds of $\mathbf{n}$ with common boundary τ.

Let us further define the sets

$$\mathbf{a}^+ = \{\gamma \in \mathbf{n}^+ \,|\, \gamma t \in \mathbf{N} \text{ for all } t \geq 0 \text{ for which } \gamma t \text{ is defined}\},$$

which represents all ingress points for which the solution does not leave $\mathbf{N}$, and

$$\mathbf{a}^- = \{\gamma \in \mathbf{n}^- \,|\, \gamma t \in \mathbf{N} \text{ for all } t \leq 0 \text{ for which } \gamma t \text{ is defined}\},$$

which represents all egress points originating from inside $\mathbf{N}$. Consequently, $\mathbf{n}^+ \setminus \mathbf{a}^+$ is the set of all points in $\mathbf{n}^+$ corresponding to solutions that enter and eventually leave $\mathbf{N}$, and $\mathbf{n}^- \setminus \mathbf{a}^-$ is the set of points in $\mathbf{n}^-$ corresponding to solutions that leave $\mathbf{N}$ after having entered it earlier some time. Also let us define the function

$$\pi : \mathbf{n}^+ \setminus \mathbf{a}^+ \to \mathbf{n}^- \setminus \mathbf{a}^-, \quad \pi(\gamma) = \gamma t^+,$$

where $t^+ = \sup\{t \geq 0 \mid \gamma t \in \mathbf{N}\}$. Notice that π is well defined and that it is a homeomorphism from $\mathbf{n}^+ \setminus \mathbf{a}^+$ to $\mathbf{n}^- \setminus \mathbf{a}^-$.

We are now in the position to define the concept of block regularization. Assume that $\mathbf{N}$ is an isolating block and that $\boldsymbol{\Delta}$ is a subset of $\mathbf{N}$. If π admits a unique extension as a homeomorphism from $\mathbf{n}^+$ to $\mathbf{n}^-$, then $\boldsymbol{\Delta}$ is called *block regularizable*. In the N-body problem, $\boldsymbol{\Delta}$ is a collision singularity, so block regularizing the collision means connecting a collision solution with an ejecting one in a unique way such that continuity of solutions with respect to initial data follows for all nearby orbits.

Regarding block regularization of the Newtonian two-body problem, Robert Easton proved the following result [E,1971].

Theorem 11.1 *(Easton, 1971). The collision solutions of the Newtonian two-body problem are block regularizable.*

The proof is based on the idea stated earlier, that the rectilinear two-body problem, in which collision occurs, is a limiting case of the elliptic orbits. In connection with block regularization of triple collisions, Richard McGehee showed the following property [Mc,1974], whose proof is too long for our introductory scope.

Theorem 11.2 *(McGehee, 1974). Except for a negligible set of masses, the triple collision solutions of the Newtonian three-body problem are not block regularizable.*

In fact, a necessary condition that collisions are block regularizable is that they are exceptional in the measure theoretical sense (which is the case in the Newtonian N-body problem, as proved by Donald Saari between 1971 and 1973; see [D,1992]). Let us prove a result in this sense.

Theorem 11.3 *Assume that for any isolating block* **N**, *the corresponding homeomorphism* $\pi : \mathbf{n}^+ \setminus \mathbf{a}^+ \to \mathbf{n}^- \setminus \mathbf{a}^-$ *is such that* $\mathbf{a}^+$ *and* $\mathbf{a}^-$ *contain sets that are open in the induced topology. Then the singularity set* Δ *is not block regularizable.*

Proof. Let **N** be an isolating block for which the corresponding homeomorphism π can be extended to a homeomorphism from $\mathbf{n}^+$ to $\mathbf{n}^-$. But $\mathbf{a}^+$ and $\mathbf{a}^-$ are closed sets in **N**, which is compact, so they are compact as well. Assume first that $\mathbf{a}^+$ and $\mathbf{a}^-$ are one-dimensional. Then at least some connected subset of them is homeomorphic to some closed interval $[a, b]$ and the boundary points are mapped to the ends of the interval. But there are infinitely many homeomorphisms from $[a, b]$ to $[a, b]$ that map a to a and b to b. So, the extension fails to be unique. For higher dimensional sets $\mathbf{a}^+$ and $\mathbf{a}^-$, we can generalize this idea using hypercubes instead of intervals. This completes the proof.

A consequence of this theorem is that binary collisions in the Manev two-body problem are not block regularizable. This is because the set of initial data leading to collisions contains an open set. It is also interesting to note that there is no relationship between block regularization and analytic regularization along the solution. For more details see [D,1992].

12   The Collision Manifold

An important tool for understanding collision singularities was provided by the American mathematician Richard McGehee [Mc,1974]. His idea was to use suitable transformations that blow up the singularity into a *collision manifold*, which extends the phase space. The flow on the collision manifold is fictitious in the sense that it has no physical meaning, but due to the continuity of the solutions with respect to initial data, its study offers information on the physical behavior of the orbits that pass close to collision. In many cases this information can be used to draw global conclusions about the qualitative properties of the flow.

In what follows we will apply this idea to the two-body problem given by a homogeneous potential $U(\mathbf{x}) = |\mathbf{x}|^{-\alpha}$, where $|\mathbf{x}|$ is the distance between two unit-mass particles and $\alpha > 0$ is a constant. Notice that if $\alpha = 1$ we recover the Newtonian potential of unit masses. (For more details on this analysis see [Mc,1981].) The equations of motion are

$$\ddot{\mathbf{x}} = -\alpha |\mathbf{x}|^{-\alpha - 2} \mathbf{x}.$$

Obviously, collisions occur only if $\mathbf{x} = (x_1, x_2) = \mathbf{0}$. The energy and angular momentum integrals are given by the equations

$$\frac{1}{2} |\mathbf{y}|^2 - |\mathbf{x}|^{-\alpha} = h \quad \text{and} \quad x_1 y_2 - x_2 y_1 = c,$$

where $\mathbf{y} = \dot{\mathbf{x}}$ represents the momentum. Let us further consider the McGehee transformation

$$(\mathbb{R}^2 \setminus \mathbf{0}) \times \mathbb{R}^2 \to (0, \infty) \times [0, 2\pi) \times \mathbb{R} \times \mathbb{R}, \quad (\mathbf{x}, \mathbf{y}) \to (r, \theta, w, v),$$

given by

$$\begin{cases} x_1 = r^\gamma \cos\theta, \\ x_2 = r^\gamma \sin\theta, \\ y_1 = r^{-\beta\gamma}(v \cos\theta - w \sin\theta), \\ y_2 = r^{-\beta\gamma}(v \sin\theta + w \cos\theta), \end{cases}$$

where $\beta = \frac{\alpha}{2}$ and $\gamma = \frac{1}{\beta+1}$. This analytic diffeomorphism leads to the new equations

$$\begin{cases} \dot{r} = (\beta + 1)v, \\ \dot{\theta} = r^{-1}w, \\ \dot{w} = r^{-1}(\beta - 1)wv, \\ \dot{v} = r^{-1}[w^2 + \beta(v^2 - 2)], \end{cases} \tag{7}$$

with the energy and angular momentum relations given by

$$w^2 + v^2 - 2 = 2hr^{\alpha\gamma} \quad \text{and} \quad r^{(1-\beta)\gamma}w = c.$$

We define the *constant energy manifold*

$$\mathbf{M}(h) = \{(r, \theta, w, v) \in \mathbb{R}^4 \,|\, r \geq 0, \ w^2 + v^2 - 2 = 2hr^{\alpha\gamma}\}$$

and the *collision manifold*

$$\mathbf{C} = \{(r, \theta, w, v) \in \mathbb{R}^4 \,|\, r = 0, \ w^2 + v^2 = 2\},$$

which is contained in $\mathbf{M}(h)$ for every value of h. Notice that $\mathbf{C}$ is homeomorphic with the two-dimensional torus $S^1 \times S^1$. System (7) is undefined on **C**. To extend system (7) to **C** (which geometrically means to paste the torus to the phase space) we rescale the time variable with the transformation

$$dt = rd\tau,$$

which is an analytic diffeomorphism that changes system (7) into

$$\begin{cases} r' = (\beta + 1)rv, \\ \theta' = w, \\ w' = (\beta - 1)wv, \\ v' = w^2 + \beta(v^2 - 2). \end{cases} \tag{8}$$

The prime denotes here differentiation with respect to the fictitious time variable τ. The energy and angular momentum relations as well as the manifolds $\mathbf{M}(h)$ and **C** maintain their form in these new variables. Notice that system (8) is now defined for $r = 0$, so the phase space extends to the collision manifold **C**. The last two equations of (8) form the independent system

$$\begin{cases} w' = (\beta - 1)wv, \\ v' = w^2 + \beta(v^2 - 2), \end{cases} \tag{9}$$

for which the circle $w^2 + v^2 = 2$ corresponds to the collision manifold **C**. Any solution of system (9) that approaches this circle tends to a collision. (Using the results of sections 3 and 4 it is easy to see that pseudocollisions cannot take place.) To check whether collisions can take place, we need to understand first the flow on the collision manifold, which has no direct physical significance. System (9) will also provide us with the global qualitative picture of the problem. This picture depends on the values of the parameter β. In what follows we will discuss each possible case.

13   The Case $\beta < 1$

In this section we will determine the global flow of system (9) for $\beta < 1$. As we will see, the qualitative behavior of the solutions in this class is similar in part to those of the Kepler problem ($\beta = \frac{1}{2}$), which is contained here.

System (9) has four equilibria: $(\pm\sqrt{\alpha}, 0)$ and $(0, \pm\sqrt{2})$. The circle $w^2 + v^2 = 2$ and the line $w = 0$ are invariant sets. The rest of the flow can be drawn by linearizing the system at each equilibrium. Linearization leads us to the conclusion that $(0, \pm\sqrt{2})$ are saddles. Using the energy integral we can also see that $(\pm\sqrt{\alpha}, 0)$ are centers. Therefore, the flow looks as in Figure 1. From the energy relation we can conclude that the circle $w^2 + v^2 = 2$ corresponds to $h = 0$, whereas the flows inside and outside the circle correspond to $h < 0$ and $h \geq 0$, respectively. Recalling that the invariant circle represents the collision manifold $\mathbf{C}$, we can now draw the following conclusions from systems (8) and (9).

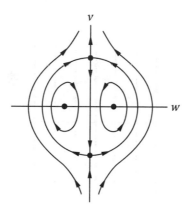

Figure 1. The flow of system (9) for $\beta < 1$.

Theorem 13.1 *(McGehee, 1981). For $h \geq 0$ (which corresponds to hyperbolic or parabolic orbits in the Kepler problem) and $w \neq 0$ (i.e., for nonzero angular momentum) all solutions are unbounded in both directions; if $w = 0$ solutions either begin or end in a collision. For $h < 0$ (which corresponds to elliptic orbits in the Kepler problem) all solutions are bounded. There are two stable circular periodic orbits corresponding to the critical points $(\pm\sqrt{\alpha}, 0)$; for $w = 0$ the orbits begin and end in a collision; all the other orbits are either periodic or quasi-periodic.*

14 The Case $\beta = 1$

In this section we will determine the global flow of system (9) for $\beta = 1$. This is a borderline case between $\beta < 1$ and $\beta > 1$, and it exhibits some interesting qualitative behavior.

The variable w becomes the angular momentum. For $h = 0$, v also becomes a first integral. From the energy relation we see that the circle $w^2 + v^2 = 2$ corresponds to $h = 0$, whereas the flows inside and outside the circle correspond to $h < 0$ and $h \geq 0$, respectively. System (9) has infinitely many equilibria, all lying on the circle $w^2 + v^2 = 2$. The other orbits are vertical in the phase plane, so they either reach the collision manifold $\mathbf{C}$ or they don't. The flow is depicted in Figure 2. This leads to the following result.

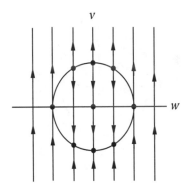

Figure 2. The flow of system (9) for $\beta = 1$.

Theorem 14.1 *(McGehee, 1981). For $h > 0$ and $|w| > \sqrt{2}$, every solution is unbounded in both time directions; for $|w| \leq \sqrt{2}$, every solution either begins or ends in a collision and is unbounded in the other direction. For $h = 0$ and $v \neq 0$ every solution also begins and ends in a collision and is unbounded in the other direction; for $v = 0$ every solution is circular. For $h < 0$ every solution begins and ends in a collision.*

15   The Case $\beta > 1$

In this section we will determine the global flow of system (9) for $\beta > 1$. The qualitative behavior of the flow is now very different from that of the Kepler problem.

System (9) again has four equilibria: $(\pm\sqrt{\alpha}, 0)$ and $(0, \pm\sqrt{2})$, but unlike in the case $\beta < 1$, the first two equilibria are now outside the circle

$w^2 + v^2 = 2$, which is an invariant set. The rest of the flow can be drawn by linearizing the system at each equilibrium. Linearization leads us to the conclusion that $(\pm\sqrt{\alpha}, 0)$ are saddles, $(0, \sqrt{2})$ is a source, and $(0, -\sqrt{2})$ is a sink, so the flow looks as in Figure 3. From the energy relation we see that the circle $w^2 + v^2 = 2$ corresponds to $h = 0$, whereas the flows inside and outside the circle correspond to $h < 0$ and $h \geq 0$, respectively. Recalling that the invariant circle represents the collision manifold $\mathbf{C}$, we can now draw the following conclusions from systems (8) and (9).

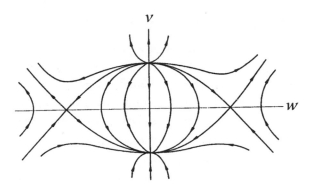

Figure 3. The flow of system (9) for $\beta > 1$.

Theorem 15.1 *(McGehee, 1981). The circular periodic orbits corresponding to the equilibria $(\pm\sqrt{\alpha}, 0)$, which now occur for positive energy, are unstable. Some of the solutions asymptotic to one of the circular orbits lead to collisions, whereas the others are unbounded. For $h > 0$ there exist unbounded collision solutions, solutions beginning and ending in collisions, and solutions unbounded in both directions. For $h \leq 0$ all solutions begin and end in collision except for those with zero energy and angular momentum, which begin or end in a collision and are unbounded in the other direction.*

16   Conclusions and Perspectives

The study of singularities is essential for understanding the qualitative behavior of solutions in particle-system dynamics. In some cases the set of initial conditions leading to singularities is negligible, but in others it can be large from the measure theoretical and from the topological point of view as well (see also [S,1975]).

Many of the above ideas can be further used and extended to potentials that are not homogeneous, for example, the *quasi-homogeneous* ones, which are of the form $\dfrac{a}{r^\alpha} + \dfrac{b}{r^\beta}$, where a, b, α, and β are constants with $0 \leq \alpha < \beta$. These generalize many of the potentials used in astronomy, physics, and chemistry, namely those of Newton, Manev, Coulomb, Van der Waals, Schwarzschild, Birkhoff, Lennard-Jones, Liboff, etc.

Also important are the anisotropic potentials, of which only two have been studied up to now in the two-body case: the anisotropic Kepler [Gu,1990] and Manev [CD,1999], [D,2000], [DS,2000] problems. The former is useful for understanding the connections existing between classical and quantum mechanics, whereas the latter lies at the intersection of classical, quantum, and relativistic mechanics. In each of these cases, as well as in the general one, singularities play a crucial role. From this point of view, the study of quasi-homogeneous potentials is far from complete.

References

[B,1887]   Bruns, H. Über die Integrale des Vielkörper-Problems, *Acta Math.* **11** (1887), 25–96.

[C,1920]   Chazy, J. Sur les singularités impossibles du problème des n corps, *C. R. Hebd. Séanc. l'Acad. Sci. Paris* **170** (1920), 575–577.

[CD,1999]   Craig, S., Diacu, F., Lacomba, E. A., and Perez, E. On the anisotropic Manev problem, *J. Math. Phys.* **40** (1999), 1359–1375.

[DD,1996]   Delgado, J., Diacu, F., Lacomba, F. N., Mingarelli, A., Mioc, V., Perez, E., and Stoica, C. The global flow of the Manev problem, *J. Math. Phys.* **37**, 6 (1996), 2748–2761.

[D,1992]   Diacu, F. *Singularities of the N-Body Problem—An Introduction to Celestial Mechanics*, Les Publications CRM, Montréal, 1992.

[D,1996]   Diacu, F. The solution of the n-body problem, *Math. Intelligencer* **18**, 3 (1996), 66–70.

[D,2000]   Diacu, F. Stability in the anisotropic Manev problem, *J. Phys. A* **33** (2000), 6573–6578.

[DH,1996]   Diacu, F., and Holmes, P. *Celestial Encounters—The Origins of Chaos and Stability*, Princeton University Press, Princeton, NJ, 1996.

[DS,2000]   Diacu, F., and Santoprete, M. Nonintegrability and chaos in the anisotropic Manev problem, *Phys. D* **156** (2001), 39–52.

[E,1971]   Easton, R. Regularization of vector fields by surgery, *J. Differential Equations* **10** (1971), 92–99.

[G,1984]   Gerver, J. A possible model for a singularity without collisions in the five-body problem, *J. Differential Equations* **52** (1984), 76–90.

[G,1991]   Gerver, J. The existence of pseudocollisions in the plane, *J. Differential Equations* **89** (1991), 1–68.

[Gu,1990]   Gutzwiller, M. C. *Chaos in Classical and Quantum Mechanics*, Springer-Verlag, New York, 1990.

[LC,1920]   Levi-Civita, T. Sur la régularisation du problème des trois corps. *Acta Math.* **42** (1920), 99–144.

[MM,1975]   Mather, J., and McGehee, R. Solutions of the collinear four-body problem which become unbounded in finite time. In *Dynamical Systems Theory and Applications*, pp. 573–587, ed. J. Moser. Lecture Notes in Physics. Springer-Verlag, New York, 1975.

[Mc,1974]   McGehee, R. Triple collision in the collinear three-body problem, *Invent. Math.* **27** (1974), 191–227.

[Mc,1981]   McGehee, R. Double collisions for a classical particle system with nongravitational interaction, *Comment. Math. Helvetici* **56** (1981), 524–557.

[Mc,1986]   McGehee, R. Von Zeipel's theorem on singularities in celestial mechanics, *Expo. Math.* **4** (1986), 335–345.

[P,1897]   Painlevé, P. *Leçons sur la théorie analytique des équations différentielles*, Hermann, Paris, 1897.

[Po,1993]   Poincaré, H. *New Methods of Celestial Mechanics* (edited and introduced by D. L. Goroff), Amer. Inst. Phys., Melville, NY, 1993.

[S,1973]   Saari, D. Singularities and collisions of Newtonian gravitational systems, *Arch. Rational Mech. Anal.* **49** (1973), 311–320.

[S,1975]   Saari, D. Collisions are of first category, *Proc. Amer. Math. Soc.* **47**, 2 (1975), 442–445.

[SD,1994]   Saari, D., and Diacu, F. Superhyperbolic expansion, noncollision singularities and symmetry configurations, *Celestial Mechanics* **60** (1994), 91–98.

[SX,1996]   Saari, D., and Xia, Z. Singularities in the Newtonian n-body problem, *Contemp. Math.* **198** (1996), 21–30.

[Si,1941]   Siegel, C. L. Der Dreierstoss, *Annals of Math.* **42**, 1 (1941), 127–168.

[SM,1971]   Siegel, C. L. and Moser, J. K. *Lectures on Celestial Mechanics*, Springer-Verlag, Berlin, Heidelberg, New York, 1971.

[S,1970]   Sperling, H.-J. On the real singularities of the N-body problem, *J. Reine Angew. Math.* **245** (1970), 15–40.

[Su,1912]   Sundman, K. Mémoire sur le problème des trois corps, *Acta Math.* **36** (1912), 105–179.

[vZ,1908]   von Zeipel, H. Sur les singularités du problème des n corps, *Ark. Mat. Astr. Fys.* **4**, 32 (1908), 1–4.

[W,1988]   Whittaker, E. T. *A Treatise on the Analytical Dynamics of Particles and Rigid Bodies*, Cambridge University Press, Cambridge, UK, 1988.

[Wi,1941]   Wintner, A. *The Analytical Foundations of Celestial Mechanics*, Princeton University Press, Princeton, NJ, 1941.

[X,1992]   Xia, Z. The existence of noncollision singularities in the N-body problem, *Annals Math.* **135** (1992), 411–468.

Lectures on the Two-Body Problem

Alain Albouy

ASD/IMC-CNRS-UMR 8028
77, AVENUE DENFERT-ROCHEREAU
F-75014 PARIS, FRANCE
e-mail: albouy@bdl.fr
TRANSLATED BY H. S. DUMAS

Introduction

It is natural to ask why we should still be interested in the classical two-body problem. Yet the founding discoveries of Kepler and Newton remain strikingly beautiful, and indeed are so familiar to us that we sometimes forget that they contain surprising features.

Certain later results, for example those associated with the names of Lambert, Hamilton, Tait, or Moser, are also surprising. The set of lectures in this chapter aims to present some of this remarkable work in a form accessible to a broad readership. Some of the results are quite old, and many are not mentioned in standard texts. We have attempted throughout to simplify, synthesize, or supplement this material with personal remarks.

Lecture 1 introduces the problem, along with the unifocal equation for conic sections, which is studied in greater depth in Lecture 4.

Lecture 2 presents the eccentricity vector and Hamilton's hodograph and explains the process of reduction-integration. In Propositions 9 and 11 we try to synthesize the traditional recipes for the Kepler problem into two "images."

Lecture 3 presents Bertrand's theorem and, through the remarks of Jacobi, Appell, Koslov, and Harin, attempts to "explain" why orbits are closed.

Lecture 4 gathers together a number of remarks on the geometry of the problem, thereby clarifying the constructions of Moser, Dziobek, Laplace, and Lagrange.

Lecture 5 establishes Lambert's theorem, discusses certain interesting parametrizations of the family of Keplerian orbits passing through two points in the plane, and presents a result of C. Simó.

For the reader wishing to know more, we recommend the texts of Whittaker and Wintner. For more recent developments, the reader may consult Souriau's interesting synthesis. Finally, sources cited in the text may be found in the bibliography, where they are arranged alphabetically by author (and where their appearance in the text is indicated inside square brackets).

Note. The term Kepler problem, which we use to designate the "$1/r^2$

central force problem," is used more classically in the sense of "the problem of solving Kepler's equation."

Acknowledgments. I wish to acknowledge the students and colleagues who attended and criticized the oral version of these lectures in Recife, and particularly H. Cabral. I am indebted to A. Chenciner for innumerable discussions, to H. S. Dumas for the elegant translation of my French draft, and to R. Cushman for some historical information and his careful reading of this text. Thanks to F. Diacu for the conception of this book and to C. Castilho, A. Santos, J. Laskar, C. Simó, D. Sauzin, F. Joutel, P. Robutel, K. Abdullah, D. Greear, M. Chapront-Touzé, J. Soudée, R. Moeckel, M. Serrero, N. Guicciardini, N. Capitaine, E. Tabacman, M. A. Jalali, and R. McKay for various contributions.

Lecture 1. Preliminaries

The Kepler Problem

The problem is to find all solutions, and to find the unique solution for a given initial condition, of the system of differential equations

$$\ddot{x} = -xr^{-3}, \qquad \ddot{y} = -yr^{-3}. \tag{1}$$

The real numbers x and y are the coordinates of the body attracted by the *fixed center* at the origin. The radius r is positive and satisfies $r^2 = x^2 + y^2$. Finally, $\dot{x} = dx/dt$ denotes the first derivative of the variable x with respect to the time variable t, and $\ddot{x}$ denotes the second derivative.

We begin with several questions that will serve to organize the discussions at the end of each lecture.

1. *Question.* Why two coordinates? One often prefers three coordinates (x, y, z) for obvious physical reasons. However, as the dimension ν is increased, whereas we clearly see the Kepler problem become richer in passing from $\nu = 1$ to $\nu = 2$, the passage to arbitrary dimension $\nu > 2$ is an exercise in geometry that should be treated separately. If some interesting feature distinguishes $\nu = 3$ from larger values of ν, we would like to bring it to light, and it is precisely for this reason that we do not set $\nu = 3$ at the outset. When we move beyond dimension 2, it will be to an unspecified

dimension ν.

2. *Question.* Why use coordinates at all? With its fixed center, our space has precisely the structure of a Euclidean vector space of dimension 2. We shall always keep in mind the more satisfying form of (1):

$$\ddot{\vec{r}} = -\|\vec{r}\|^{-3}\vec{r}. \tag{2}$$

This form is independent of the dimension ν of the space.

3. *Question.* Why treat t as a separate coordinate? We have, among others, the following option. First, write $p = \dot{x}$ and $q = \dot{y}$. The system (1) becomes

$$\dot{x} = p, \quad \dot{y} = q, \quad \dot{p} = -xr^{-3}, \quad \dot{q} = -yr^{-3}. \tag{3}$$

We then change our point of view and pose the Kepler problem with a "Pfaff system," defining a "direction field" in (x, y, p, q, t)-space:

$$dx = p\,dt, \quad dy = q\,dt, \quad dp = -xr^{-3}dt, \quad dq = -yr^{-3}dt. \tag{4}$$

4. *Question.* Why write a system of differential equations when we can define the problem using a Lagrangian or a Hamiltonian? The notions of Lagrangian or Hamiltonian mechanics will be used only when they provide some computational or conceptual simplification. One of the messages we wish to transmit is that these formalisms do not simplify the reduction process in any way.

The Trajectories: Conic Sections

All the methods of solution presented here lead us to the same characterization of conic sections embodied in the remarkable equation of Lemma 5, or the polar equation of Lemma 6.

5. *Lemma.* Let α, β, and γ be real numbers with $\gamma > 0$. Consider the plane $\mathbb{R}^2$ with coordinates (x, y) and $r = \sqrt{x^2 + y^2}$. The curve described by the equation $r = \alpha x + \beta y + \gamma$ is a branch of the conic section with eccentricity $k = \sqrt{\alpha^2 + \beta^2}$, with focus at the origin, and with directrix the straight line given by $\alpha x + \beta y + \gamma = 0$. This curve is more precisely the branch of the conic section located on the same side of the directrix as the focus, in other words it is an ellipse, a parabola, or half of a hyperbola.

Proof. At any point on the curve, the ratio of the distance from the focus to the distance from the directrix is $\sqrt{\alpha^2 + \beta^2}$.

The well-known equation for conic sections in polar coordinates centered at the focus is obtained by dividing the equation $r = \alpha x + \beta y + \gamma$ by r. We introduce the polar angle θ such that $x = r\cos\theta$ and $y = r\sin\theta$, and the angle θ_0 such that $\alpha = -k\cos\theta_0$ and $\beta = -k\sin\theta_0$.

6. *Lemma.* The plane curve described by the polar equation $\gamma/r = 1 + k\cos(\theta-\theta_0)$ is a branch of a conic section with semiparameter γ, eccentricity k, and angle of the pericenter θ_0.

7. *Exercise.* Find the maximal number of points of intersection of two curves with respective equations $r = \alpha x + \beta y + \gamma$ and $r = \alpha' x + \beta' y + \gamma'$. Find the maximal number of points of intersection of two conic sections having the same focus. Answers: two, four.

8. *Definition.* In order to organize discussions in the remainder of the lecture, we shall use the term *trigonometric parametrizations* of a conic section with center to designate the elementary parametrizations by trigonometric functions. Thus an ellipse, which in an orthonormal coordinate frame may always be written in the form $x^2/a^2 + y^2/b^2 - 1$, admits the trigonometric parametrization $x = a\cos\theta$, $y = b\sin\theta$. The other case is the hyperbola, the equation of which takes the form $x^2/a^2 - y^2/b^2 - 1$, and which admits the trigonometric parametrization $x = a\cosh\theta$, $y = b\sinh\theta$. Later we will define the parametrization of an ellipse in Lemma 5 by the "eccentric anomaly" as a trigonometric parametrization.

Brief Review of Steps Leading to the Kepler Problem in the Plane

Our point of departure is the *two-body problem.* We may denote the vector joining the two bodies, or a vector joining the center of mass to one of the bodies, by $\vec{r}$. For a suitable choice of units, this vector satisfies the equation

$$\ddot{\vec{r}} = -\|\vec{r}\|^{-3}\vec{r}. \tag{2}$$

Let E be the space of motions and F the plane generated by $\vec{r}_0$ and $\vec{v}_0$ at the initial time $t = 0$. The first thing to show is that the motion is planar, and more precisely that $\vec{r} \in F$ for all t. We shall work in the *state space* (also called the *phase space*) of pairs of vectors $(\vec{r}, \vec{v})$. The assertion then becomes: Any solution of the differential equation $(\dot{\vec{r}}, \dot{\vec{v}}) = (\vec{v}, -r^{-3}\vec{r})$ with initial condition in $F \times F \subset E \times E$ continues in the vector subspace $F \times F$. To prove the assertion, it is enough to note that $(\vec{v}, -r^{-3}\vec{r}) \in F \times F$ if $(\vec{r}, \vec{v}) \in F \times F$ and to use the following result from the theory of ordinary differential equations, which we state in geometrical language: *A submanifold is invariant under the flow of a vector field if and only if the vector field is everywhere tangent to the submanifold.* The proof of this general result uses the existence-uniqueness theorem for solutions of a differential equation.

Remark. A number of authors use the invariance of angular momentum to prove the result above. Their arguments are correct and avoid use of the existence and uniqueness theorem, but such arguments do not extend very far. For example, they do not serve to prove that if initially $\vec{r} \wedge \vec{v} = 0$, then the motion will continue along the straight line containing $\vec{r}$ and $\vec{v}$. Arguments relying on invariance of angular momentum show simply that $C = \vec{r} \wedge \vec{v}$ remains zero, in other words, that $\vec{r}$ and $\vec{v}$ remain dependent.

Lecture 2. Two Solutions by Reduction

The Kepler problem was solved long ago. Among "modern" notions that may be brought to bear on its solution, the most useful is perhaps the quotient space. In essence, a reduction is simply a series of restrictions and quotients. With this in mind, we may view what others have seen as a multiplicity of procedures as variations of a single procedure.

The Solution of the Kepler Problem by Hermann and Lagrange

We consider the system

$$\ddot{x} = -xr^{-3}, \qquad \ddot{y} = -yr^{-3}. \tag{1}$$

Elementary calculations establish the following three properties.

(i) The quantity $C = x\dot{y} - y\dot{x}$, called the angular momentum or areal constant, is a first integral of motion. In other words, $\dot{C} = 0$ or, in still other terms, C is a conserved quantity.

(ii) The quantity $\alpha = xr^{-1} - C\dot{y}$ is a first integral of motion.

(iii) The quantity $\beta = yr^{-1} + C\dot{x}$ is a first integral of motion.

We have three functionally independent first integrals. In the space of states $(x, y, \dot{x}, \dot{y})$ of the system, which is also the space of possible initial conditions, a given triple (C, α, β) completely determines an *integral curve*. To describe this curve, we may project it onto the configuration space (x, y) and onto the velocity space $(\dot{x}, \dot{y})$. The first projection is effected by eliminating the velocities. We multiply (ii) by x, (iii) by y, and add. We find

$$xa + y\beta = r - C^2, \tag{5}$$

the equation of a conic section from Lemma 5. The second projection, called the hodograph, is obtained for $C \neq 0$ by writing $\dot{x} = (\beta - yr^{-1})/C$, $\dot{y} = (-\alpha + xr^{-1})/C$. The hodograph is thus contained in a circle with center $(\beta/C, -\alpha/C)$ and radius $1/C$, as was noticed by Hamilton.

Time. A complete description of the trajectories must give the state of the system as a function of time. What we find explicitly is the elapsed time $t - t_0$ since passage through the pericenter (the point on the orbit closest to the fixed center) as a function of position on the orbit. For this we use the "law of equal areas," the geometric interpretation of the first integral (i). This relation may be written $C\,dt = 2dA$ or, upon integration, $C(t - t_0) = 2A$, where A is the area "swept out" by the radius vector $\vec{r}$ starting at the pericenter.

In the case of "conic sections with center" – circles, ellipses, or hyperbolas – this swept-out area may easily be calculated as a function of the trigonometric parameter (Definition 8). In fact, the area swept out from the center of the conic section is proportional to this parameter. The area A swept out from a focus of the conic section is then obtained by adding the oriented area of a triangle. The formula thus obtained for the elapsed time $t - t_0$, called Kepler's equation, will be obtained below by other means.

A parabola has neither a center nor a trigonometric parametrization, thus precluding use of the same method. But if we consider the parabolic orbit corresponding to the values $\alpha = 1$, $\beta = 0$, we easily find $12C^2A = -y(y^2 + 3C^4)$.

Remarks on the Reduction Process

System (1) is autonomous (time does not appear explicitly in the right-hand side) and of order 4 (where "order" means the dimension of the state space). In the preceding section we reduced this system using three functionally independent first integrals. We obtained integral curves as intersections of the level surfaces of these first integrals. Yet the system was still not altogether solved: it remained to find the time parametrization of each integral curve, in other words, to *solve* (1) *restricted to an integral curve*.

"To reduce" means to reduce the order of the system of differential equations; in our case we transform the autonomous system of order 4 to an autonomous system of order 1. Such a system is locally nothing other than a differential equation of the form $\dot{w} = f(w)$, where f is a real function. It is to this form that one arrives by *eliminating* three of the four variables of (1) with the help of the first integrals. The general solution procedure is to write this equation as $dt = dw/f(w)$. In this form, the system is "reduced to quadrature," since it suffices to find a primitive of $1/f$ to make the time t an explicit function of w.

The procedure using the "swept-out area" described above seems to differ from the general procedure. Yet it also expresses the time as a function of location on the orbit, rather than the reverse, and also requires a quadrature (i.e., an integration).

The reduction process of the previous section uses only "restrictions" that are expressed analytically by elimination of variables. The process described in the following section is more general: it makes use of restrictions and of a passage to the quotient.

The Radial Potential Problem and Newton's Method

Here we treat a problem more general than Kepler's problem, defined instead by the system

$$\ddot{x} = -\varphi(I)x, \qquad \ddot{y} = -\varphi(I)y, \tag{6}$$

where $I = x^2 + y^2$ and φ is an arbitrary real function. The Kepler problem corresponds to the particular case $\varphi(I) = I^{-3/2}$.

We shall reduce this problem to quadrature. The specific features of

Kepler's problem will appear only in the process of quadrature.

In order to justify our use of the term "radial potential problem," we note that the "force field" that defines the system (6) may be characterized by any two of the following three properties, which in turn imply the third property:

(i) The force field is invariant under isometries of the plane that fix the origin.

(ii) The force is central.

(iii) The force field is derived from a potential.

Property (ii) signifies that the force is directed toward (or away from) the origin. Property (iii) signifies that there exists a function whose field of gradient vectors is the force field. When the force field is that of system (6), this function may be written as $-\frac{1}{2}\Phi(I)$, where Φ denotes a primitive of the function φ. In the Kepler problem, we then have $\Phi(I) = -2I^{-1/2}$.

In order to solve (6), we note the following three properties.

(i) The isometries of the plane fixing $(0,0)$ form a symmetry group of (6). That is, the transformation

$$\begin{pmatrix} x & \dot{x} \\ y & \dot{y} \end{pmatrix} \longmapsto \begin{pmatrix} a & b \\ a' & b' \end{pmatrix} \begin{pmatrix} x & \dot{x} \\ y & \dot{y} \end{pmatrix}$$

sends any solution of (6) onto a solution of (6) provided $a^2+b^2 = a'^2+b'^2 = 1$ and $aa' + bb' = 0$.

(ii) The quantity $C = x\dot{y} - y\dot{x}$ is a first integral of system (6).

(iii) The quantity $H = \frac{1}{2}(\dot{x}^2 + \dot{y}^2 + \Phi(I))$, the energy, is a first integral of system (6).

We also note that the first integrals H and C^2 are invariant under the isometries (i) and are functionally independent. We can already foresee that the reduction process we are about to use will lead us to an autonomous system of order 1, as was the case for system (1) using the method of Hermann and Lagrange.

The reduction takes place in three stages, one for each of the ingredients (i), (ii), and (iii). We may choose the order of the stages. We begin by using

(i) to reduce the order by one. We choose three independent functions of $(x, y, \dot{x}, \dot{y}) \in \mathbb{R}^4$ invariant under the action of the group. The most natural choices are $I = x^2 + y^2$, $J = x\dot{x} + y\dot{y}$, and $K = \dot{x}^2 + \dot{y}^2$. Upon calculating the derivative and replacing $\ddot{x}$ and $\ddot{y}$ by the right-hand sides of (6), we obtain:

$$\dot{I} = 2J, \quad \dot{J} = K - I\varphi(I), \quad \dot{K} = -2J\varphi(I). \tag{7}$$

This is the system of order 3 that we sought. Some remarks will serve to clarify its status.

The classical treatises on Lagrangian mechanics describe the preceding reduction in terms of *ignorable variables*. Consider the angle θ such that $x = \sqrt{I}\cos\theta$ and $y = \sqrt{I}\sin\theta$. On suitable open sets, the transformation $(x, y, \dot{x}, \dot{y}) \mapsto (I, J, K, \theta)$ is a smooth change of variables. System (6) is transformed into the three equations (7) and the equation

$$\dot{\theta} = \frac{C}{I}. \tag{8}$$

The angle θ is "ignorable" because it does not appear in the right-hand sides above; we may study system (7) while ignoring θ. Once an explicit solution of (7) is found, we may deduce a family of solutions of (6) from it by integrating $d\theta = C\,dt/I$, thereby introducing a constant of integration.

To this classical presentation, we prefer an explanation in terms of the *quotient space*, which has the advantage of not introducing an angle θ that is subsequently ignored. In $(x, y, \dot{x}, \dot{y})$-space, we represent the *orbits of the action of the group of isometries* (i), not to be confused with the orbits of the system that we ultimately seek to describe. Except for degeneracies, these are topologically pairs of circles in the space $\mathbb{R}^4$, because the group of plane isometries is topologically a pair of circles. We may check that C is positive on one circle and negative on the other and that the degeneracies are characterized precisely by the equation $C = 0$. To give meaning to system (7), it is enough to note that a triple (I, J, K) characterizes one of these orbits. The system thus describes motion in a new, three-dimensional space whose "points" are the pairs of circles above. This is the quotient space.

We continue to reduce the order. The quantities H and C^2 are first integrals of system (7). We have

$$2H = K + \Phi(I) \qquad \text{and} \qquad C^2 = IK - J^2. \tag{9}$$

From the geometric standpoint, this remark achieves the reduction to order 1: these two equations define curves in (I, J, K)-space that are invariant under system (7) and on which this system induces an autonomous differential equation of order 1.

In preparation for quadrature, one ordinarily writes this differential equation explicitly by eliminating J and K. By eliminating K from (9), we obtain

$$C^2 + J^2 = -I\Phi(I) + 2HI, \tag{10}$$

which may be solved for J. Then system (7) becomes the autonomous differential equation

$$\dot{I} = \pm 2\sqrt{-I\Phi(I) + 2HI - C^2}. \tag{11}$$

This is the form of the equation we expected. However, we shall put (11) aside to avoid arguing about the appropriate sign in front of the square root. We prefer to work with (10), since it more faithfully represents the geometry and topology of curves in (I, J, K)-space: a closed curve does not lend itself to parametrization by one of the three coordinates. Having made this remark, we are ready to present a good algorithm for integrating system (6).

(i) Choose a quadruple $(x_0, y_0, \dot{x}_0, \dot{y}_0)$ of initial conditions and an initial time t_0. Compute H and C. Calculate the initial values I_0, J_0, and θ_0 of the variables I, J, and the angle θ.

(ii) Consider the curve (10) in the (I, J)-plane. It contains the initial point (I_0, J_0). Now *move* along this curve; in other words, choose a parameter w and a path (I_w, J_w) that describes the curve. We take $w = 0$ at the initial point.

(iii) Integrate along this path the following two differential forms (which are smooth on the curves (10) thanks to the symmetry $J \mapsto -J$ of the curves):

$$dt = \frac{dr}{\dot{r}} = \frac{r\,dr}{r\dot{r}} = \frac{dI}{2J} \qquad \text{and} \qquad d\theta = \dot{\theta}\,dt = \frac{C\,dt}{r^2} = \frac{C\,dI}{2IJ}. \tag{12}$$

In this way we obtain the functions t_w and θ_w of the parameter w.

(iv) Now return to the original variables. The quintuple $(x_w, y_w, \dot{x}_w, \dot{y}_w, t_w)$ is a function of the parameter w. We may consider w as a function of t since dt does not vanish.

Bernoulli's Method for the Particular Case of $1/r^2$-Attraction

When $\varphi(I) = I^{-3/2}$, equation (10) becomes

$$C^2 + J^2 = 2\sqrt{I} + 2HI. \tag{13}$$

We arrive at a rational equation by setting $r = \sqrt{I}$ or $\rho = 1/\sqrt{I}$.

(i) *The variable r and the eccentric anomaly u.* We call the curve described by $J^2 = 2Hr^2 + 2r - C^2$ the *first auxiliary conic*. We set

$$k^2 = 1 + 2HC^2. \tag{14}$$

The condition $k^2 \geq 0$ is equivalent to the existence of a real point on this conic section or to the existence of a real point in the intersection of the conic section and the half-plane $r > 0$. We distinguish three possibilities: $H < 0$, $H = 0$, or $H > 0$. The particular case $C = 0$ is notable only for the following reason: the conic section contains a point such that $r = 0$. We restrict ourselves to the case $H < 0$, where the auxiliary conic is an ellipse entirely contained in the half-plane $r \geq 0$. We set

$$a = -\frac{1}{2H} \tag{15}$$

and we introduce the trigonometric parametrization (Definition 8) of the auxiliary ellipse by the angle u, called the *eccentric anomaly*:

$$r = a(1 - k \cos u), \qquad J = k\sqrt{a} \sin u. \tag{16}$$

We obtain

$$dt = r\sqrt{a}\, du = a^{3/2}(1 - k \cos u)du \quad \text{and} \quad d\theta = \frac{C\, du}{\sqrt{a}(1 - k \cos u)}. \tag{17}$$

The first equation immediately gives *Kepler's equation*

$$a^{-3/2}(t - t_0) = u - k \sin u. \tag{18}$$

To integrate the second equation, it is enough to use a rational parametrization of the ellipse, in other words, to take as the new variable the tangent of $u/2$, which leads us to introduce a new angle v, depending on u through the formula

$$\tan \frac{u}{2} = \sqrt{\frac{1-k}{1+k}} \tan \frac{v}{2}.$$

We shall be led to this angle by our preference of the variable $\rho = 1/r$ over the variable r at the outset. The hypothesis $H < 0$ will not be necessary.

(ii) *The variable ρ and the true anomaly v.* We multiply both sides of (13) by ρ^2 and we note that it is convenient to use the variable $\dot{r}$ as the ordinate rather than $J = r\dot{r}$. There remains $\dot{r}^2 = 2H + 2\rho - C^2\rho^2$. This *second auxiliary conic* possesses a real point if and only if $k^2 = 1 + 2HC^2 \geq 0$. It is an ellipse provided C is nonzero, with trigonometric parametrization:

$$\rho = \frac{1 + k\cos v}{C^2}, \qquad \dot{r} = \frac{k\sin v}{C}. \tag{19}$$

We obtain

$$dt = \frac{C^3}{(1+k\cos v)^2} dv \qquad \text{and} \qquad d\theta = dv.$$

The first quadrature is difficult this time, but the second is miraculously simple:

$$\theta - \theta_0 = v. \tag{20}$$

(iii) *Synthesis.* Equation (20) makes v into a polar angle. Following Lemma 6, the first equation (19) is thus the equation of a conic section with eccentricity k. The angle θ_0 indicates the direction of the pericenter: r is minimal for $\theta = \theta_0$.

Classically, one introduces three *anomalies*, defined as angles starting from the pericenter, and distinguished from the *longitudes*, which are angles starting from a fixed direction. Thus v is the *true anomaly* and θ the *true longitude*. The angle $l = a^{-3/2}(t - t_0)$, appearing on the left-hand side of Kepler's equation (18), is called the *mean anomaly*. We have finally introduced the *eccentric anomaly u*, the trigonometric parametrization of the first auxiliary ellipse. It remains to see that it is also a trigonometric parametrization of the trajectory ellipse. This is easily deduced from the following proposition.

9. *Proposition.* Suppose the eccentricity k is nonzero. Let $\xi = r \cos v$ and $\eta = r \sin v$ be the coordinates of the body in a frame having the fixed center as its origin and the direction of the pericenter as the ξ-axis. The affine transformation $(r, J) \mapsto (\xi, \eta) = k^{-1}(C^2 - r, CJ)$ takes the first auxiliary conic onto the trajectory while preserving the time parametrization.

Proof. We put the first equation (19) in the form of Lemma 5, as $r = -k\xi + C^2$, and we multiply the second by r to obtain $J = k\eta/C$.

To give the trigonometric parametrization explicitly in the case of an ellipse of semimajor axis a, we use the system (16), taking into account the relation $C^2 = a(1 - k^2)$, which we deduce from (14) and (15). We obtain

$$\xi = a(\cos u - k), \qquad \eta = a\sqrt{1 - k^2} \sin u. \tag{21}$$

Choosing the pericenter as the origin, rather than the apocenter, is justified – for example, in Gauss's *Theoria Motus* – by the absence of an apocenter in the case of hyperbolic motion. The choice of the pericenter is thus required for the true anomaly and is extended to the other anomalies. The eccentric anomaly is specific to the ellipse and does not continue through $H = 0$. However, it has an exact analog for $H > 0$. In the formulas, it suffices to replace the sin and cos by sinh and cosh, the $\sqrt{a}$ by $\sqrt{-a}$, and finally the $a^{3/2}$ by $a\sqrt{-a}$.

Comparison of the Two Solution Methods for the Kepler Problem

The Hermann–Lagrange method is much quicker. Nevertheless, it requires knowledge of an expression for the first integrals as a function of the coordinates $(x, y, \dot{x}, \dot{y})$; the simplicity of this expression remains surprising, even after the deep study that constitutes the second method.

The approach of Newton and Bernoulli has the advantage of being very general. In his *Mécanique analytique*, Lagrange treated the problems of rigid bodies, of the spherical pendulum, and of two fixed centers, by means of the same algorithm of reduction-integration.

The polar angle and the true anomaly are not the variables that appear naturally in the first method. The second method has the advantage of bringing them to light and providing formulas for them. We note that from points (i) and (ii), which allowed the synthesis (iii), we may surmise a useful principle: when an integration is difficult using the eccentric anomaly, it may be easy using the true anomaly and viceversa. The following exercise illustrates this rule.

10. *Exercise.* Consider a planet in an elliptic orbit around the sun. Take the time average of the heat received by the planet. How does this quantity depend on the semimajor axis a and the eccentricity k? Also calculate the mean distance between the two bodies over one revolution. What is to be made of the discussion of climate at the end of Appendix 8 in V. I. Arnold's *Mathematical Methods of Classical Mechanics*?

The formulas for the eccentric anomaly may be extracted from the first method even more naturally than from the second. This requires defining the eccentric anomaly as the trigonometric parametrization of the trajectory ellipse, as Kepler did. This amounts to writing system (21). We could then continue with $k\xi = -\alpha x - \beta y = C^2 - r$, $k\eta = -\alpha y + \beta x = CJ$, equivalent to Proposition 9, arriving finally at (16) and (18). The following result, "dual" to Proposition 9, is obtained in this way:

11. *Proposition.* Suppose the eccentricity k is nonzero. Let $\dot{\xi} = -(\alpha \dot{x} + \beta \dot{y})/k$ and $\dot{\eta} = (-\alpha \dot{y} + \beta \dot{x})/k$ be the coordinates of the velocity of the body in a frame having the fixed center as its origin and the direction of the pericenter as the ξ-axis. If k is nonzero, the affine transformation $(\rho, \dot{r}) \mapsto (\dot{\xi}, \dot{\eta}) = k^{-1}(-\dot{r}, 2HC + C\rho)$ sends the second auxiliary conic onto the hodograph while preserving the time parametrization.

Recall that the hodograph is a truncated circle in the case $H > 0$. The second auxiliary conic, which may be transformed into a circle by taking $C\rho$ as the ordinate instead of ρ, is thus truncated in the same way by the condition $\rho > 0$.

Concerning Question 3. In (12), we wrote expressions for dt and $d\theta$ as Newton did in his Proposition 41. We cannot avoid being struck by seeing the same role played by two variables introduced in such asymmetric ways. We now show how to correct the defect in our presentation.

Question 3 presents a space of five dimensions foliated into solution curves of a Pfaff system. Nothing obliges us to consider these curves as parametrized by the time t. We may think of two symmetry groups acting on this space: the translations $t \mapsto t + \Delta t$ and the rotations $\theta \mapsto \theta + \Delta\theta$. The reduction thus includes two passages to the quotient, one ignoring t and the other ignoring θ. The symmetry is thus perfectly reestablished.

Concerning Question 4. System (1) is naturally associated to the *Hamiltonian H* and to the *symplectic form $dp\wedge dx + dq\wedge dy$*. System (4) is the *characteristic system* of the differential form $pdx + qdy - Hdt = \dot{r}dr + Cd\theta - Hdt$. These important notions are explained in Arnold's book, which also treats the Lagrangian formalism. The role of these structures in the reduction procedure is to associate a first integral to the action of "good" symmetry groups. Thus the rotation group is associated to C, and time translation is associated to H. Why did we not need these structures to carry out the reduction? Simply because, in the radial potential problem, we already knew the first integrals H and C. We close this lecture by showing an example of reduction by a "bad" symmetry group, which serves just as well for the reduction procedure.

12. *Exercise.* Reduce the radial potential problem in the homogeneous case $\Phi(I) = \kappa^{-1}I^\kappa$ using three ingredients: the first integral $|H|^{-(\kappa+1)/2\kappa}C$, the isometries, and the one-parameter group of transformations $(x, y, p, q) \mapsto (\lambda x, \lambda y, \lambda^\kappa p, \lambda^\kappa q)$ (with parameter $\lambda > 0$).

Lecture 3. Why Are Keplerian Orbits Closed?

The Kepler problem is singular among other problems with radial potential. All of its bounded orbits are periodic. By contrast, the general bounded orbits of the other problems are only quasi-periodic. This is made precise by Bertrand's theorem.

The proof of this theorem provides very little "explanation" of the phenomenon. The statement itself may be criticized, in the sense that it imposes a priori radial potential problems as the generalization to consider.

We shall seek to explain the closure of the orbits by trying, in spite of Bertrand's theorem, to generalize the framework of the problem. In the literature, we find two results, which we combine here: one due to Jacobi, the other due to Appell and supplemented by Koslov and Harin.

Clairaut's Solution of the Kepler Problem

Since time does not enter into the question of the closure of the orbits, we shall eliminate it at the outset. It turns out that there is a solution method for Kepler's problem in which one begins by eliminating time and which is highly effective when one wishes to solve generalizations of the problem. In this lecture we shall use this method and no other. This technical choice will cause us to stray occasionally from classical methods of proof, but will not complicate things in any way.

We begin by deriving a remarkable form of the reduced system that may also be deduced from the formulas established in the preceding lecture. We identify the plane of motion with $\mathbb{C}$, and we set $\vec{r} = re^{i\theta}$. The first derivative of $\vec{r}$ is $(\dot{r}+ir\dot{\theta})e^{i\theta}$; the second derivative is $(\ddot{r}-r\dot{\theta}^2+2i\dot{r}\dot{\theta}+ir\ddot{\theta})e^{i\theta}$. As usual, we denote the angular momentum by $C = r^2\dot{\theta}$. Once again we point out that the acceleration is central if and only if $\dot{C} = r(2\dot{r}\dot{\theta}+r\ddot{\theta}) = 0$.

We consider once again equations (6) for the radial potential problem. By identifying the acceleration with the force, we find

$$\ddot{r} - \frac{C^2}{r^3} = -r\varphi(r^2). \tag{22}$$

We then use $\rho = r^{-1}$ in place of the radius r, and the polar angle θ in place of the time t. From $C = r^2\dot{\theta}$ we arrive at

$$\frac{d\rho}{d\theta} = -\frac{\dot{r}}{C} \quad \text{and} \quad \frac{d^2\rho}{d\theta^2} = -\frac{r^2\ddot{r}}{C^2} = -\rho + \frac{\varphi(\rho^{-2})}{C^2\rho^3}, \tag{23}$$

which, upon setting $\varphi(I) = I^{-3/2}$, yields

$$\frac{d^2\rho}{d\theta^2} + \rho = \frac{1}{C^2}.$$

The general solution of this linear equation is $C^2\rho - 1 = k\cos(\theta - \theta_0)$, which is the first of equations (19). We have thus proved again that the solution trajectories of the Kepler problem are the conic sections of Lemma 6.

A Theorem of Bertrand

13. *Theorem.* Consider a radial potential problem defined by the equation $\ddot{\vec{r}} = -\varphi(I)\vec{r}$, where $I = \|\vec{r}\|^2 = r^2$ and where $\varphi : (0,+\infty) \longmapsto \mathbb{R}$ is an *analytic* function. Assume that all the *bounded* nonrectilinear orbits are

closed orbits. Then either the central force is everywhere repulsive, that is, $\varphi(I) \leq 0$ for all values of I and there exist no bounded orbits, or there is a $\mathcal{G} > 0$ such that $\varphi(I) = \mathcal{G}I^{-3/2}$ or $\varphi(I) = \mathcal{G}$.

Proof. The differential equation (23) will be our tool. It contains a parameter C. By setting $d^2\rho/d\theta^2$ equal to zero we see that the radii r_0 of circular orbits with angular momentum C are such that the $\rho_0 = 1/r_0$ are solutions of the equation $C^2 = \rho_0^{-4}\varphi(\rho_0^{-2})$. For the moment this equation cannot be solved explicitly. It is convenient to study orbits having the same angular momentum as the circular orbit of radius r_0 and to use $\rho_0 = 1/r_0$ as a new local parameter of the differential equation (23). This amounts to using the explicit substitution $C^2 = \rho_0^{-4}\varphi(\rho_0^{-2})$.

We set $\psi(\rho) = \rho^{-3}\varphi(\rho^{-2})$, and we have $C^2 = \psi(\rho_0)/\rho_0$. Finally, we set $\Psi(\rho_0, \rho) = \rho - \rho_0\psi(\rho)/\psi(\rho_0)$. Equation (23) then becomes the system

$$\frac{d\rho}{d\theta} = \rho', \qquad \frac{d\rho'}{d\theta} = -\rho + C^{-2}\psi(\rho) = -\Psi(\rho_0, \rho). \tag{24}$$

The form of system (24) gives us the first integral

$$2\mathcal{E} = \rho'^2 + \rho^2 + C^{-2}\Phi(\rho^{-2}). \tag{25}$$

We easily verify that $\mathcal{E} = C^{-2}H$. We shall call $\mathcal{E}$ the *angular energy* to distinguish it from the energy H. Whenever a smooth level curve of $\mathcal{E}$, traced out in the (ρ, ρ')-half-plane defined by $\rho > 0$, is closed, an important quantity appears: the period of the solution $\big(\rho(\theta), \rho'(\theta)\big)$ of the system (24) which describes it. We call this the *angular period* and denote it by Θ. This period may be interpreted as follows: it is the angle separating two successive passages of the body through pericenters. Finally, we call the ratio $2\pi/\Theta$ the *orbital ratio*, and we denote its square by $\sigma = (2\pi/\Theta)^2$. The following lemma is self-evident, but fundamental.

14. *Lemma.* A solution of the system $\ddot{\vec{r}} = -\varphi(I)\vec{r}$ associated to a periodic solution $\big(\rho(\theta), \rho'(\theta)\big)$ of system (24) is periodic if and only if the orbital ratio is rational.

There is a circular orbit of radius $r_0 = \sqrt{I_0} > 0$ if and only if $\varphi(I_0) > 0$. The first step in the proof of Bertrand's theorem is a first-order study of orbits such that r remains perpetually close to r_0. But the existence of such orbits must first be guaranteed. The following lemma shows that when they do not exist, the hypothesis of Bertrand's theorem is not satisfied.

15. *Lemma.* Consider φ as in Theorem 13 and $I_0 > 0$ such that $\varphi(I_0) > 0$, and let C be the angular momentum of the direct circular orbit of radius $r_0 = \sqrt{I_0} = 1/\rho_0$. The function $\rho \mapsto \Psi(\rho_0, \rho)$ vanishes at ρ_0. If its derivative is strictly negative at ρ_0, then there exists a bounded orbit asymptotic to the circular orbit, which is hence nonclosed.

Proof. By hypothesis, the function $\rho \mapsto \rho^2 + C^{-2}\Phi(\rho^{-2})$, which is a primitive of the function in the lemma, takes its maximal value $2\mathcal{E}_0$ at ρ_0. Thus $(\rho, \rho') = (\rho_0, 0)$ is a critical point of saddle type for the function $\mathcal{E}$. The level curve $\mathcal{E} = \mathcal{E}_0$ has a branch in the sector $\rho > \rho_0$, $\rho' > 0$, which continues in this sector, unless it happens to cross the axis $\rho' = 0$, in which case it continues symmetrically. In any case we have $\rho > \rho_0$ and thus $r < r_0$ on this branch: the orbits that follow it are bounded and asymptotic to the circular orbit $r = r_0$.

The hypotheses of the theorem thus imply that $\partial\Psi/\partial\rho \geq 0$ at $\rho = \rho_0$. We shall need strict inequality, and for this purpose we first dispose of the intermediate case.

16. *Lemma.* If the function $\rho \mapsto \Psi(\rho_0, \rho)$ has a degenerate zero at $\rho = \rho_0$, and if this remains true as the parameter ρ_0 is varied, then there exists a constant $\mathcal{G}$ such that $\varphi(I) = \mathcal{G}I^{-2}$.

Proof. We take the derivative of the function to be zero at ρ_0, which gives $\rho_0\psi'(\rho_0) = \psi(\rho_0)$ in a neighborhood of ρ_0 or $\psi(\rho_0) = \mathcal{G}\rho_0$ in this neighborhood and thus, by analyticity, throughout its domain of definition. The desired value of φ follows.

The law of attraction $\varphi(I) = \mathcal{G}I^{-2}$ has been well known since the time of Newton and Cotes. It is known that for negative values of the energy, the orbit remains bounded but spirals toward the origin. We leave the verification of this fact as an exercise in the use of Newton's reduction, and we note that the formulas are slightly simpler if one works with the variable $I = r^2$ rather than r.

We may continue our search for laws of attraction such that all bounded orbits are closed. The case where $\partial\Psi/\partial\rho < 0$ at $\rho = \rho_0$ is excluded by Lemma 15. In the case where $\partial\Psi/\partial\rho = 0$ at $\rho = \rho_0$, we may, according to

Lemmas 15 and 16, change the value of ρ_0 slightly to obtain

$$\frac{\partial \Psi}{\partial \rho}\Big|_{\rho=\rho_0} > 0.$$

In what follows we shall assume that ρ_0 is chosen so as to satisfy this inequality. In the (ρ, ρ')-plane, the level curves of $\mathcal{E}$ form closed curves, at least in the neighborhood of the point $(\rho_0, 0)$.

The orbital ratio associated to a closed, smooth level curve of $\mathcal{E}$ must be rational but must also vary continuously as either $\mathcal{E}$ or C are varied. It must thus remain constant in a neighborhood of the chosen curve. The following lemma applies this idea to curves near the point $(\rho, \rho') = (\rho_0, 0)$.

17. *Lemma.* Under the hypotheses of Theorem 13, there exists $\mathcal{G} > 0$ such that $\varphi(I) = \mathcal{G}I^{-2+\sigma/2}$, where σ is the square of the orbital ratio.

Corollary. Newton's law is the only law that satisfies Bertrand's hypotheses, if in addition one demands that the orbital ratio equal unity.

Proof. Let ρ_0 be such that $A = \partial \Psi / \partial \rho|_{\rho=\rho_0} = 1 - \rho_0 \psi' \psi^{-1} > 0$. Upon taking C equal to the angular momentum of the circular orbit of radius $1/\rho_0$, and upon setting $\varepsilon = \rho - \rho_0$, equation (23) becomes

$$\frac{d^2\varepsilon}{d\theta^2} + \varepsilon = -\rho_0 + \frac{\rho_0 \psi(\rho_0 + \varepsilon)}{\psi(\rho_0)} = \frac{\rho_0 \psi'(\rho_0)}{\psi(\rho_0)} \varepsilon + O(\varepsilon^2).$$

The limit of the period Θ of the function $\varepsilon(\theta)$ for $\varepsilon \to 0$ is $2\pi/\sqrt{A}$. We thus have $A = \sigma$, and this equality is maintained as ρ_0 is varied, in other words as C is varied. Upon replacing A by its expression above, we arrive at the differential equation $(1 - \sigma)\psi(\rho_0) = \rho_0 \psi'(\rho_0)$, leading to $\psi(\rho_0) = \mathcal{G}\rho_0^{1-\sigma}$, which is equivalent to the desired expression.

Remark. If $\varphi(I) = \mathcal{G}I^b$, then $\sigma = 2(2 + b)$ and the limiting angular period is $2\pi/\sqrt{2(2+b)}$. On a quasi-circular orbit, this is the angle between two consecutive pericenters. This angle was calculated by Newton.

We have reduced the number of possible force laws considerably, but we must go further. There are two classical strategies. The first is to calculate the limiting period above to second order. This calculation was performed by Lagrange and by a host of other authors afterward. The second strategy, due to Bertrand, is to calculate the limiting period for orbits starting with higher and higher velocities from a given point. We present both calculations here, which the reader may compare. We first summarize our formulas, setting $\psi(\rho) = \rho^{1-\sigma}$:

$$\frac{d\rho}{d\theta} = \rho', \quad \frac{d\rho'}{d\theta} = -\rho + C^{-2}\rho^{1-\sigma}, \quad 2\mathcal{E} = \rho'^2 + \rho^2 - \frac{2}{2-\sigma}C^{-2}\rho^{2-\sigma}, \quad (26)$$

where the last term must be replaced by $-C^{-2}\ln\rho$ when $\sigma = 2$. We note that $\sqrt{\sigma}$ is not rational in this latter case.

First End of Proof. We set $\rho_0 = 1$ and $C = 1$. Because of the homogeneity of φ, these choices do not restrict the class of solutions studied. Upon setting $\rho = 1 + \varepsilon$, equation (23) becomes

$$\frac{d^2\varepsilon}{d\theta^2} = -(1+\varepsilon)+(1+\varepsilon)^{1-\sigma} = -\sigma\varepsilon+\frac{(\sigma-1)\sigma}{2}\varepsilon^2-\frac{(\sigma-1)\sigma(\sigma+1)}{6}\varepsilon^3+O(\varepsilon^4).$$

We may eliminate a power of σ by setting $\tau = \theta\sqrt{\sigma}$:

$$\frac{d^2\varepsilon}{d\tau^2} + \varepsilon = \frac{\sigma-1}{2}\varepsilon^2 - \frac{(\sigma-1)(\sigma+1)}{6}\varepsilon^3 + O(\varepsilon^4). \quad (27)$$

We know that solutions close to the trivial solution $\varepsilon = 0$ trace out closed curves in the $(\varepsilon, d\varepsilon/d\tau)$-plane. We wish to know how the period of these solutions varies as we move away from the origin. This fundamental problem in perturbation theory is not so easy to solve.

The key idea is to introduce a change of variables that reduces the equation to a simple model. The chief difficulty is to choose a class of equations large enough to contain such a model. This is the class of Hamiltonian equations, in other words those of the form

$$\frac{d\varepsilon}{d\tau} = \frac{\partial G}{\partial z}, \quad \frac{dz}{d\tau} = -\frac{\partial G}{\partial \varepsilon},$$

where G, the "Hamiltonian," is a function of ε and of the so-called "conjugate" variable z. Equation (27) corresponds to

$$G = \frac{1}{2}(z^2 + \varepsilon^2) - \frac{\sigma-1}{6}\varepsilon^3 + \frac{(\sigma-1)(\sigma+1)}{24}\varepsilon^4 + O(\varepsilon^5). \quad (28)$$

By direct identification, we find that the near-identity variables change $(\varepsilon, z) \mapsto (\xi, \eta)$, which simultaneously preserves the area form $d\varepsilon \wedge d\eta$, and the symmetry $z \mapsto -z$ are of the form

$$\varepsilon = \xi + a\xi^2 + b\eta^2 + (3c - 4ab)\xi\eta^2 + d\xi^3 + \cdots,$$

$$z = \eta - 2a\xi\eta + (4a^2 - 3d)\xi^2\eta - c\eta^3 + \cdots,$$

where a, b, c, d, are four real parameters. We abbreviate by writing $m = 1 - \sigma$. As we begin substitution, we notice that the terms of order 3 in the new Hamiltonian are eliminated if and only if $b = 2a = -m/3$. Completing the substitution, we find

$$G = \frac{1}{2}(\eta^2 + \xi^2) + \left(d - \frac{5m^2}{72} + \frac{m(m-2)}{24}\right)\xi^4$$

$$+ 3\left(c - d - \frac{m^2}{18}\right)\xi^2\eta^2 + \left(\frac{m^2}{18} - c\right)\eta^4 + \cdots \quad .$$

Setting $c = m^2/18 - \gamma$, then $d = m(m+3)/24 + \gamma$, and finally $\gamma = -m(m+3)/96$, we find, on returning to the parameter σ, the "normal form"

$$G = \frac{1}{2}(\eta^2 + \xi^2) - \frac{(1-\sigma)(4-\sigma)}{96}(\eta^2 + \xi^2)^2 + c\ldots \quad . \tag{29}$$

It is now clear that the period can remain constant only if $\sigma = 1$ or $\sigma = 4$. We thus arrive at the exponents appearing in the statement of Bertrand's theorem, which is now proved.

Second End of Proof. We consider the laws of attraction of Lemma 17, and we imagine the body starting from an arbitrary point with higher and higher velocity. If $0 < \sigma < 2$, there is an "escape velocity" above which orbits will not be bounded. We shall calculate the limiting period Θ as the velocity approaches this limit. On the other hand, if $2 \leq \sigma$, the body will always return. We shall calculate the limiting period Θ for an infinite velocity.

First Case. If $0 < \sigma < 2$, we normalize by setting $C = 1$. We see that the curve $\mathcal{E} = 0$ is the boundary of the region where the level sets of $\mathcal{E}$ are closed. We make the substitution $\rho' = -J\rho$, so that the equation of this level curve may be written

$$J^2 + 1 = \frac{2}{2-\sigma}\rho^{-\sigma}.$$

We see that the curve is parametrized by J. Differentiation gives

$$J\,dJ = -\frac{\sigma}{2-\sigma}\rho^{-\sigma-1}d\rho = -\sigma\frac{d\rho}{2\rho}(1 + J^2).$$

The total variation of the angle is thus

$$\Theta = \int d\theta = \int \frac{d\rho}{\rho'} = \frac{2}{\sigma}\int_{-\infty}^{+\infty} \frac{dJ}{1 + J^2} = \frac{2\pi}{\sigma}.$$

But Lemma 17 gives $\Theta = 2\pi\sigma^{-1/2}$. Thus $\sigma = 1$.

Second Case. If $2 < \sigma$, the energy is always positive. We consider the level curve of $\mathcal{E}$, denoted by $\mathcal{N}$, which passes through $(\rho, \rho') = (1, 0)$. Its equation is

$$1 - \frac{2}{2-\sigma} C^{-2} = \rho'^2 + \rho^2 - \frac{2}{2-\sigma} C^{-2} \rho^{2-\sigma}. \tag{30}$$

When $C^{-2} \to 0$, the closed curve $\mathcal{N}$ approaches the boundary of the half-disk $0 < \rho$, $\rho'^2 + \rho^2 < 1$. To evaluate the period Θ in the limit, we cut this curve into two pieces, one approaching the semicircle $\rho'^2 + \rho^2 = 1$, $\rho > 0$, the other approaching the segment $-1 < \rho' < 1$, $\rho = 0$. It is convenient to use two cuts, one at a point A near $(0, 1)$ and satisfying $\rho' = 1 - \sqrt{\rho}$, the other at a point B near $(0, -1)$ and satisfying $\rho' = -1 + \sqrt{\rho}$.

The way in which the curve $\mathcal{N}$ is traversed is determined by the *vector field* tangent to $\mathcal{N}$ with components $(\rho', -\rho + C^{-2}\rho^{1-\sigma})$. In fact, these components make up the right-hand side of system (24). In the limit $C^{-2} \to 0$, the interval (A, B) of the curve $\mathcal{N}$ becomes the semicircle $\rho'^2 + \rho^2 = 1$, $\rho > 0$, equipped with the vector field with components $(\rho', -\rho)$, which defines a uniform motion. The contribution to Θ of this interval tends to π.

The contribution of the interval (B, A) is more difficult to evaluate, since $\rho^{1-\sigma} \to +\infty$ in the limit. To remove the indeterminacy in $C^{-2}\rho^{1-\sigma}$, we write (30) in the form

$$C^{-2} = \frac{2-\sigma}{2} \times \frac{1 - \rho^2 - \rho'^2}{1 - \rho^{2-\sigma}}.$$

In the zone of interest, we have $C^{-2} \sim (\sigma/2 - 1)(1 - \rho'^2)\rho^{\sigma-2}$. We have $|\rho'| < 1 - \sqrt{\rho}$ and thus $1 - \rho'^2 > 2\sqrt{\rho} - \rho$. The term $C^{-2}\rho^{1-\sigma}$ on the right-hand side of the system is thus bounded from below by $(\sigma/2 - 1)\rho^{-1/2}$. It therefore becomes infinite, so that the segment $\rho = 0$, $-1 < \rho' < 1$, is traversed instantaneously in the limit. In sum, $\Theta \to \pi$ and the orbital ratio is 2, so that $\sigma = 4$. Bertrand's theorem is proved again.

Jacobi's Generalization of the Kepler Problem

We consider the following system, which generalizes (1):

$$\ddot{\vec{r}} = -f\vec{r}. \tag{31}$$

The vector $\vec{r}$ belongs to the (not necessarily Euclidean) vector space E, and f is a real-valued function of $\vec{r}$ and $\dot{\vec{r}} = \vec{v}$. We say that we are dealing with a *central force problem*. The quantity $C = \vec{r} \wedge \vec{v}$ is a first integral. The motion takes place in a plane.

Jacobi assumes that f is a homogeneous function of degree -3 of the position $\vec{r} \in E \setminus \{0\}$, and he proceeds to integrate. His solution resembles a translation of Hermann's and Lagrange's method into polar coordinates. Here we give a slightly different solution using Clairaut's method.

Consider an arbitrary linear identification of the plane of motion with the complex plane. We write $\vec{r} = r e^{i\theta}$, then $\zeta(\theta) = f(e^{i\theta})$, to get $f(\vec{r}) = r^{-3}\zeta(\theta)$. Setting $\rho = 1/r$, equation (31) becomes

$$\frac{d^2\rho}{d\theta^2} + \rho = C^{-2}\zeta(\theta).$$

To solve this equation, we expand ζ in a Fourier series in θ. The particular case $\zeta(\theta) = C^2 e^{ik\theta}$ gives the particular solution $\rho = \theta e^{ik\theta}$ if $k = \pm 1$, and $\rho = (1 - k^2)^{-1} e^{ik\theta}$ otherwise. The general case

$$\zeta(\theta) = \sum_k z_k e^{i\theta}$$

results in

$$\rho(\theta) = C^{-2}\left(z_0 + Ae^{i\theta} + Be^{-i\theta} + z_1\theta e^{i\theta} + z_{-1}\theta e^{-i\theta} + \sum_{|k|\geq 2} \frac{z_k}{1 - k^2} e^{ik\theta}\right), \quad (32)$$

where A is an arbitrary complex number and B its conjugate.

18. *Proposition.* In a central force problem (31), where f is a homogeneous function of $\vec{r}$ of degree -3, all the nonrectilinear bounded orbits are closed.

We note that the implication of the proposition is often empty, in the sense that the problem may not, in general, admit any bounded nonrectilinear solutions. There is, however, an interesting class of problems containing the Kepler problem and characterized by D'Alembert's hypothesis below.

19. *Definition.* We say that f satisfies D'Alembert's hypothesis if the Fourier coefficient $z_1 = \bar{z}_{-1}$ vanishes.

20. *Proposition.* D'Alembert's hypothesis is a necessary condition for the existence of bounded nonrectilinear orbits.

The proofs are obtained immediately from expression (32) for $\rho(\theta)$, the inverse of the distance. We note that the convergence of the Fourier series for ρ is "better" than that of the Fourier series for ζ.

21. *Remark.* The above presentation is not very satisfactory in the sense that the identification of the plane of motion with $\mathbb{C}$ furnishes this plane with an arbitrary Euclidean form. In particular, it's not clear why d'Alembert's hypothesis is invariant under a change of this Euclidean form. Here we shall be content to indicate that this hypothesis amounts to the vanishing, for any linear function l, of the integral of the closed form $l\,f\,(x\,dy - y\,dx)$ over a cycle encircling the origin.

Radial Potential Problems in Non-Euclidean Geometry

P. Serret, P. Appell, then H. Liebmann, then later V. Koslov and A. Harin remarked that the "Keplerian miracle" – the fact that the orbits are conic sections – persists in non-Euclidean geometry. In truth, all the "miracles" of central forces presented in this lecture survive when we bend the space of motions, as we shall see.

We consider a vector space F and a vector $Z \in F$. We denote by $[Z]$ the socalled "vertical" line generated by Z. We choose a "horizontal" subspace E such that $F = [Z] \oplus E$. Finally, we denote by $\omega \in F^*$ the linear form such that $\ker \omega = E$ and $\langle \omega, Z \rangle = 1$.

We consider the differential equation $\ddot{X} = AX + BZ$, where A and B are two real functions of the position $X \in F$ and of the velocity $\dot{X}$. Let $h = \langle \omega, X \rangle$ be the "height" of X. A brief calculation will bring out a remarkable property of the evolution of $h^{-1}X$, the central projection of X onto the horizontal plane containing Z. We write $X_E = h^{-1}X - Z \in E$, and we have

$$h^2 \dot{X}_E = h\dot{X} - \dot{h}X, \qquad \frac{d}{dt}(h^2\dot{X}_E) = h\ddot{X} - \ddot{h}X.$$

But $\ddot{h} = Ah + B$, according to the differential equation and the definition of h. Thus

$$\frac{d}{dt}(h^2\dot{X}_E) = hBZ - BX = -hBX_E, \quad \text{or} \quad \frac{d^2}{d\tau^2}X_E = -h^3BX_E,$$

upon setting $d\tau = h^{-2}dt$. Remarkably, the function A has disappeared from the equation. Nevertheless, we must not forget that knowledge of X_E and $\dot{X}_E$ is not sufficient to evaluate h and B.

We now suppose that F is Euclidean, that $\|Z\| = 1$, and that Z and E are orthogonal. Let S be the unit sphere of F. The motions of X restricted to the unit sphere are such that the normal acceleration $\ddot{X}_N = -\|\dot{X}\|^2 X$ and such that the tangential acceleration $\ddot{X}_T$ is arbitrary. To verify this, it suffices to twice differentiate the constraint $\|X\|^2 = 1$.

If $\ddot{X}_T = aX + BZ$, where a and B are two functions of X and $\dot{X}$, we say that we are dealing with a *central force problem on the sphere* S. This amounts to assuming that $\ddot{X}_T$, or equivalently $\ddot{X}$, remains in the plane generated by X and Z. Using the fact that $\ddot{X}_T$ is tangent to the sphere, we find $a = -Bh$, which gives $\ddot{X} = -(Bh + \|\dot{X}\|^2)X + BZ$.

From the short calculation above, we obtain

$$\frac{d^2}{d\tau^2} X_E = -h^3 B X_E, \quad \text{with} \quad h^2(1 + \|X_E\|^2) = 1. \tag{33}$$

The quantity h is determined up to a choice of sign by the second equation. We may arrange to study only the hemisphere $h > 0$, for which a "projective" chart is the space E. In our examples, B will be a function of X only, calculated via h and X_E. The central force problem on the sphere S is thus reduced to a central force problem on the space E.

For example, the classical problem of the (inverted) spherical pendulum is defined by $B = g > 0$, the constant acceleration of gravity. It corresponds, after introducing a new time variable τ, to the radial potential problem

$$\frac{d^2}{d\tau^2} X_E = -\frac{g}{(1 + \|X_E\|^2)^{3/2}} X_E.$$

This system defines a kind of interpolation between the two radial potential problems whose closed orbits are bounded.

Now going backwards, if we seek a problem on the sphere that is sent to the Kepler problem, it suffices to take $B = \mathcal{G}(1 - h^2)^{-3/2}$.

By "non-Euclidean geometries" in the title above, we mean geometries of spaces of constant curvature. We have just treated the case of constant positive curvature by reducing the study of central force problems on the sphere to ordinary central force problems. It remains to treat the case of constant negative curvature. The calculations are exactly the same. It suffices to take the "unit sphere" of a Lorentz space as our model of hyperbolic space.

We assume that $\langle Z, Z \rangle = -1$, that Z is orthogonal to E, and that the Lorentz form restricted to E is positive definite. The "unit sphere" H is the set of X with $\langle X, X \rangle = -1$. Recall that the hyperbolic space H is not topologically a sphere, but is instead contractible (homeomorphic to $\mathbb{R}^\nu$). The equations are the same. Only the relation between h and X_E is changed, as follows:

$$h^2(1 - \langle X_E, X_E \rangle) = 1, \qquad 0 \le \langle X_E, X_E \rangle < 1.$$

To finish, and to prepare for concluding remarks, we calculate, for the example of the sphere S, the *divergence* δ_S of the field of tangent accelerations $\ddot{X}_T = B(-hX + Z)$. The useful formula is $d(\ddot{X}_T \lrcorner \mu_S) = \delta_S \mu_S$, where d designates the exterior differential, μ_S the standard area form on the sphere S, and $\lrcorner$ the contracted product. For the calculation we use the projective chart for the hemisphere described above. The chart E is equipped with the standard area form μ_E. Let $\nu + 1$ be the dimension of F and μ_F be the standard volume form on this space. We note that in S, or in the hyperplane $h = 1$, we may measure the area of a domain $\mathcal{A}$ by dividing the volume of the cone with base $\mathcal{A}$ and vertex at the origin of F by $\nu + 1$. We thus check that in the chart, μ_S becomes $h^{\nu+1}\mu_E$. We also check that the tangent vector $-hX + Z$ becomes $-h^{-1}X_E$. In the chart the formula for δ_S becomes

$$d(-h^\nu B X_E \lrcorner \mu_E) = h^{\nu+1}\delta_S \mu_E. \tag{34}$$

The divergence δ_E of the acceleration field comprising the right-hand side of (33) satisfies

$$d(-h^3 B X_E \lrcorner \mu_E) = \delta_E \mu_E. \tag{35}$$

It is quite remarkable that when $\nu = 3$ the two divergences δ_E and δ_S vanish simultaneously.

Concerning Question 4. We conclude by making several important remarks. The works of Hamilton, Jacobi, Cauchy, Liouville, and others signal the beginning of a kind of dogma which holds that a classical mechanical system is a Hamiltonian system. In our investigation into the closure properties of Keplerian orbits, we encountered an interesting *non-Hamiltonian* generalization of the Kepler problem. The author of this generalization, Jacobi, stressed that the integrability of the problem is explained outside of the Hamiltonian framework. Our presentation shows the extent to which the closure property of the orbits is alien to this framework. Our conclusion is that a mechanical system is more than a Hamiltonian system. In practice, working directly with the equations of motion, rather than with the Hamiltonian, may simplify calculations considerably.

Concerning Question 1. A fundamental question tied to the one asked in the title of this lecture is; Why is Newton's law of attraction inversely proportional to the square of the distance? Laplace made two distinct remarks on this subject. One remark is that a homothety of ratio λ sends a trajectory of the n-body problem onto another trajectory, provided care is taken to multiply all the masses by λ^3, thus transforming them as one transforms three-dimensional volumes. The other remark is that in a Euclidean space of dimension three, the $1/r$-potential satisfies Laplace's equation $\Delta(1/r) = 0$. We shall examine this second remark more closely.

The Laplacian of the potential is nothing other than the divergence of the force field. Let us *assume that physical space is of dimension $\nu = 3$* and reexamine, one by one, all the examples of this lecture where the bounded orbits are closed. We shall simply calculate the divergence δ of the force field. For the two laws of force appearing in Bertrand's theorem, we find, respectively, $\delta = 0$ and $\delta = -3\mathcal{G}$. Let us focus on the first law, corresponding to the Kepler problem. A short calculation shows that the central force fields studied by Jacobi are *characterized* by the condition $\delta = 0$. We now transport these fields onto the sphere S^3 or onto the hyperbolic space H^3 using the projection described above. The bounded orbits (which are not asymptotic to the poles of S^3) remain closed, and the condition $\delta = 0$ still holds.

These somewhat strange remarks indicate a distinguished role for dimension 3 and a dynamical role for the hypothesis of vanishing divergence.

Lecture 4. Concerning the Eccentricity Vector

On the unifocal equation for conic sections

The equation $r = \alpha x + \beta y + \gamma$ for conic sections appearing in Lemma 5 is not well known. In essence, it is simply a translation into symbols of the description of a conic section in terms of the focus and directrix; its invention goes back at least to Pappus of Alexandria. This description is also similar to the more widely known polar equation for conic sections.

But we wish to stress, as Gauss did in his *Theoria Motus*, that there are advantages to considering precisely this former equation as the primary and best description of Keplerian conics. For this reason we give the equation a name: the *unifocal equation*.

Exercise 7 is very instructive. In order to find the maximal number of points of intersection of curves described by the equations $r = \alpha x + \beta y + \gamma$ and $r = \alpha'x + \beta'y + \gamma'$, we take the difference $0 = (\alpha - \alpha')x + (\beta - \beta')y + (\gamma - \gamma')$. This is the equation of a line whose intersection with one of the conic sections contains at most two points. It's difficult to see how this could be obtained in any simpler way. One of the very few other efficient methods is the Levi–Civita transformation.

Let us try to understand the advantages of the unifocal equation over other equations for conic sections. In the first place, it gives a very simple *parametrization* of the family of nonrectilinear Keplerian orbits. Each triple $(\alpha, \beta, \gamma) \in \mathbb{R}^3$, $\gamma > 0$, is associated with such an orbit and vice-versa. We must acknowledge that the behavior of the parametrization when $\gamma = C^2$ tends to zero is somewhat curious: all the rectilinear orbits are sent onto the circle $\gamma = 0$, $k = \sqrt{\alpha^2 + \beta^2} = 1$. But observe the solution of our unifocal equation when $\alpha^2 + \beta^2 > 1$ and $\gamma = 0$. It is a pair of lines. This degenerate conic section is not devoid of interpretation; it corresponds to a limit of hyperbolic trajectories for which the angular momentum is zero and the energy is infinite.

We may think of the unifocal equation as lying between the bipolar equation and the Cartesian equation for conic sections. Consider two foci in the plane with coordinates (x_1, y_1) and (x_2, y_2). The bipolar equation of a conic section with semimajor axis a is

$$\pm\sqrt{(x-x_1)^2 + (y-y_1)^2} \pm \sqrt{(x-x_2)^2 + (y-y_2)^2} = 2a.$$

For simplicity we set $x_2 = y_2 = 0$. We might expect elimination of the radicals to give a quartic equation, but in fact the first step gives

$$\pm 4a\sqrt{x^2 + y^2} = 2xx_1 + 2yy_1 + 4a^2 - x_1^2 - y_1^2.$$

The squares in x and y have disappeared, thus the second step, eliminating the other radical, gives a quadratic equation, the Cartesian equation of the conic section. It is still necessary to divide by $4a$ to arrive at the unifocal equation. The condition $a \neq 0$ gives our equation an advantage over the bipolar equation; namely, our equation admits parabolas.

The advantage of the equations with radicals over the Cartesian equation is that the choice of $\pm$ signs allows us to distinguish branches of hyperbolas. From equation (1) defining the Kepler problem, we can change the sign in front of the radical and thus define the *repulsive* problem. Our solution methods in this case proceed in the same way, and in the end designate the convex branch of the hyperbola as the only possible trajectory.

The Eccentricity Vector and Tensor Analysis

Hamilton used the eccentricity vector to illustrate his method of vector analysis based on quaternions. Much later, Gibbs developed a vector analysis based on the notion of vector product, and he too wrote down a formula expressing the eccentricity vector in the so-called Gibbs-Heaviside system. In keeping with the spirit of our Question 1, we reject these systems, since they are confined to dimension 3. We shall instead write multidimensional formulas using tensor analysis and exterior calculus (a special case of the former).

To our way of thinking, all the quantities entering into tensor analysis are tensors, and the fundamental operations consist only of sums, tensor products, contractions, and permutations. Tensor analysis contains linear algebra: the operations of evaluation and composition of linear maps must always be translated in terms of contracted products of tensors, the linear maps being "encoded" by the tensors with which they are associated.

These theoretical demands may run into problems of notation. The conventions conceived for writing *all* the formulas of tensor analysis may seem overly cumbersome and ill-suited to replacing the conventions of abstract linear algebra. But this is not the real issue. The problem is not to find a universal notation but instead to find conventions adapted to tensor analysis, and in particular conventions that give contraction the stature of a fundamental operation, something not afforded by traditional abstract linear algebra.

We shall not need a very elaborate system of notation. We will be content to use well-known symbols in accord with the principles just described. We thus write

$$\vec{v} = \dot{\vec{r}}, \qquad C = \vec{r} \wedge \vec{v},$$

which at first glance appears to introduce nothing new. A closer look reveals that the angular momentum is a bivector. We have performed an exterior rather than a vector product. The latter operation is only possible in dimension 3, and even then it first requires the introduction of an arbitrary orientation of space, so that bivectors may be considered as vectors. Physicists often point out that C is not a "true" vector, since it does not change sign when the system is reflected in the plane $(\vec{r}, \vec{v})$.

We then write

$$\vec{L} = \frac{\vec{r}}{r} + \vec{v} \lrcorner C.$$

This is the eccentricity vector. We have denoted the inner product by $\lrcorner$, which is a special *contracted product*. A contracted product is a tensor product followed by a *contraction*. If E is a finite-dimensional vector space, elementary contraction associates a scalar to an element of $E \otimes E^*$. If this element is $X \otimes \xi$, $X \in E$, $\xi \in E^*$, then the scalar is the duality bracket $\langle X, \xi \rangle$. If this element is $\omega \in E \otimes E^*$, the tensor which "encodes" a unique endomorphism of E, the contraction is the trace of the endomorphism.

To carry out the contraction implicit in our symbol $\lrcorner$, an identification is necessary: the identification of our ν-dimensional physical space E with its dual E^* via the Euclidean structure g. In symbols, we decree $E = E^*$, $g = \text{Id}$. We do not change notation in any way. Henceforth we shall simply consider the contraction of an element of $E \otimes E$ to be legitimate. Thus, with this convention, the duality bracket, which is simply a notation for the most elementary contracted product, becomes the scalar product.

We now know how to calculate

$$\vec{v}\lrcorner C = \vec{v}\lrcorner(\vec{r}\wedge\vec{v}) = \vec{v}\lrcorner(\vec{r}\otimes\vec{v} - \vec{v}\otimes\vec{r}) = \langle\vec{v},\vec{r}\rangle\vec{v} - \langle\vec{v},\vec{v}\rangle\vec{r}.$$

One quickly learns to skip the third step by remembering Leibniz's rules for exterior algebra. Here we have the analog of the vector triple product, and this formula is more general and simpler to remember. It is now clear that our $\vec{L}$ is truly a vector and not a bivector, as is C.

An Exercise in Geometry. To try out the above calculation, we are going to deduce the expression for $\vec{L}$ from the bifocal properties of the conic sections. We begin as Newton did in his Proposition 17. Let $\vec{r}$ and $\vec{v}$ be position and velocity. Let us find the Keplerian orbit, for which we note that it is enough to find the second focus. It is situated on the ray of light that starts from the first focus, at the origin, and reflects off the tangent line passing through the point at position $\vec{r}$ with direction $\vec{v}$. The first exercise in vector geometry is thus to calculate the vector $\vec{s}$, the image of $\vec{r}$ under orthogonal reflection with respect to the tangent line. With a little cleverness, the expression for $\vec{s}$ may be deduced immediately from the formula

$$\vec{r} + \vec{s} = 2\frac{\langle\vec{r},\vec{v}\rangle}{\langle\vec{v},\vec{v}\rangle}\vec{v}.$$

We need instead another formula, which may be deduced from this one using Leibniz's rule mentioned above, but which we wish to obtain directly:

$$-\vec{r} + \vec{s} = \frac{2}{\langle\vec{v},\vec{v}\rangle}\vec{v}\lrcorner(\vec{r}\wedge\vec{v}).$$

A Short Recipe. We may write down the above formula directly once we are in the habit of considering $\vec{w} = \vec{u}\lrcorner(\vec{r}\wedge\vec{v})$ each time we need a vector orthogonal to $\vec{u}$ in the $(\vec{r},\vec{v})$-plane. If this plane is oriented by the basis $(\vec{r},\vec{v})$, then the basis $(\vec{u},\vec{w})$ defines the same orientation. In other words, we have the following proposition.

22. *Proposition.* After suitable normalization, the operation $\vec{u}\longmapsto\vec{w} = \vec{u}\lrcorner(\vec{r}\wedge\vec{v})$ is a rotation by $+\pi/2$ in the plane generated and oriented by the basis $(\vec{r},\vec{v})$.

Proof. It suffices to use the recipe

$$\langle \vec{u}, \vec{w} \rangle = \vec{u} \lrcorner \big(\vec{u} \lrcorner (\vec{r} \wedge \vec{v}) \big) = \langle \vec{u} \wedge \vec{u} | \vec{r} \wedge \vec{v} \rangle = 0,$$

$$\vec{u} \wedge \vec{w} = \vec{u} \wedge \big(\vec{u} \lrcorner (\vec{r} \wedge \vec{v}) \big) = \vec{u} \lrcorner (\vec{u} \wedge \vec{r} \wedge \vec{v}) + \langle u, u \rangle \vec{r} \wedge \vec{v} = \langle u, u \rangle \vec{r} \wedge \vec{v}.$$

In order to understand the second equality above, we must assign a precise convention to the duality bracket, or scalar product, for multivectors. Most authors do not bother to distinguish between the convention used in exterior algebra and the convention used for general tensors. We use the notation $\langle a | b \rangle$ for what we call the "simplified bracket," when a and b are p-vectors or alternating p-forms. The same convention must be used for symmetric tensors. The general bracket is denoted $\langle a, b \rangle$. We have $\langle a, b \rangle = p! \langle a | b \rangle$ in the case of our p-vectors. There's no need for further explanation here; the reader may simply use the formula $\langle \vec{u} \wedge \vec{w} | \vec{r} \wedge \vec{v} \rangle = \langle \vec{u}, \vec{r} \rangle \langle \vec{w}, \vec{v} \rangle - \langle \vec{u}, \vec{v} \rangle \langle \vec{w}, \vec{r} \rangle$.

Commentary. Standard formulas for the vector product are taught all over the world. Our formulas are no more complicated, and have the advantage of working in any dimension. It is true that in dimension 3, when no spatial orientation is furnished by the problem, we must distinguish between vectors and bivectors, but this is a very pertinent distinction.

Back to the Calculation of $\vec{L}$. We have determined the direction $\vec{s}$ of the reflected ray of light. It remains to locate the second focus on this ray. Suppose that we are dealing with an ellipse of semimajor axis a. Let $\vec{f}$ be the difference vector between the two foci. We have

$$\vec{f} = \vec{r} + \frac{2a - r}{r} \vec{s}.$$

It is easy to determine a using the expression for the energy

$$-\frac{1}{2a} - \frac{1}{2} v^2 - \frac{1}{r} \qquad \text{or} \qquad \frac{2a - r}{r} = a v^2.$$

We find

$$\vec{f} = \vec{r} + a v^2 \Big(\vec{r} + \frac{2}{\langle \vec{v}, \vec{v} \rangle} \vec{v} \lrcorner (\vec{r} \wedge \vec{v}) \Big) = 2a \Big(\frac{\vec{r}}{r} + \vec{v} \lrcorner (\vec{r} \wedge \vec{v}) \Big).$$

After division by $2a$, which allows parabolas to be treated with the same formula, we find the expression for the eccentricity vector, along with its geometric interpretation. This concludes our exercise in geometry.

23. *Exercise.* Compute the duality bracket $\langle \vec{L}, \vec{r} \rangle$. Verify the result by comparing with the method of Hermann and Lagrange.

One More Dimension

Here we gather together some interesting remarks on the Kepler problem due to Moser, Dziobek, Laplace, and Lagrange; these remarks shed light on the status of the eccentricity vector (α, β) or of the unifocal equation, and they have the common feature that an extra dimension is added to the physical space.

Consider the vector leading from the *center* of a Keplerian ellipse of semimajor axis a to a particle on the ellipse, and denote this vector by $2a\vec{\xi}$. It is given by a remarkable formula obtained from the vectors $\vec{f}$ and $\vec{s}$ described in the previous paragraph:

$$2a\vec{\xi} = 2\vec{r} - \vec{f} = \vec{r} - av^2\vec{s} = \vec{r} - av^2\left(-\vec{r} + 2\frac{\langle \vec{r}, \vec{v} \rangle}{v^2}\vec{v}\right),$$

or, setting $J = \langle \vec{r}, \vec{v} \rangle$,

$$\vec{\xi} = \frac{\vec{r}}{r} - J\vec{v}.$$

The vector $\vec{\xi}$ describes an ellipse centered at the origin with unit semimajor axis. It is natural to think of it as the projection onto physical space E of a vector $X = (h, \vec{\xi})$, which describes the great circles of the unit sphere in $\mathbb{R} \oplus E$. We seek the "vertical" component h of X. We have

$$\|\vec{\xi}\|^2 = \left\|\frac{\vec{r}}{r} - J\vec{v}\right\|^2 = 1 - \frac{2}{r}J^2 + v^2 J^2 = 1 - \frac{1}{a}J^2.$$

It is thus convenient to set $h = a^{-1/2}J$, so that $\|X\|^2 = h^2 + \|\vec{\xi}\|^2 = 1$. We recopy these formulas, this time dispensing with the intermediate notation: the vector

$$X = \left(a^{-1/2}\langle \vec{r}, \vec{v} \rangle, a^{-1}(\vec{r} - a\vec{L})\right)$$

describes the great circles of the unit sphere in $\mathbb{R} \oplus E$. Its derivative Y with respect to time may be written as

$$Y = \dot{X} = \left(a^{-1/2}(v^2 - r^{-1}), a^{-1}\vec{v}\right).$$

It of course satisfies $\langle X, Y \rangle = 0$. We have $\|Y\|^2 = a^{-1}(v^4 - 2v^2r^{-1} + r^{-2} + v^2(2r^{-1} - v^2)) = a^{-1}r^{-2}$. The vector

$$\bar{Y} = \frac{r}{a}Y = \left(a^{-3/2}(rv^2 - 1), a^{-2}r\vec{v}\right)$$

thus satisfies $\|\bar{Y}\|^2 = a^{-3}$. Replacing the velocity vector Y by $\bar{Y}$ therefore amounts to uniformizing the motion on the great circles and to regularizing the motion on the vertical great circles. The regularized motion is geodesic motion on the sphere. Its dynamics is clearly poorer, that is, more symmetric than the dynamics defined by Y.

Hyperbolas. The construction may be generalized to hyperbolic orbits by changing $\sqrt{a}$ to $\sqrt{-a}$ as usual. Geometrically it's enough to change the unit sphere of $\mathbb{R} \oplus E$ into the hyperboloid given by $-h^2 + \|\vec{\xi}\|^2 = 1$. As in Lecture 3, this amounts to changing from Euclidean to Lorentzian geometry. It must be pointed out that the hyperbolic space H is the *other* hyperboloid, given by $-h^2 + \|\vec{\xi}\|^2 = -1$.

24. *Remark.* Moser's construction, and the extensions to positive energy obtained later, begin by exhibiting the hodograph of Kepler's problem as the stereographic projection of geodesic motion on the sphere or on the hyperbolic space H. Our construction is simpler. To see that the two constructions are compatible, it's enough to recall that if X describes a geodesic on the sphere, so does its derivative X'. And if X describes a geodesic on H, X' describes a geodesic, not on H, but on the other hyperboloid.

To conclude, we turn to the fate of the first integrals $\vec{L}$ and C. For this purpose, we call the vertical unit vector Z, and we compute

$$X \wedge \bar{Y} = \left(a^{-1/2}\langle \vec{r}, \vec{v} \rangle Z + r^{-1}\vec{r} - \langle \vec{r}, \vec{v} \rangle \vec{v}\right) \wedge \left(a^{-3/2}(rv^2 - 1)Z + a^{-2}r\vec{v}\right)$$
$$= a^{-3/2}Z \wedge \vec{L} + a^{-2}\vec{r} \wedge \vec{v}.$$

We thus see the angular momentum and the eccentricity vector consolidated in a single bivector. This interesting consolidation also occurs in the following construction.

25. *Dziobek's Construction.* Dziobek proposed the following method for solving the Kepler problem. Consider $J = \langle \vec{r}, \vec{v} \rangle$. We have $\dot{J} = v^2 - r^{-1}$ and $\ddot{J} = -r^{-3}J$. This equation is clearly close to system (1): by setting

$\mathcal{X} = (J, \vec{r}) \in \mathbb{R} \oplus E$, we have

$$\ddot{\mathcal{X}} = -r^{-3}\mathcal{X}.$$

The motion in $\mathbb{R} \oplus E$ is therefore governed by a central force. The angular momentum $\mathcal{X} \wedge \dot{\mathcal{X}}$ is constant. This bivector may be written

$$(JZ + \vec{r}) \wedge (\dot{J}Z + \vec{v}) = Z \wedge (J\vec{v} - (v^2 - r^{-1})\vec{r}) + \vec{r} \wedge \vec{v} = Z \wedge \vec{L} + \vec{r} \wedge \vec{v}.$$

In this way Dziobek "discovers" the expression for the eccentricity vector and concludes the way we did using the first method of Lecture 2. There is evidently a strong connection between this and the preceding construction. To bring them still closer, it's enough, when $a > 0$, to start with $\sqrt{a}J$ instead of J, which does not change the chain of reasoning. The orbits of the differential equation for $\mathcal{X}$ which correspond to Keplerian ellipses are now all circles. To form Moser's sphere, it remains only to translate these circles horizontally until their centers coincide with the origin.

26. *Laplace's Method.* Laplace proposed a method for integrating the Kepler problem, which, as R. Weinstock remarked in "Inverse Square Orbits" uses no first integrals. This method is no doubt at the root of Dziobek's construction. Let us subject the equation $\ddot{J} = -r^{-3}J$ to a few mysterious manipulations. We have $2J = d(r^2)/dt$ and thus $2\ddot{J} = d^3(r^2)/dt^3$. Using on one hand the relation $J = r\dot{r}$, and the remarkable identity $2(r^3 r'')' = r^2(r^2)'''$ on the other, we arrive at

$$r^2 \ddot{J} = -\frac{dr}{dt} = \frac{d}{dt}\left(r^3 \frac{d^2 r}{dt^2}\right).$$

Now consider the linear differential equation of third order $-\dot{f} = d(r^3 \ddot{f})/dt$. Its coefficients depend on time through the intermediary of the function $r(t)$ corresponding to an orbit of the Kepler problem. We immediately find four solutions: $f = r$, by means of the relation above; $f = x$ and $f = y$, by way of the equations of motion; and finally $f = 1$. There must therefore exist constants α, β, and γ along with a linear dependency $r = \alpha x + \beta y + \gamma$. The trajectory is a branch of a conic section according to Lemma 5.

27. *Lagrange's Method.* The starting point is (22) from Lecture 3 in the particular case of Newtonian attraction:

$$\ddot{r} = \frac{C^2}{r^3} - \frac{1}{r^2}.$$

Lagrange very subtly sets $w = C^2 - r$. We thus have $\ddot{w} = -wr^{-3}$, an equation analogous to those of system (1). The second-order linear differential equation $\ddot{f} = -fr^{-3}$, with coefficient r^{-3} depending on time, possesses three solutions x, y, and w, which are therefore linearly dependent. We thus have $w + \alpha x + \beta y = 0$, the equation of Lemma 5.

28. *Short Synopsis.* In each case, we sought to put a solution of the equation $\ddot{f} = -f/r^3$ into a $(\nu + 1)$st dimension. Of the two candidates $C^2 - r$ and J, the second is the more interesting above rectilinear orbits of the Kepler problem: it causes the fictitious body to describe a smooth curve, which reminds us of the first auxiliary conic from Lecture 2. We note that the integral of the angular momentum in Dziobek's construction is connected with the Wronskian in the approaches of Lagrange and Laplace.

Concerning Question 1. No particular feature of dimension $\nu = 3$ has yet appeared in this lecture. Nonetheless we note that when $\nu = 3$, Moser's construction makes use of the unit sphere in four-dimensional space, whose geometry is very special. Let us focus our attention, for example, on the space of Keplerian orbits of semi-major axis $a = 1$. Moser's construction gives this space a beautiful geometry by identifying it with the so-called *Grassmannian* space of oriented planes in $\mathbb{R} \oplus E$. When $\nu = 3$, this latter space may be identified with the product of two spheres S^2. Following Souriau, we may directly construct the map that associates two points on the unit sphere in E to a Keplerian orbit. It suffices to compute the vector $\vec{L}$ and the bivector C which characterize the orbit, to identify C with a vector by choosing an orientation of E, to divide this vector by $\sqrt{a}$, and to let $\vec{S}$ denote the result. The two desired points are $\vec{S} + \vec{L}$ and $\vec{S} - \vec{L}$. The parametrization of Keplerian ellipses obtained in this way is quite remarkable.

Lecture 5. Lambert's Theorem

Introduction

We begin with a configuration of three points A_1, A_2 and O in the Euclidean plane, with $A_1 \neq A_2$, and we seek to connect A_1 and A_2 by an arc of a Keplerian orbit with "attractive focus" O (we use this term rather than our usual "fixed center" because in what follows it is convenient to think of O as an arbitrary point in the plane, rather than as fixed at the origin). This may be done in infinitely many ways. There is in fact a three-parameter family of planar Keplerian orbits with attractive focus O. The condition of passing through a point determines one of the parameters. There is thus a one-parameter family of orbits passing through A_1 and A_2.

We shall use the words "arc," "orbit," and "configuration" in very precise senses. We may associate to an arc the orbit which traces it out, and also the configuration formed by the first extremity A_1, the second extremity A_2, and the attractive focus O. We consider an arc of an orbit to trace out *less than one full turn* in joining its two extremities. An orbit does not have extremities, but does have an attractive focus O. We consider orbits to be oriented, in other words to have a sense of circulation. Arcs of orbits are thus also oriented, though not necessarily from A_1 toward A_2.

The quantities $r_1 = \|OA_1\|$, $r_2 = \|OA_2\|$, and $c = \|A_1A_2\|$ are functions on the space of arcs that depend only on the configuration. The energy H is a function on the space of arcs that depends only on the orbit. The time of travel Δt is a function of the arc, which may be positive or negative, but never zero. We may now state Lambert's theorem.

29. *Theorem.* The four functions c, $r_1 + r_2$, H, and Δt are functionally dependent.

A more traditional statement is as follows: *The time of travel Δt depends only on the energy H, the sum $r_1 + r_2$, and the length c of the chord.* This statement may be deduced from ours by means of the implicit function theorem. It is only correct locally. In fact, it is clear that since H, c, and $r_1 + r_2$ depend only on the orbit, they cannot determine Δt, which also depends on the arc. We point out that to a single elliptic orbit, there correspond two distinct arcs. We shall also see shortly that to a configuration and to a given energy there correspond two non-oriented orbits (in other words, four oriented orbits).

Lambert's result is surprising. One expects a priori a functional relation among r_1, r_2, c, H, and Δt. In fact, once the first three quantities are known, the configuration is known (up to isometry), and there remains as we said only a single one-parameter family of arcs. There must therefore be a relation of functional dependence between the two parameters H and Δt. The surprise is that r_1 and r_2 enter this relation only by way of their sum.

Lambert's result is useful. It tells us that if we want to express Δt as a function of H or H as a function of Δt for a given configuration, we may reduce to the case of another configuration provided that $r_1 + r_2$ and c remain the same. We may therefore deform the configuration by displacing O along an ellipse with foci A_1 and A_2, and, in particular, we may reduce to the case where O, A_2, and A_1 are aligned in that order and where the orbit is rectilinear. In short, the relation between H and Δt is the same in the case of a rectilinear orbit and in the case of a nonrectilinear orbit. We shall illustrate this idea by reproving a result of C. Simó. Lambert himself used his result to simplify tables giving Δt as a function of the orbital parameters; these tables were useful at the time for the problem of determining the orbits of comets.

Lambert's theorem aroused considerable attention. We shall cite only the best-known names. Before Lambert, Euler had obtained the particular case for parabolic orbits, which is also found in Newton's work in a slightly different form. After Lambert's proof in 1761, using the geometric "synthesis," Lagrange was the first to publish an analytic proof in 1766, followed by three other such proofs in 1778. Laplace, Gauss, Hamilton, Jacobi, Cayley, Sylvester, Adams, Catalan, and more recently Souriau have all published proofs in turn, by more or less modifying those existing beforehand.

What makes the result remarkable is that it gives a property that is useful and extremely difficult to see, in a context where one might instead think that everything is clear: the relations between H and Δt are obtained by elementary means, through the eccentric anomaly and Kepler's equation.

Lambert Parameters, and First Among Them, Energy

We might think that Lambert's theorem describes a property specific to the time-energy pair. To disabuse ourselves of this way of thinking and to clarify the theorem's proof, we introduce the notion of Lambert parameters.

30. *Definition.* Lambert parameters are real analytic functions whose domain of definition contains an open set in the space of arcs, and which satisfy the following axioms:

  (i) A Lambert parameter is not constant on a family of arcs with the same configuration;

  (ii) If f and g are two Lambert parameters, the four functions c, $r_1 + r_2$, f, and g are functionally dependent;

(iii) The energy H is a Lambert parameter.

    We shall successively introduce various Lambert parameters. At the same time we shall establish the formulas by which they are interrelated, all of which are very simple. At the end, we will have established the following result, which contains Lambert's theorem.

31. *Proposition.* The energy H, the normalized ordinate $\hat{\beta}$ of the eccentricity vector, the increment ΔJ of the scalar product $J = \langle \vec{r}, \vec{v} \rangle$, the increment Δu of the eccentric anomaly, the time increment Δt, and the action integral ΔS are all Lambert parameters.

    Let us now try to understand how the energy parametrizes the orbits passing through A_1 and A_2. The simplest means is by way of the following geometric construction. We first concentrate on the elliptic case, where the semimajor axis $a = -(2H)^{-1}$ is positive. Given O, A_1, A_2, and a, we must find the orbit, which amounts to finding the second focus F of the ellipse. Since $\|OA_1\| + \|A_1F\| = 2a$, it is located on the circle $\mathcal{C}_1$ with center A_1 and radius $2a - r_1$, and also on the circle $\mathcal{C}_2$ with center A_2 and radius $2a - r_2$. The intersection $\mathcal{C}_1 \cap \mathcal{C}_2$ may be empty or may consist of two points, or of a single point. In the latter case, the segment A_1A_2 contains F, and we have $4a = r_1 + r_2 + c$. It is in this case that the semimajor axis is minimal; in other words, the energy is minimal. In the family of conic sections passing through A_1 and A_2, the smallest possible energy is thus $H_{\min} = -2(r_1 + r_2 + c)^{-1}$. To an energy $H > H_{\min}$ there correspond exactly two orbits. We have just proved this for negative energies only, but the result easily extends to $H \geq 0$.

Parameters Arising from the Unifocal Equation

Let x and y be orthogonal coordinates in the plane. Let $A_1 = (x_1, y_1)$ and $A_2 = (x_2, y_2)$. If the Keplerian orbit $r = \alpha x + \beta y + \gamma$ passes through the two points, the triple $(\alpha, \beta, \gamma) \in \mathbb{R}^3$ satisfies the two constraints $r_1 = \alpha x_1 + \beta y_1 + \gamma$ and $r_2 = \alpha x_2 + \beta y_2 + \gamma$. These equations determine a line in $\mathbb{R}^3$, whose points on which γ is positive "parametrize" our family of orbits.

We use the following conventions: the axis x is parallel to the line $A_1 A_2$, which amounts to $y_1 = y_2$, and the two inequalities $y_1 \geq 0$ and $x_2 < x_1$ are satisfied. The length of the chord is thus $c = x_1 - x_2 > 0$. Subtracting the two constraints above, we find $r_1 - r_2 = \alpha c$. The abscissa α of the eccentricity vector is thus the same for all the Keplerian orbits passing through A_1 and A_2.

There remains the linear relation $r_1 - \alpha x_1 = \beta y_1 + \gamma$, which allows us to eliminate β or γ. We would like to retain $\gamma = C^2$, which is a more natural parameter than β. Unfortunately, in the interesting particular case $y_1 = 0$, γ is not a parameter. We therefore choose to keep β.

32. *Remark.* Let us be more explicit about the particular case $y_1 = 0$. Suppose A_2, O, and A_1 are aligned in that order. For all orbits joining A_1 and A_2, the norm of the angular momentum is the same: we have $\gamma = C^2 = 2r_1 r_2 (r_1 + r_2)^{-1}$. It is perhaps interesting to compare this odd fact with another peculiarity of the situation. Suppose for a moment that the space of motions is three-dimensional. These configurations are then the only ones for which the plane of the orbit is undetermined. We remark lastly that Lambert's theorem reduces these particular configurations to the rectilinear case $O = A_2$.

We now seek to give a symmetric form to the expression for γ as a function of β. Our triangle was determined by the three numbers x_1, x_2, and $y_1 = y_2$. We express these numbers as a function of the new variables r_1, r_2, and c:

$$x_1 = \frac{r_1^2 + c^2 - r_2^2}{2c}, \qquad x_2 = \frac{r_1^2 - c^2 - r_2^2}{2c}, \qquad y_1 = y_2 = \hat{y} \sin \phi,$$

with

$$\hat{y} = \frac{1}{2}\sqrt{(r_1 + r_2)^2 - c^2}, \qquad \sin \phi = \frac{1}{c}\sqrt{c^2 - (r_1 - r_2)^2}.$$

It is convenient to write the last quantity as the sine of an angle ϕ, since we have already encountered $\cos\phi = \alpha = (r_1 - r_2)/c$. The triangle inequality guarantees that $|\alpha| \le 1$. The triple (α, β, γ) is now given as a function of β by the formulas

$$\alpha = \cos\phi,$$

$$\gamma = \bar{r}\sin^2\phi - \beta\hat{y}\sin\phi,$$

$$\bar{r} = \tfrac{1}{2}(r_1 + r_2).$$

Finally, we prescribe an acceptable interval for β by writing the condition $\gamma > 0$. The interval is $(-\infty, \beta_0)$, with $\beta_0 = \bar{r}\sin\phi/\hat{y}$ in the case $\hat{y} > 0$, and $(-\infty, +\infty)$ if $\hat{y} = 0$.

The variable β conveniently parametrizes the family, except in the particular case $\sin\phi = 0$, which corresponds to the alignment of the three points in the order OA_2A_1. The following observation integrates this exception into the general case. We express the energy H as a function of β. We have

$$2H = \frac{k^2 - 1}{C^2} = \frac{\alpha^2 + \beta^2 - 1}{\gamma} = \frac{\beta^2 - \sin^2\phi}{\bar{r}\sin^2\phi - \beta\hat{y}\sin\phi},$$

or, setting

$$\hat{\beta} = \frac{\beta}{\sin\phi}, \qquad 2H = \frac{\hat{\beta}^2 - 1}{\bar{r} - \hat{\beta}\hat{y}}. \tag{36}$$

In this way the normalized ordinate $\hat{\beta}$ of the eccentricity vector is connected to H by a formula in which only $\bar{r}$ and c appear. *It is thus a Lambert parameter.*

The parameter β was zero for all rectilinear orbits. Because of the indeterminacy in the formula that defines it, $\hat{\beta}$ is by contrast a good local parameter in this situation also, since the energy is a good local parameter. We shall return later to this parametrization.

We now study the function H of the variable $\hat{\beta}$. It vanishes at $\hat{\beta} = \pm 1$. Its derivative vanishes if and only if $\hat{y}\hat{\beta}^2 - 2\hat{\beta}\bar{r} + \hat{y} = 0$, that is, at $\hat{\beta} = 0$ if $\hat{y} = 0$, and at $\hat{\beta} = (r_1 + r_2 \pm c)/2\hat{y}$ if $\hat{y} > 0$. We must choose the negative sign to remain in the interval $(-\infty, \hat{\beta}_0)$, where $\hat{\beta}_0 = (r_1 + r_2)/2\hat{y}$. Then $\bar{r} - \hat{\beta}\hat{y} = c/2$ and $H = -2(r_1 + r_2 + c)^{-1}$, which is indeed the value $H_{\min}$ computed earlier.

A Third Lambert Parameter

We write $J_1 = \langle \vec{r}_1, \vec{v}_1 \rangle$ and $J_2 = \langle \vec{r}_2, \vec{v}_2 \rangle$. We shall show that $\Delta J = J_2 - J_1$ is a Lambert parameter.

We begin with the formula $CJ = -\alpha y + \beta x$, which we encountered in Lecture 2 just before the statement of Proposition 11. By subtracting what it gives at the points A_2 and A_1, we find $C\Delta J = -\beta c$. But $C = \pm\sqrt{\gamma}$. We thus set $\hat{C} = C/\sin\phi = \pm(\bar{r} - \hat{\beta}\hat{y})^{1/2}$. The formula $\hat{C}\Delta J = -\hat{\beta}c$ expresses ΔJ as a function of $\bar{r}$, c, and the Lambert parameter $\hat{\beta}$, from which the desired result follows.

We note that H and ΔJ have a simple and explicit expression in terms of the variable $\hat{\beta}$. The relation between H and ΔJ obtained by eliminating $\hat{\beta}$ is more complicated. We also note that the appearance of a choice of sign in the expression for ΔJ was foreseeable, since this quantity depends on the sense of circulation, whereas neither H nor $\hat{\beta}$ does.

The Eccentric Anomaly and the Variable s

Consider a circle of radius R. We may relate the length l of a chord to the angle θ subtended by the chord by means of the elementary formula $l = 2R\sin(\theta/2)$. What remains of this formula when we replace the circle by an elliptic Keplerian orbit with chord $c = \|A_1 A_2\|$ allows us to relate the increment Δu of the eccentric anomaly to the Lambert parameter ΔJ.

To calculate the length of the chord c, we may use Proposition 9 to place ourselves in the (r, J)-plane:

$$k^2 c^2 = (r_2 - r_1)^2 + C^2(J_2 - J_1)^2.$$

But in the variables (r, Y), where $Y = \sqrt{a}J$, the first auxiliary conic is a circle of radius $R = ak$. Thus

$$\left(2ak\sin(\Delta u/2)\right)^2 = (r_1 - r_2)^2 + a(J_2 - J_1)^2.$$

Upon subtracting, using the formula $C^2 = a(1 - k^2)$, and dividing by k^2, we find

$$4a^2 \sin^2(\Delta u/2) - c^2 = a(\Delta J)^2. \tag{37}$$

This expression shows that Δu is a Lambert parameter.

In comparison with H, $\hat{\beta}$, or ΔJ, the parameter Δu has several novel features. The expression relating it to the other parameters is not algebraic, since it contains a trigonometric function. In addition, Δu depends not only on the orbit, but also on the arc of the elliptic orbit chosen to connect A_1 to A_2. Finally, in contrast to the previous Lambert parameters, Δu is not defined for all orbits in the family, but rather only for elliptic orbits.

We may extend the formula to parabolic and hyperbolic orbits. We introduce the entire function σ such that $\sinh x = x\sigma(x^2)$, or in other words $\sin x = x\sigma(-x^2)$. We then have $\sigma(y) = 1 + y/6 + y^2/120 + \cdots$.

We also introduce a variable s, the "regularizing time" of Levi-Civita, which is similar to the eccentric anomaly, but which remains defined for all parabolic and hyperbolic orbits. In order to define it, we modify formula (17), $du = r^{-1}a^{-1/2}dt$. We define s by the formula $ds = r^{-1}dt$. In this way, if $s = 0$ at $u = u_0$, we have $s = \sqrt{a}(u - u_0)$. The formula becomes

$$\left(\sqrt{a}\,\Delta s\,\sigma\left(-\frac{(\Delta s)^2}{4a}\right)\right)^2 - c^2 = a(\Delta J)^2,$$

or, since $H = -1/2a$,

$$\left(\Delta s\,\sigma\left(\frac{H}{2}(\Delta s)^2\right)\right)^2 + 2Hc^2 = (\Delta J)^2. \tag{38}$$

In this way the Lambert parameter Δs is related to the Lambert parameters H and ΔJ.

Time and Action

We recall the formula $\dot{J} = v^2 - r^{-1}$, or $dJ = (v^2 - r^{-1})dt$. Let us relate this expression to $ds = r^{-1}dt$, and to $2H dt = (v^2 - 2r^{-1})dt$. We have

$$dJ = ds + 2H dt \quad \text{or} \quad \Delta t = \frac{1}{2H}(\Delta J - \Delta s). \tag{39}$$

The quantity Δt is thus related to the Lambert parameters ΔJ, Δs, and H, which makes it a Lambert parameter. Lambert's classical result is therefore proved.

The Actions. We write $\Delta S = \int v^2 dt$. This is Hamilton's "characteristic function." The formula

$$\Delta S = \Delta J + \Delta s \tag{40}$$

shows that ΔS is a Lambert parameter. There is another action integral, the integral of the Lagrangian $\int (v^2/2 + r^{-1})dt$, called the "principal function" by Hamilton, which gives a Lambert parameter in the same way.

33. *Remark.* We mention here a remark due to Tait on the action integral ΔS. *In the case of an elliptic orbit, this quantity is proportional to the area swept out by the segment connecting the body to the second focus of the ellipse.* The proof proceeds in much the same way as the geometric proof of Kepler's equation sketched at the beginning of Lecture 2. In fact, the passage from the first to the second focus exactly translates the change of sign between the expressions $2H\Delta t = \Delta J - \Delta s$ and $\Delta S = \Delta J + \Delta s$.

Euler's Formula

The formulas used in the proof above are

$$\hat{C}^2 = \bar{r} - \hat{\beta}\hat{y}, \qquad 2H\hat{C}^2 = \hat{\beta}^2 - 1, \qquad \Delta J = -\frac{c\hat{\beta}}{\hat{C}} = \Delta s + 2H\Delta t,$$

and the trigonometric formula (38) which gives Δs. These formulas allow us to obtain expressions for Δt, which we now wish to simplify in the parabolic case $H = 0$. Expression (38) becomes $(\Delta s)^2 = (\Delta J)^2$, but also the above relation becomes $\Delta s - \Delta J$, which is more precise. In fact, the integral formulas that led us to the expression for Δt contain very natural implicit sign conventions. Thus, the sign of Δs is the same as that of $\Delta t = t_2 - t_1$. The form of relation (38) "oriented" by these conventions in the neighborhood of $H = 0$ is

$$\Delta J = \Delta s \sqrt{\sigma^2 \left(\frac{H}{2}(\Delta s)^2 \right) + 2H \left(\frac{c}{\Delta s} \right)^2}.$$

Expanding, we find

$$2\Delta t = \frac{\Delta J - \Delta s}{H} \bigg|_{H=0} = \Delta s \left(\frac{(\Delta s)^2}{12} + \frac{c^2}{(\Delta s)^2} \right),$$

an expression in which we may set $\Delta s = \Delta J$. The hypothesis $H = 0$ leads to $\hat{\beta}^2 = 1$. We write $\hat{\beta} = \epsilon = \pm 1$. Other notation will be useful below:

$$x_1 = \frac{1}{2}(r_1 + r_2 + c), \qquad x_2 = \frac{1}{2}(r_1 + r_2 - c),$$

which gives

$$\bar{r} = \frac{1}{2}(x_1 + x_2), \qquad \hat{y} = \sqrt{x_1 x_2}, \qquad c = x_1 - x_2.$$

We thus have

$$\hat{C}^2 = \frac{1}{2}(x_1 + x_2) - \epsilon\sqrt{x_1 x_2} = \left(\sqrt{\frac{x_1}{2}} - \epsilon\sqrt{\frac{x_2}{2}}\right)^2.$$

We choose

$$\hat{C} = \sqrt{\frac{x_1}{2}} - \epsilon\sqrt{\frac{x_2}{2}} > 0,$$

a choice that corresponds to a sense of circulation of the orbit. Substituting $\Delta s = \Delta J = -\epsilon c/\hat{C}$ in the expression for Δt, and simplifying, we find

$$\Delta t = -\frac{4}{3}\epsilon\left(\left(\frac{x_1}{2}\right)^{3/2} + \epsilon\left(\frac{x_2}{2}\right)^{3/2}\right). \tag{41}$$

This is Euler's formula. We quickly see that the calculations just presented are unnecessary. In fact, for a *rectilinear* parabolic motion, the time required to join the origin to the point with abscissa x_1 is

$$T_1 = \int dt = \int_0^{x_1} \frac{dx}{\dot{x}} = \int_0^{x_1} \frac{dx}{\sqrt{2x^{-1}}} = \frac{4}{3}\left(\frac{x_1}{2}\right)^{3/2}.$$

In the rectilinear case we may thus foresee an expression of the form $\Delta t = \pm T_1 \pm T_2$. If, instead of a recipe, we retain from Lambert's theorem that the expression for Δt is the same in the case of a general arc as it is in the case of a rectilinear arc, and if we can determine this rectilinear arc with enough precision so as to choose the signs, then we will know how to write Euler's formula directly in its simplified form.

Determining the Rectilinear Arc

To any general arc (we use the adjective "general" in contradistinction to the particular case "rectilinear"), corresponding to a configuration of sides r_1, r_2, and c, we may associate a unique rectilinear arc having the same c, the same sum $r_1 + r_2$, and the same Lambert parameters, as follows.

Choice of the Rectilinear Configuration. This choice is obtained by moving O along the ellipse passing through O with foci A_1 and A_2 in such a way as to achieve the alignment O, A_2, A_1, in that order. The respective abscissas of these three points are then 0, $x_2 = (r_1+r_2-c)/2$, and $x_1 = (r_1+r_2+c)/2$.

Choice of the Rectilinear Orbit. We choose the rectilinear orbit having the same $\hat{\beta}$ as the general orbit.

Choice and Description of the Rectilinear Arc. An arc may be either "direct" or "indirect." It is indirect if it "goes around" O, in other words, if O is in its convex hull. An indirect rectilinear arc is an arc with "bounce" at O. We choose the direct arc if and only if the general arc was direct.

34. *Cayley's Criterion and Description of the Rectilinear Orbit.* Two values of $\hat{\beta}$ give the same value of H. The two rectilinear orbits associated with these two values are similar, but we shall distinguish between them. Consider first the general case. Cayley gave a qualitative criterion for distinguishing the two ellipses with semimajor axis a passing through the two points A_1 and A_2, which may be deduced from the compass construction described at the beginning of this lecture: *For only one of these two ellipses, the line A_1A_2 separates the two foci O and F.* To determine which one, we list the intervals where $\hat{\beta}$ takes its values:

$$H' = (-\infty, -1), \quad E' = \left(-1, \sqrt{\frac{x_2}{x_1}}\right),$$

$$E'' = \left(\sqrt{\frac{x_2}{x_1}}, 1\right), \quad H'' = \left(1, \frac{x_1 + x_2}{2\sqrt{x_1 x_2}}\right).$$

A little care shows that if $\hat{\beta} \in E'$, the line A_1A_2 does not separate the two foci, while it does separate them if $\hat{\beta} \in E''$. Passing to the limit of rectilinear orbits, this criterion becomes the following: *We consider a rectilinear orbit to be composed of the superposition of two "branches," and we may pass from one branch to the other by "bouncing" from the point O, or, in the elliptic case, when the velocity vanishes at the "top" of the orbit; if $\hat{\beta} < \sqrt{x_2/x_1}$, the two points A_1 and A_2 are on the same branch; if $\hat{\beta} > \sqrt{x_2/x_1}$, the two points are on two distinct branches.*

35. *Determination of the Signs in Euler's Formula.* Once again we let T_1 and T_2 denote the times required to join the origin to x_1 and x_2, respectively, by means of a rectilinear parabolic motion. We assumed above the hypothesis $\hat{C} > 0$, which fixed the sense of circulation of the orbit in order

to establish the formula $\Delta t = -\epsilon(T_1 + \epsilon T_2)$. We wish to rederive this formula, for which it is enough to determine the signs. We see that $\hat{C} > 0$ is equivalent to $v_2 < 0$ in the case of a rectilinear orbit. If $\hat{\beta} = \epsilon = -1$, the two points A_1 and A_2 are on the same branch of the rectilinear parabolic orbit. The initial velocity v_1 is therefore negative, so that $\Delta t = T_1 - T_2$. If $\hat{\beta} = \epsilon = 1$, the two points are on distinct branches, hence $v_1 > 0$ and x_1 is attained after one bounce. Thus $\Delta t = -T_1 - T_2$.

A Result of C. Simó

36. *Proposition.* Given any configuration OA_1A_2 with $A_1 \neq A_2$, and given $\Delta t > 0$, there exist a unique direct arc and a unique indirect arc joining A_1 to A_2 in a time Δt.

Recall that we exclude arcs that execute more than one turn around O. We note that in the particular case where A_1, O, and A_2 are aligned in that order, the two arcs are mutually symmetric. The distinction between the direct and the indirect arc is a matter of convention only. If in addition A_2 coincides with O, the two arcs are rectilinear and thus identical.

Proof. According to Lambert's theorem, it is enough to treat the rectilinear case. We will need some auxiliary results.

37. *A Sequence of Orbits.* Having fixed the configuration, we have up to now considered $\hat{\beta}$ as the natural parameter of the family of orbits. We now show that ΔJ, or in the rectilinear case v_1, are monotone functions of $\hat{\beta}$. The conversion formulas are very simple:

$$\Delta J = -\frac{c\hat{\beta}}{\hat{C}}, \qquad v_1 = \frac{\hat{\beta} - \sqrt{x_2/x_1}}{\hat{C}}. \tag{42}$$

We assume that $\hat{C} > 0$, that is, that $v_2 < 0$. To show that v_1 is an increasing function of $\hat{\beta}$, we compute the derivative of the above expression. It is $\hat{C}^{-3}(\bar{r} - x_2 + \hat{C}^2)/2$, which is strictly positive. Computing in turn the derivative of ΔJ, we find $-c\,\hat{C}^{-3}(\bar{r} + \hat{C}^2)/2$, which is strictly negative.

38. *The Sequence of the Arcs.* The direct arc exists and is continuously deformed as $\hat{\beta}$ moves through the interval $(-\infty, 1)$. When $\hat{\beta} = -\infty$, it is a straight arc traversed instantaneously. As $\hat{\beta} \to 1$, it is a piece of an ellipse stretching to infinity. The indirect arc exists and is continuously deformed in the interval $(-1, (x_1+x_2)/2\sqrt{x_1 x_2})$. At the end of the interval,

it becomes the union of the two segments A_1O and OA_2, also traversed instantaneously. If we choose a sense of circulation of the orbit such that $\Delta t > 0$ for the direct arc, and if this choice is maintained as $\hat{\beta}$ is varied, then the indirect arc will be traversed in a time $\Delta t < 0$.

Heuristic Proof of Simó's Result. We first consider the direct arc. It is enough to show that the derivative of Δt with respect to $\hat{\beta}$ is strictly positive. But this is very intuitive upon replacing $\hat{\beta}$ by the parameter v_1. The more negative the initial velocity v_1, the more quickly we reach A_2. The same argument applies to the indirect arc.

A Rigorous Proof. We write $H = v_1^2/2 - 1/x_1$ and we consider the formula

$$\Delta t = \int_{x_2}^{x_1} \frac{dx}{\sqrt{2H + 2x^{-1}}}.$$

Upon differentiating this formula with respect to H, we immediately find that Δt is a decreasing function of H, and thus an increasing function of $\hat{\beta}$ on the intervals H' and E'. For the indirect arc, the same argument shows that Δt is decreasing (here we choose the sense of circulation so that this number is positive) on the intervals E'' and H''. We now return to the direct arc on the interval E''. To show that Δt is increasing, it is enough to note that $\Delta t_{\text{direct}} = T - \Delta t_{\text{indirect}}$, where T is the period of the elliptic motion, that T is an increasing function, and $\Delta t_{\text{indirect}}$ a decreasing function. The end of the proof for the indirect arc is analogous.

39. *Supplement.* In order to show that the derivative of Δt with respect to $\hat{\beta}$ is everywhere strictly positive, it remains to evaluate this derivative at $\hat{\beta} = \sqrt{x_2/x_1}$. We restrict ourselves to the direct arc for simplicity. We find that at this point

$$\hat{C}^2 = \frac{c}{2}, \qquad \frac{d\Delta J}{d\hat{\beta}} = -x_1\sqrt{\frac{2}{c}}, \qquad \frac{d\Delta t}{d\Delta J} = \frac{1}{H},$$

which completes the proof. This last formula is quite remarkable. It is very naturally deduced from

$$\frac{d\Delta S}{dH} = \Delta t, \tag{43}$$

which is taken from Hamilton's theories; it is usually derived by the calculus of variations and is valid for all problems in Lagrangian mechanics involving families of orbits connecting two fixed points in configuration space. From this formula we deduce that ΔS, considered as a function of $\hat{\beta}$, possesses a critical point at $\hat{\beta} = \sqrt{x_2/x_1}$. Formula (40) thus shows that $d(\Delta s)/d(\Delta J) = -1$, which may be used with expression (39) for Δt.

Concerning Question 4. Several authors have presented Lambert's theorem in the Hamiltonian framework. Dziobek went so far as to say that this framework was the "true source" of the theorem. However, the calculations in such presentations do not seem simpler than those presented here, or than those of Laplace, for example. It is nonetheless interesting to recall the classical formula (43), which gives a differential relation among three Lambert parameters. Since it suffices to know two of the parameters ΔJ, Δs, ΔS and Δt in order to know them all, we could be satisfied with the expression for ΔJ, and from this relation establish Lambert's theorem. Relation (37) may be omitted in this way. But the proof would lose some of its elementary character, without a corresponding gain in insight or brevity.

In these lectures, we only used Hamiltonian notions in §39 and in one of the two proofs of Bertrand's theorem, during a perturbation calculation. It was perturbation theory that was the basis for these notions' success in the 19th century, and they remain of capital importance today. Nonetheless, we leave the final words of conclusion to G. D. Birkhoff: "I believe that the whole apparatus of transformation which is used in connection with the Hamiltonian equations has been overrated."

March, 2000

Bibliography and Author Index

J. C. Adams, *On a simple proof of Lambert's Theorem*, Messenger of Math. 7 (1877) pp. 97–100 [L5].

J. le Rond d'Alembert, *Recherches sur differens points importans du système du monde*, Livre premier, Theorie de la lune (1754) [L3] p. 34.

P. Appell, *Sur les lois de forces centrales faisant décrire à leur point d'application une conique quelles que soient les conditions initiales*, American Journal of Mathematics 13 (1891) pp. 153–158, [L3] p. 156.

V. I. Arnold, *Mathematical Methods of Classical Mechanics*, Springer-Verlag, New York 1978, 1989, [L2].

J. Bernoulli: see Herman.

J. Bertrand, *Théorème relatif au mouvement d'un point attiré vers un centre fixe*, Comptes Rendus 77 (1873) pp. 849–853 [L3].

G. D. Birkhoff, *A mathematical critique of some physical theories*, Bull. Amer. Math. Soc. 33 (1927) pp. 165–181; reprint ibid. (New Series) 37/1 pp. 65–74 [L5] p. 70.

E. Catalan, *Note sur le théorème de Lambert*, Nouv. Annal. de Math. 3/3 (1884) pp. 506–513 [L5].

A. Cayley, *Note on Lambert's theorem for elliptic motion*, Month. Notices of the R.A.S. 29 (1868) pp. 318–320; Math. Papers 7 pp. 387–389, [L5 §34].

A. Cayley, *On Lambert's theorem for elliptic motion*, Monthly Notices of the Royal Astronomical Society 22 (1862) pp. 238–242; Math. Papers 3, Cambridge, pp. 562–565 [L5].

A. C. Clairaut, *Théorie de la lune* (1750), [L3].

R. Cotes, *Logometria*, Philosophical Transactions 29 (1714) pp. 5–45, and *Harmonia Mensurarum*, pp. 31, 98, [L3].

O. Dziobek, *Mathematical Theories of Planetary Motions* (German original, 1888, translation 1892, Dover 1962), [L4] p. 6, [L5] p. 136.

L. Euler, *Determinatio orbitae cometae qui mense martio huius anni 1742 potissimum fuit observatus*, Miscellanea Berolinensia 7 (1743) pp. 1–90; opera omnia 2/28 pp. 28–104, [L5]; opera p. 40.

K. F. Gauss, *Theoria Motus Corporum Coelestium in sectionibus conicis solem ambientium* (1809), translation by C.H. Davis (1857), Dover (1963), [L2] p. 6, [L4] p. 3, [L5] p. 145.

J. W. Gibbs, E.B. Wilson, *Vector Analysis, a Text-Book for the Use of Students of Mathematics and Physics*, Scribners, New York (1901), Dover (1960), [L4] p. 135.

W. R. Hamilton, *On the application of the method of quaternions to some dynamical questions*, Proc. Roy. Irish Acad. 3 (1847) July 1845; Math. Papers 3, pp. 441–448, [L4] p. 442.

W. R. Hamilton, *Elements of Quaternions*, (1866, 1899), Chelsea (1969), vol. II, [L2] p. 300, [L5] p. 314.

W. R. Hamilton, *On a general method in dynamics by which the study of the motions of all free systems of attracting or repelling points is reduced to the search and differentiation of one central relation, or characteristic function*, Philosophical Transactions (1834) pp. 247–308; Mathematical Papers, vol. 1, pp. 103–161, [L5] pp. 280–286, p. 251.

J. Herman, J. Bernoulli, *Extrait d'une lettre & extrait de la réponse*, Histoires de L'Académie Royale des Sciences avec les Mémoires de Mathématique et Physique, Paris, (1710) pp. 102–103 and 519–544; reprint (1713), Amsterdam, pp. 682–703; Œuvres de Johann Bernoulli, Lausanneae et Genevae, vol. I, p. 470, [L2].

C. G. J. Jacobi, *De motu puncti singularis*, Crelle Journal für die reine und angewandte Mathematik, Bd. 24 (1842) pp. 5–27; Werke Bd. 4, pp. 263–288, [L3] p. 282.

C. G. J. Jacobi, *Über die Reduction der Integration der partiellen Differentialgleichungen erster Ordnung zwischen irgend einer Zahl Variabeln auf die Integration eines einzigen Systemes gewöhnlicher Differentialgleichungen*, Crelle Journal, Bd. 17, pp. 97–162; Werke Bd. 4, (1836) pp. 57–127, [L5] p. 98. See also Wintner, §248.

J. L. Lagrange, *Théorie des variations séculaires des éléments des planètes*, Œuvres, Gauthier-Villars, Paris, vol. 5, (1781) pp. 125–207, [L2] p. 132.

J. L. Lagrange, *Mécanique analytique*, (1788) quatrième édition, tome second, Œuvres, vol. 12 (1888), [L2], [L5] p. 30.

J. L. Lagrange, *Solution de différents problèmes de calcul intégral*, (1762–1765) Miscellanea Taurinensia 3; Œuvres, vol. 1, pp. 469–668, [L3] p. 573.

J. L. Lagrange, *Recherches sur la théorie des perturbations que les comètes peuvent éprouver par l'action des planètes*, Œuvres, vol. 6, (1778) pp. 403–503, [L4] p. 419.

J. L. Lagrange, *Recherches sur le mouvement d'un corps qui est attiré vers deux centres fixes*, Œuvres, vol. 2, (1766–1769) pp. 67–121, [L5] p. 84.

J. L. Lagrange, *Sur une manière particulière d'exprimer le temps dans les sections coniques, décrites par des forces tendantes au foyer et réciproquement proportionnelles aux carrés des distances*, Œuvres, vol. 4, (1778) pp. 559–582, [L5].

J. H. Lambert, *Insigniores Orbitæ Cometarum Proprietates*, Augusta Vindelicorum, Augsburg (1761), [L5]. See also Cayley and Ostwald's *Klassiker der exakten Wissenschaften*, Verlag von Willen Engelmann, no. 133, (1902).

P. S. Laplace, *Sur le principe de la gravitation universelle et sur les inégalités séculaires des planètes qui en dépendent*, (1773), Œuvres Complètes, Gauthier-Villars, Paris, tome 8, pp. 199–275, [L3 first remark] p. 214.

P. S. Laplace, *Traité de Mécanique Céleste, livre II*, (1798), Œuvres Complètes, tome 1, [L3 second remark] §11 p. 153, [L4] §17 p. 175, [L5] §27 p. 214.

T. Levi-Civita, *Sur la résolution qualitative du problème restreint des trois corps*, Acta Math. 30 (1906) pp. 305–327, [L4], [L5] pp. 311–313.

H. Liebmann, *Geometrische Theorie der Differentialgleichungen*, Enc. d. Math. Wiss. 3/3 (1914) pp. 526–528, [L3] p. 528.

V. V. Koslov, A. O. Harin, *Kepler's problem in constant curvature spaces*, Celestial Mechanics and Dynamical Astronomy 54 (1992) pp. 393–399, [L3].

J. Moser, *Regularization of Kepler's problem and the averaging method on a manifold*, Communications on Pure and Applied Mathematics 23 (1970) pp. 609–636, [L4].

I. Newton, *Mathematical Principles of Natural Philosophy*, (1687–1713–1726), Motte's translation revised by F. Cajori, University of California Press, Berkeley (1934), or a new translation by I. B. Cohen and A. Whitman, University of California Press (1999), [L2] book 1, prop. 41, [L3] prop. 9, prop. 45, [L4] prop. 17, [L5] book 3, lemma 10.

Pappus of Alexandria: see Euclid, *The thirteen books of the elements*, translated with introduction and commentary by Sir Thomas L. Heath, Second Ed., Cambridge University Press (1925), Dover (1956), [L4] p. 15.

P. Serret, *Théorie nouvelle géométrique et mécanique des lignes à double courbure*, Mallet-Bachelier, Paris (1860), [L3] p. 204.

C. Simó, *Solución del problema de Lambert mediante regularización*, Collectanea Mathematica 24 (1974) pp. 231–247, [L5].

J. M. Souriau, *Géométrie globale du problème à deux corps*, Proceedings of the IUTAM–ISIMM, Symposium on Modern Developments in Analytical Mechanics, Academy of Sciences of Turin, Turin (1983) pp. 369–418, [Introduction], [L5] p. 376.

J. M. Souriau, *Structure des systèmes dynamiques*, Dunod, Paris (1969), translation: *Structure of Dynamical Systems: A Symplectic View of Physics*, by C. H. Cushman-de Vries, Birkhäuser, (1997), [L4] §12.154.

J. J. Sylvester, *Astronomical propulsions: Commencing with an instantaneous proof of Lambert's and Euler's Theorems, and modulating through a construction of the orbit of a heavenly body from two heliocentric distances, the subtended chord, and the periodic time, and the focal property of cartesian ovals, into a discussion of motion in a circle and its relation to planetary motion*, Phil. Magazine 31 (1866) pp. 52–76; Math. Papers 2, Chelsea, pp. 519–541, [L5].

P. G. Tait, *On the application of Hamilton's characteristic function to special cases of constraint*, Trans. of the Royal Soc. of Edinburg 24 (1865); Scientific Papers 1, Cambridge, pp. 54–73 and *Note on the action in an elliptic orbit*, Quart. J. Math. 7 (1866) p. 45, [L5 §33].

R. Weinstock, *Inverse-square orbits: Three little-known solutions and a novel integration technique*, Am. J. Phys. 60 (1992) pp. 615–619, [L4].

E. T. Whittaker, *A Treatise on the Analytical Dynamics of Particles and Rigid Bodies, with an introduction to the problem of three bodies*, Fourth Edition, Cambridge University Press (1937), [Introduction].

A. Wintner, *The Analytical Foundations of Celestial Mechanics*, Princeton Math. Series 5, Princeton University Press, Princeton, NJ (1941), [Introduction].

Photograph: Antônio Tenório

Normal Forms of Hamiltonian Systems

and Stability of Equilibria

Hildeberto Eulalio Cabral

DEPARTAMENTO DE MATEMÁTICA
UNIVERSIDADE FEDERAL DE PERNAMBUCO
AV. PROF. LUIS FREIRE, S/N
CIDADE UNIVERSITÁRIA, RECIFE, PE, BRASIL
e-mail: hild@dmat.ufpe.br

1   Introduction

These notes on stability of equilibria of Hamiltonian systems are part of the material on stability theory included in a course on ordinary differential equations given at the Mathematics Department of the Federal University of Pernambuco at Recife, in 1998. It has been updated with part of the material in reference [6].

The main body of the notes consists of sections five through eight, but we have included three initial sections on Hamiltonian systems which make the exposition self-contained. In section 9 we include an application to the restricted three-body problem. Section 10 gives an exposition of the normalization technique of Deprit and Hori with its application to the proof of Theorem 6.1 on normal forms.

The texts that exerted the main influence on the preparation of these notes were the books by Markeev [20] and Meyer and Hall [21], both of which were used in courses on celestial mechanics at Recife.

I would like to thank Alain Albouy for his reading of these notes and for his suggestions.

2   Hamiltonian Systems

Consider a system of n mass particles in space with position vectors $\mathbf{r}_1, \ldots, \mathbf{r}_n$ and masses $m_1, \ldots, m_n$ moving under the action of a force field deriving from a potential function $U(r) = U(r_1, \ldots, r_n)$. Newton's equation of motion

$$m_j \ddot{r}_j = -\nabla_{r_j} U(r) \qquad (j = 1, \ldots, n),$$

where $\nabla_{r_i} U$ is the gradient of U with respect to the vector r_i, can be written as a first-order system of differential equations by setting $v_j = m_j \dot{r}_j$. We get

$$\dot{r}_j = \frac{1}{m_j} v_j, \quad \dot{v}_j = -\nabla_{r_j} U(r) \qquad (j = 1, \ldots, n).$$

Introducing the function

$$H(r, v) - \sum_{j=1}^{n} \frac{1}{2m_j} |v_j|^2 + U(r),$$

this system assumes the form

$$\dot{r}_j = H_{v_j}, \quad \dot{v}_j = -H_{r_j} \qquad (j = 1, \ldots, n).$$

An important particular case of this is the Newtonian n-body problem, in which n particles of masses $m_1, \ldots, m_n$ move in space subject only to the action of their mutual gravitational attraction; the potential function of this problem is

$$U(r) = -\sum_{i<j} \frac{m_i m_j}{|r_i - r_j|}.$$

As another example, consider the motion of a particle in the plane in a rotating coordinate system with constant angular speed ω. Introducing the moving orthonormal frame

$$e_1 = (\cos \omega t, \sin \omega t), \quad e_2 = (-\sin \omega t, \cos \omega t)$$

and using the relations $\dot{e}_1 = \omega e_2$, $\dot{e}_2 - -\omega c_1$, we see that the acceleration of the position vector

$$r = \xi \, e_1 + \eta \, e_2$$

is given by

$$\ddot{r} = (\ddot{\xi} - 2\omega\dot{\eta} - \omega^2 \xi)e_1 + (\ddot{\eta} + 2\omega\dot{\xi} - \omega^2\eta)e_2.$$

Now, the expression for the gradient of the potential V,

$$\nabla V = \frac{\partial V}{\partial \xi} \, e_1 + \frac{\partial V}{\partial \eta} \, e_2,$$

shows that the equation of motion

$$\ddot{r} = \nabla V(r)$$

can be written as the system of differential equations

$$\ddot{\xi} - 2\omega\dot{\eta} - \omega^2 \xi = \frac{\partial V}{\partial \xi}, \quad \ddot{\eta} + 2\omega\dot{\xi} - \omega^2\eta = \frac{\partial V}{\partial \eta}.$$

Setting

$$x_1 = \xi, \quad x_2 = \eta, \quad y_1 = \dot{\xi} - \omega\eta, \quad y_2 = \dot{\eta} + \omega\xi,$$

this system can be written as

$$\dot{x}_j = H_{y_j}, \quad \dot{y}_j = -H_{x_j} \qquad (j = 1, 2), \tag{1}$$

where

$$H = \frac{1}{2}(y_1^2 + y_2^2) + \omega(x_2 y_1 - x_1 y_2) - V(x_1, x_2, t).$$

An important particular case is the planar circular restricted three-body problem: two particles of masses m_1 and m_2, the primaries, move in circular orbits around their center of mass with angular speed ω, and a third particle of infinitesimal mass moves under the action of the primaries' attraction without disturbing their motion. Normalizing the units of mass and length so that the sum of the masses equals 1 and the distance between the primaries also equals 1, we have $\omega = 1$, and setting $m_1 = \mu$, $m_2 = 1 - \mu$, the potential is given by

$$V(r) = \frac{1 - \mu}{\rho_1} + \frac{\mu}{\rho_2},$$

with

$$\rho_1^2 = (\xi + \mu)^2 + \eta^2, \quad \rho_2^2 = (\xi - 1 + \mu)^2 + \eta^2.$$

A third particular example is the isosceles three-body problem in space with a symmetry axis. Here two particles of equal masses $m_1 = m_2 = m$ move symmetrically with respect to an axis, and a third particle, of mass m_3, moves always along this axis, the only forces on them coming from the Newtonian law of attraction. The symmetry and the center of mass fixed at the origin show that the motion of one of the particles determines the motion of the others. If r, θ, z are the cylindrical coordinates of one of the particles of mass m, the equations of motion are

$$\ddot{\rho} - \rho\dot{\theta}^2 = -\frac{m}{4\rho^2} - \frac{m_3}{[\rho^2 + (\frac{M}{m_3})z^2]^{3/2}},$$

$$\ddot{z} = -\frac{Mz}{[\rho^2 + (\frac{M}{m_3})z^2]^{3/2}}, \qquad \rho\ddot{\theta} + 2\dot{\rho}\dot{\theta} = 0,$$

where $M = 2m + m_3$ is the total mass of the system.

The third equation shows that $\rho^2\dot\theta = c$ is a constant of the motion. Using this relation to eliminate θ from the first two equations and setting $u = \sqrt{M/m_3}\, z$, we obtain

$$\ddot\rho = -\frac{\partial V}{\partial\rho}, \qquad \ddot u = -\frac{\partial V}{\partial u},$$

where

$$V(\rho, u) = \frac{c^2}{2\rho^2} - \frac{m}{4\rho} - \frac{m_3}{\sqrt{\rho^2 + (M/m_3)u^2}}.$$

Setting $x_1 = \rho$, $x_2 = u$, $y_1 = \dot\rho$, $y_2 = \dot u$ and taking

$$H(x_1, x_2, y_1, y_2) = \frac{1}{2}(y_1^2 + y_2^2) + V(x_1, x_2),$$

the equations for ρ and u can be written as in (1).

Systems of ordinary differential equations of the form

$$\dot x_j = H_{y_j}, \quad \dot y_j = -H_{x_j} \qquad (j = 1, \ldots, n) \qquad (2)$$

with $H = H(x_1, \ldots, x_n, y_1, \ldots, y_n)$ are called *Hamiltonian* or *canonical* systems of ordinary differential equations. The function H is called the *Hamiltonian function*, or simply, the *Hamiltonian* of the system. The variables $x_1, \ldots, x_n$ and $y_1, \ldots, y_n$ are said to be *conjugate* to each other. The former are the *positions*, whereas the latter are the *momenta*. The number n of variables $x_1, \ldots, x_n$ is called the number of degrees of freedom of the system.

Introducing the vector $z = (x_1, \ldots, x_n; y_1, \ldots, y_n)^T$, where ^{T} stands for transpose and considering the standard symplectic matrix

$$J = \begin{pmatrix} O & I \\ -I & O \end{pmatrix},$$

where I is the $n \times n$ identity matrix, system (2) can be written more concisely as

$$\dot z = J\nabla H(z), \qquad (3)$$

where $\nabla H(z)$ is the gradient of H at z.

The relations $J^{-1} = J^T = -J$, $J^2 = -I$ will frequently be used in the subsequent sections.

We close this section by noticing that *the Hamiltonian function H is a first integral of the system.* Indeed, if $z(t)$ is a solution of (3), then

$$\frac{d}{dt}H(z(t)) = \nabla H(z(t))^T J \nabla H(z(t)) = 0;$$

therefore, H remains constant along the solutions, and so the result follows.

3   Symplectic Changes of Coordinates and Generating Functions

We ask for the coordinate changes

$$\zeta = \phi(z), \tag{4}$$

which leave invariant the form (3) of the equations, that is, such that in the new variable ζ,

$$\dot{\zeta} = J \nabla \mathcal{H}(\zeta)$$

for some function $\mathcal{H}(\zeta) = \mathcal{H}(\xi_1, \dots, \xi_n; \eta_1, \dots, \eta_n)$.

Setting $\tilde{H}(\zeta) = H(z)$ and denoting by $Q = Q(z)$ the Jacobian matrix of $\phi(z)$, we get $\nabla H(z) = Q^T \nabla \tilde{H}(\zeta)$, so that (3) becomes

$$\dot{\zeta} = QJQ^T \, \nabla \tilde{H}(\zeta).$$

Therefore, if the matrix Q satisfies the relation

$$QJQ^T = \mu J \tag{5}$$

for some number $\mu \neq 0$, we get

$$\dot{\zeta} = J \nabla \mathcal{H}(z), \tag{6}$$

with

$$\mathcal{H}(\zeta) = \mu H(\phi^{-1}(\zeta)). \tag{7}$$

Conversely, if the coordinate transformation (4) has the property of transforming *every* Hamiltonian system (3) into another Hamiltonian system (6), then its Jacobian matrix satisfies equation (5) for some μ and the Hamiltonians are related by the expression (7) (see [21] for a proof).

A $2n \times 2n$ matrix Q that satisfies (5) is called a *symplectic matrix with multiplier* μ, or simply, a μ-symplectic matrix. If $\mu = 1$, it is called a *symplectic matrix*. The set of all symplectic matrices forms a group, called the *symplectic group*.

Equation (5) is equivalent to the equation

$$Q^T J Q = \mu J, \tag{8}$$

as can be seen by substituting in one of them the expression for Q^T obtained from the other, namely, $Q^T = -\mu J Q^{-1} J$.

Since the determinant of J is 1, it follows from (5) that the absolute value of the determinant of a symplectic matrix is equal to 1. In fact, the determinant is equal to 1 but the proof is more complicated (see [21, Chapter 2]).

If Q is represented in block form in terms of four $n \times n$ matrices,

$$Q = \begin{pmatrix} A & B \\ C & D \end{pmatrix},$$

then Q is μ-symplectic if and only if A, B, C, D have the properties

$$A^T C \quad \text{and} \quad B^T D \quad \text{are symmetric and} \quad A^T D - C^T B = \mu I. \tag{9}$$

A transformation (4) is said to be μ-*symplectic* if its Jacobian matrix $Q(z)$ is μ-symplectic for each z in the domain of ϕ. It is called a *symplectic transformation* if $Q(z)$ is symplectic.

For example, the mapping

$$(I_1, \ldots, I_n; \theta_1, \ldots, \theta_n) \mapsto (x_1, \ldots, x_n; y_1, \ldots, y_n)$$

defined by

$$x_j = \sqrt{2I_j} \cos \theta_j, \quad y_j = \sqrt{2I_j} \sin \theta_j \qquad (j = 1, \ldots, n) \tag{10}$$

is a symplectic transformation since its Jacobian matrix consists of four diagonal blocks

$$A = \mathrm{diag} \left[\frac{1}{\sqrt{2I_1}} \cos \theta_1, \ldots, \frac{1}{\sqrt{2I_n}} \cos \theta_n \right],$$

$$B = \text{diag}\left[-\sqrt{2I_1}\sin\theta_1, \ldots, -\sqrt{2I_n}\sin\theta_n\right],$$

$$C = \text{diag}\left[\frac{1}{\sqrt{2I_1}}\sin\theta_1, \ldots, \frac{1}{\sqrt{2I_n}}\sin\theta_n\right],$$

$$D = \text{diag}\left[\sqrt{2I_1}\cos\theta_1, \ldots, \sqrt{2I_n}\cos\theta_n\right].$$

The variables I_j, θ_j $(j = 1, \ldots, n)$ are called *action-angle* variables.

The question is, How does one find symplectic transformations? An answer is: by means of generating functions. To see how this is done, let us first give another characterization of symplectic transformations.

We recall that the exterior product of the differentials of two functions $f(x_1, \ldots, x_n)$ and $g(x_1, \ldots, x_n)$ is given by

$$df \wedge dg = \sum_{j<k}\left(\frac{\partial f}{\partial x_j}\frac{\partial g}{\partial x_k} - \frac{\partial f}{\partial x_k}\frac{\partial g}{\partial x_j}\right) dx_j \wedge dx_k.$$

Therefore, given a coordinate transformation

$$x_j = x_j(\xi, \eta), \quad y_j = y_j(\xi, \eta) \qquad (j = 1, \ldots, n), \tag{11}$$

we compute

$$
\begin{aligned}
dx_i \wedge dy_i = &\sum_{j<k}\left(\frac{\partial x_i}{\partial \xi_j}\frac{\partial y_i}{\partial \xi_k} - \frac{\partial x_i}{\partial \xi_k}\frac{\partial y_i}{\partial \xi_j}\right) d\xi_j \wedge d\xi_k \\
&+ \sum_{j<k}\left(\frac{\partial x_i}{\partial \eta_j}\frac{\partial y_i}{\partial \eta_k} - \frac{\partial x_i}{\partial \eta_k}\frac{\partial y_i}{\partial \eta_j}\right) d\eta_j \wedge d\eta_k \\
&+ \sum_{j,k}\left(\frac{\partial x_i}{\partial \xi_j}\frac{\partial y_i}{\partial \eta_k} - \frac{\partial x_i}{\partial \eta_k}\frac{\partial y_i}{\partial \xi_j}\right) d\xi_j \wedge d\eta_k
\end{aligned}
$$

and get

$$
\begin{aligned}
\sum dx_i \wedge dy_i = &\sum_{j<k}\left\{\left(\frac{\partial x}{\partial \xi}\right)^T\left(\frac{\partial y}{\partial \xi}\right) - \left(\frac{\partial y}{\partial \xi}\right)^T\left(\frac{\partial x}{\partial \xi}\right)\right\}_{jk} d\xi_j \wedge d\xi_k \\
&+ \sum_{j<k}\left\{\left(\frac{\partial x}{\partial \eta}\right)^T\left(\frac{\partial y}{\partial \eta}\right) - \left(\frac{\partial y}{\partial \eta}\right)^T\left(\frac{\partial x}{\partial \eta}\right)\right\}_{jk} d\eta_j \wedge d\eta_k \\
&+ \sum_{j,k}\left\{\left(\frac{\partial x}{\partial \xi}\right)^T\left(\frac{\partial y}{\partial \eta}\right) - \left(\frac{\partial y}{\partial \xi}\right)^T\left(\frac{\partial x}{\partial \eta}\right)\right\}_{jk} d\xi_j \wedge d\eta_k,
\end{aligned}
$$

where $\left(\frac{\partial x}{\partial \xi}\right)$ is the Jacobian matrix of $x = x(\xi, \eta)$ with respect to ξ, and so on.

From the block characterization (9) of symplectic transformations, we see that the transformation (11) is μ-symplectic if and only if

$$\sum_{i=1} dx_i \wedge dy_i = \mu \sum_{i=1} d\xi_i \wedge d\eta_i. \tag{12}$$

Now, given a function $S(x, \eta) = S(x_1, \ldots, x_n; \eta_1, \ldots, \eta_n)$ of class C^2, consider the matrix of partial derivatives $S_{x\eta}(x, \eta) = \left(\frac{\partial^2 S}{\partial x_i \partial \eta_j}\right)$. If $\det S_{x\eta}(x^*, \eta^*) \neq 0$, then introducing the variables $y_1, \ldots, y_n; \xi_1, \ldots, \xi_n$ by means of the equations

$$y_i = S_{x_i}(x, \eta), \quad \xi_i = S_{\eta_i}(x, \eta) \qquad (i = 1, \ldots, n), \tag{13}$$

the implicit function theorem guarantees that the second set of equations can be solved implicitly for x, giving $x_i = x_i(\xi, \eta)$ $(i = 1, \ldots, n)$ as functions of $\xi_1, \ldots, \xi_n; \eta_1, \ldots, \eta_n$, which, when inserted in the first set of equations provides a transformation of the form (11).

This transformation is symplectic. Indeed, since $y_i = \frac{\partial S}{\partial x_i}(x, \eta)$, we have

$$\sum_{i=1}^n dx_i \wedge dy_i = \sum_{i=1}^n dx_i \wedge \left[\sum_{j=1}^n \frac{\partial^2 S}{\partial x_j \partial x_i} dx_j + \frac{\partial^2 S}{\partial \eta_j \partial x_i} d\eta_j\right]$$

$$= \sum_i \sum_j \frac{\partial^2 S}{\partial \eta_j \partial x_i} dx_i \wedge d\eta_j,$$

because

$$\sum_i \sum_j \frac{\partial^2 S}{\partial x_j \partial x_i} dx_i \wedge dx_j = 0.$$

Analogously, from $\xi_j = \frac{\partial S}{\partial \eta_j}(x, \eta)$, we get

$$d\xi_j \wedge d\eta_j = \sum_i \sum_j \frac{\partial^2 S}{\partial x_i \partial \eta_j} dx_i \wedge d\eta_j,$$

and since S is of class C^2, we have the equality of the mixed derivatives. We conclude that $\sum_i dx_i \wedge dy_i = \sum_i d\xi_i \wedge d\eta_i$, so the transformation is symplectic.

The function $S = S(x, \eta)$ is called the *generating function* of the transformation (11) defined implicitly by (13).

If S is given as a function of the variables $x_1, \ldots, x_n; \xi_1, \ldots, \xi_n$, say, $S = S(x, \xi)$, with $\det S_{x\xi}(x, \xi) \neq 0$, then setting

$$y_i = S_{x_i}(x, \xi), \quad \eta_i = S_{\xi_i}(x, \xi) \qquad (i = 1, \ldots, n),$$

we have for the resulting transformation (11),

$$\sum_i \sum_j dx_i \wedge dy_i = \sum_i \sum_j \frac{\partial^2 S}{\partial \xi_j \partial x_i} dx_i \wedge d\xi_j \qquad \text{and}$$

$$\sum_i \sum_j d\xi_j \wedge d\eta_j = \sum_i \sum_j \frac{\partial^2 S}{\partial x_i \partial \xi_j} d\xi_j \wedge dx_i,$$

from which we see that $\sum_i dx_i \wedge dy_i = -\sum_i d\xi_i \wedge d\eta_i$. Therefore, the transformation is symplectic with multiplier -1. If we want to get a true symplectic transformation we should define it implicitly by the equations

$$y_i = S_{x_i}(x, \xi), \quad \eta_i = -S_{\xi_i}(x, \xi) \quad \text{or by}$$
$$y_i = -S_{x_i}(x, \xi), \quad \eta_i = S_{\xi_i}(x, \xi) \qquad (i = 1, \ldots, n).$$

We leave as an exercise for the reader to write down the equations that implicitly define a symplectic transformation when $S = S(y, \xi)$ and $S = S(y, \eta)$, to find the condition on the determinants that allows us to do that, and to prove that the transformation is symplectic.

Once we know how to find symplectic transformations, the problem is to obtain a function $S = S(x, \eta)$ that generates a tranformation by means of (13) and that carries the Hamiltonian $H(z)$ to a simpler one, $\mathcal{H}(\zeta)$. Since $H(x, y) = \mathcal{H}(\xi, \eta)$, we must solve the partial differential equation

$$H(x, S_x) = \mathcal{H}(S_\eta, \eta)$$

known as the *Hamilton–Jacobi* equation in order to get S. This formulation is used to get the Birkhoff normal form as done in [23] or [26].

4   Hamiltonian Flows

Let $\phi(t, z)$ be the flow of the Hamiltonian system (3) defined on a region $\Omega \subset \mathbb{R}^{2n}$.

Theorem 4.1 *For each t, the mapping $z \mapsto \phi_t(z) = \phi(t,z)$ is symplectic.*

Proof. Let $M(t,z) = D_z\phi(t,z)$. We must show that M is a symplectic matrix for each $z \in \Omega$. Denoting by H_z the gradient of $H(z)$ we have, by definition of flow,

$$\frac{\partial}{\partial t}\phi(t,z) = JH_z(\phi(t,z)).$$

Differentiating with respect to z and interchanging the order of differentiation, we get

$$\dot{M} = \Phi M,$$

where $\Phi = JH_{zz}$; here, H_{zz} stands for the Hessian of H, a symmetric matrix. It follows that $J\Phi$ is symmetric, hence, so is $M^T J\dot{M} = M^T J\Phi M$. Consequently, $M^T J\dot{M} = -\dot{M}^T JM$ and this gives

$$\frac{d}{dt}(M^T JM) = \dot{M}^T JM + M^T J\dot{M} = 0.$$

It follows that $M^T JM$ is a constant matrix, and since $\phi(0,z) = z$ for all z, we have $M(0) = I$, hence, $M^T JM = J$, showing that $M = M(t,z)$ is a symplectic matrix. ∎

Since the determinant of a symplectic matrix is equal to 1, Liouville's theorem on the volume-preserving character of a Hamiltonian flow follows as a corollary.

Corollary 4.1 *(Liouville). The flow $\phi(t,z)$ of a Hamiltonian system defines a volume-preserving diffeomorphism ϕ_t for each t.*

The following local converse to the above theorem is also true.

Theorem 4.2 *Let $\phi(t,z)$ be the flow of a differential equation $\dot{z} = f(z)$ defined on a region $\Omega \subset \mathbb{R}^{2n}$. Suppose that for each t, the transformation $z \mapsto \phi_t(z)$ is symplectic. Then, for each $z_0 \in \Omega$ there exists a neighborhood V of z_0 and a function H_V such that $f(z) = J\nabla H_V(z)$ on V; that is, the equation is, locally, a Hamiltonian system.*

Proof. By hypothesis, the matrix $M = M(t,z) = D_z\phi(t,z)$ is symplectic, that is, $M^T JM = J$. Differentiating this equation with respect to t, we have $\dot{M}^T J M + M^T J \dot{M} = 0$, from which we get $J \dot{M} M^{-1} = -M^{-T} \dot{M} J = (J \dot{M} M^{-1})^T$, showing that $J \dot{M} M^{-1}$ is symmetric. Since $\dot{M} = Df(\phi(t,z))M$, it follows that $-J Df(\phi(t,z)) = -J \dot{M} M^{-1}$ is symmetric, hence the Jacobian matrix of the vector field $-Jf(z)$ is symmetric. Therefore, in a star-shaped neighborhood V of each point, this is the gradient field of some function H_V, that is, $-Jf(z) = \nabla H_V(z)$, hence $f(z) = J \nabla H_V(z)$ on V. This completes the proof.   ∎

Remark. Notice that if the domain Ω is star-shaped, for example, if it is convex, then the equation $\dot{z} = f(z)$ is globally Hamiltonian.

Given two differentiable functions $f, g : \Omega \to \mathbb{R}$ defined on an open subset Ω of $\mathbb{R}^{2n}$, the *Poisson bracket* of f and g is the function $\{f, g\}$ defined by

$$\{f, g\}(z) = \nabla f(z)^T J \nabla g(z). \tag{14}$$

In local coordinates $z = (x_1, \ldots, x_n; y_1, \ldots, y_n)$, this bracket is given by

$$\{f, g\}(z) = \sum_{i=1}^{n} \left(\frac{\partial f}{\partial x_i} \frac{\partial g}{\partial y_i} - \frac{\partial f}{\partial y_i} \frac{\partial g}{\partial x_i} \right),$$

with the derivatives evaluated at z.

Theorem 4.3 *Let $z = \phi(\zeta)$ be a symplectic transformation. Given functions $f(z)$ and $g(z)$, let $F(\zeta) = f(\phi(\zeta))$ and $G(\zeta) = g(\phi(\zeta))$. Then,*

$$\{F, G\}(\zeta) = \{f, g\}(z),$$

that is, the Poisson bracket is invariant under symplectic transformations.

Proof. This is an immediate consequence of the following relations among the gradients

$$\nabla F(\zeta) = D\phi(z)^T \nabla f(z) \quad \text{and} \quad \nabla G(\zeta) = D\phi(z)^T \nabla g(z)$$

and the fact that $D\phi(\zeta)$ is symplectic.   ∎

The next theorem gives a characterization of symplectic transformation in terms of Poisson brackets.

Theorem 4.4   (a) *The transformation $\phi(\zeta) = (\phi_1(\zeta), \ldots, \phi_{2n}(\zeta))$ in $\mathbb{R}^{2n}$ is symplectic if and only if $\{\phi_i, \phi_j\} = 0$ for $j \neq i+n$ and $\{\phi_i, \phi_{i+n}\} = 1$.*

(b) *The transformation (11) is symplectic if and only if $\{x_i, x_j\} = 0$, $\{y_i, y_j\} = 0$, and $\{x_i, y_j\} = \delta_{ij}$, where δ_{ij} is the Kronecker symbol, that is, $\delta_{ij} = 0$ for $i \neq j$ and $\delta_{ii} = 1$.*

Proof. Clearly, (b) follows from (a). To prove (a), just notice that

$$[D\phi(\zeta)JD\phi(\zeta)^T]_{ij} = \nabla\phi_i(\zeta)^T J \nabla\phi_j(\zeta) = \{\phi_i, \phi_j\}(\zeta).$$

So, $D\phi(\zeta)$ symplectic is equivalent to the condition stated in (a).   ∎

5   Stability of Equilibria

Let x^* be an equilibrium of an autonomous system of ordinary differential equations

$$\dot{x} = f(x), \qquad x \in \Omega \subset \mathbb{R}^n, \tag{15}$$

that is, $f(x^*) = 0$. Denote by $\phi(t, x)$ the solution of (15) such that $\phi(0, x) = x$. For $r > 0$, denote by $B_r(x)$ the open ball in $\mathbb{R}^n$ with center x and radius r.

We say that the equilibrium solution x^* is stable (in the future) if, given $\epsilon > 0$, there exists $\delta > 0$ such that for all $x \in B_\delta(x^*)$ the solution $\phi(t, x)$ is defined for all real $t \geq 0$ and $\phi(t, x) \in B_\epsilon(x^*)$ for all $t \geq 0$.

Before stating the following theorem, let us recall that a function ψ is said to be positive definite (resp., negative definite) at a point x^* if $f(x^*) = 0$ and if there exists a neighborhood V of x^* such that $f(x) > 0$ (resp., $f(x) < 0$) for all $x \neq x^*$ in V.

Theorem 5.1 *(Dirichlet). If there exists a positive (or negative) definite integral ψ of the system (15) in a neighborhood of the equilibrium x^*, then x^* is stable.*

Proof. Indeed, let $\epsilon > 0$ and take a positive number $r < \epsilon$ such that the integral is defined and, say, positive definite on the closed ball $\overline{B_r(x^*)}$.

Let m be the minimum value of ψ on the boundary of this ball; hence, $m > 0$. Since $\psi(x^*) = 0$, the continuity of ψ assures the existence of a positive number $\delta < r$ such that $\psi(x) < m$ for all $x \in B_\delta(x^*)$. Then, for $x \in B_\delta(x^*)$, the solution $\phi(t, x)$ must remain inside the ball $B_r(x^*)$ for all time where it is defined; otherwise it would cross the boundary of this ball, and we should get $\psi(\phi(t, x)) \geq m$ for some t. But this is a contradiction because ψ, being an integral, it remains constant along the solution $\phi(t, x)$, whereas at $t = 0$ we have $\psi(x) < m$. This shows that $\phi(t, x)$ remains in the compact set $\overline{B_r(x^*)}$ for all t in its domain of definition which, therefore, must be all of $\mathbb{R}$, and since $r < \epsilon$, we have $\phi(t, x) \in B_\epsilon(x^*)$ for all t. This proves that x^* is a stable equilibrium. The proof is analogous in the negative definite case.

If f is twice differentiable in a neighborhood of the equilibrium x^*, the system (15) can be written as

$$\dot{x} = Df(x^*) \cdot (x - x^*) + O(|x - x^*|^2). \tag{16}$$

The associated linearized system is

$$\dot{u} = Df(x^*)u. \tag{17}$$

A theorem of A. M. Lyapunov [18] says that: *If all the eigenvalues of the matrix $Df(x^*)$ have negative real parts, then the equilibrium x^* of (16) is stable* (in the future, $t \geq 0$). In fact, it says more as it guarantees that the equilibrium is *asymptotically stable*, that is, that there exists $\delta_1 < \delta$ such that for $x \in B_{\delta_1}(x^*)$ not only does $\phi(t, x)$ remain inside the ball $B_\epsilon(x^*)$, but also $\phi(t, x)$ approaches x^* as t approaches infinity.

Another theorem of Lyapunov says that *if some eigenvalue of $Df(x^*)$ has positive real part, then the equilibrium is not stable.*

For a proof of these results, see Hirsch and Smale [14].

What do these theorems imply in the case of an equilibrium of a Hamiltonian system? Let A be the matrix of the linearized part of the system (3) in a neighborhood of the equilibrium z^*.

Theorem 5.2 *The characteristic polynomial of A is even.*

Proof. Notice that $A = JS$, where S is the symmetric matrix $H_{zz}(z^*)$, the Hessian of H at z^*. Since

$$\det (JS \quad \lambda I) - \det (JS - \lambda I)^T = \det (-SJ - \lambda I)$$
$$= \det [J(JS + \lambda I)J] = \det (JS + \lambda I),$$

the conclusion of the theorem follows. ∎

By this theorem, if $A = JS$ has an eigenvalue with nonzero real part, it has one with positive real part. Consequently, a theorem of Liapunov implies that *if an equilibrium of a Hamiltonian system is stable, then all the eigenvalues of the linearized system are purely imaginary numbers.*

By means of a translation we can always bring the equilibrium to the origin, so we will assume that $z^* = 0$. Then, if the Hamiltonian $H(z)$ is analytic in a neighborhood of z^*, it has a power series expansion

$$H(z) = H_2(z) + H_3(z) + \cdots, \tag{18}$$

where $H_k(z)$ is a homogeneous polynomial of degree k in z. We have omitted the constant term $H(z^*)$ as it has no effect on the dynamics, and the linear term does not appear because $\nabla H(z^*) = 0$.

Since H is a first integral of the system, Dirichlet's theorem implies that *if the quadratic form $H_2(z)$ is positive (or negative) definite, then the equilibrium is stable.*

Therefore, the interesting stability case for the equilibrium of a Hamiltonian system is that when all the eigenvalues of the linearized system are purely imaginary numbers and the quadratic part of the Hamiltonian is an indefinite form.

For non-Hamiltonian systems $\dot{x} = f(x) = Ax + \ldots$, the stability of the linearized system in the case when all the eigenvalues of A have negative real parts guarantees, by Lyapunov's theorem, the (future) stability of the equilibrium solution of the full system; in fact, it assures the asymptotic stability of the equilibrium. When some of the eigenvalues are purely imaginary numbers, the equilibrium may be stable for the linear system and yet unstable for the nonlinear system, as the simple example,

$$\dot{x}_1 = -x_2 + x_1(x_1^2 + x_2^2), \qquad \dot{x}_2 = x_1 + x_2(x_1^2 + x_2^2),$$

shows. For Hamiltonian systems, the stability of the linearized system implies that all the eigenvalues of the linearized system are purely imaginary numbers. Although the above system is not Hamiltonian, T. M. Cherry [7] provided an example of a nonstable equilibrium of a Hamiltonian system for which the linearized system is stable (see section 8). Therefore, the linear terms alone cannot decide the stability of the equilibrium solution of the full system.

The main result on stability of equilibria of Hamiltonian systems is a theorem of V. I. Arnold for systems with two degrees of freedom. We will state this result below. Little is known about stability of equilibria for Hamiltonian systems with three or more degrees of freedom.

Consider an analytic Hamiltonian system defined in a neighborhood of the origin in $\mathbb{R}^4$ and suppose that in the symplectic coordinates x_1, x_2, y_1, y_2 the Hamiltonian has the form

$$H = H_2 + H_4 + \cdots + H_{2N} + H^*, \tag{19}$$

where each $H_{2j}(I_1, I_2)$ is a homogeneous polynomial of degree j in the action variables $I_j = \frac{1}{2}(x_j^2 + y_j^2)$ $(j = 1, 2)$ and the remainder term H^* has a series expansion that begins with terms of degree at least $2N + 1$ in x_1, x_2, y_1, y_2. Moreover, suppose that the quadratic part is

$$H_2 = \omega_1 I_1 - \omega_2 I_2,$$

where ω_1 and ω_2 are positive constants. Then, we have the following result.

Theorem 5.3 *(Arnold's stability theorem).* *For a Hamiltonian system with H as above, the origin is a stable equilibrium solution provided that for some $j = 2, \ldots, N$, $D_{2j} = H_{2j}(\omega_2, \omega_1) \neq 0$, or, equivalently, provided that H_2 does not divide H_{2j}.*

The proof of this theorem uses the invariant curve theorem of Moser. In Siegel and Moser [26] the proof is given for the case when $N = 2$ in the expansion (19), and in the above formulation the proof was given by Meyer and Schmidt [22]. The theorem was originally proved independently of the invariant curve theorem; see Arnold [3, 4].

We notice that in case of stability we can never have asymptotic stability of the equilibrium solution of a Hamiltonian system since, as we have seen in section 4, the flow of such a system is volume preserving (Liouville's theorem).

When Arnold's criterion fails, or when we have resonances in systems with two or more degrees of freedom, we use normal forms of the Hamiltonian to construct Lyapunov's functions or Chetaev's functions in order to give some information about the stability of the equilibrium.

In section 9, we will apply some of the results to the stability of the Lagrangean solutions of the planar, circular, restricted three-body problem, work that has been done by several researchers: Leontovich [17], Markeev [19, 20], Deprit and Deprit-Bartholomé [9, 10], Alfriend [1, 2], Sokol'skii [27], Meyer and Schmidt [22].

A main reference on the stability of the Lagrangean solutions of the restricted three-body problem is Markeev's book [20]. We will not discuss the case of time-dependent Hamiltonian systems, so we do not consider here the elliptic restricted three-body problem. For normal forms of time-dependent Hamiltonians, see [21], and for the stability of the Lagrangean solutions in the elliptic restricted three-body problem, see [20].

6   Normal Forms

Consider an analytic Hamiltonian system in a neighborhood of the origin of $\mathbb{R}^{2n}$:

$$\dot{x} = J\nabla H(x). \tag{20}$$

Suppose that the origin is an equilibrium, so the Hamiltonian has a power series expansion

$$H(x) = \sum_{i=2}^{\infty} H_i(x), \tag{21}$$

where $H_i(x)$ is a homogeneous polynomial of degree i in the symplectic coordinates $x_1, \ldots, x_{2n}$.

The linearized system is given by

$$\dot{x} = Ax, \qquad A = JD^2 H_2(0), \tag{22}$$

with flow $\phi_0(t, x) = e^{tA}x$.

In section 10, we will prove the following theorem.

Theorem 6.1 *Suppose that the matrix A is diagonalizable in the complex field. Then there exists a formal symplectic transformation $x = \Phi(y) = y + \cdots$, which carries $H(x)$ to the Hamiltonian $H^*(y) = H(\Phi(y))$. This Hamiltonian has the expansion*

$$H^*(y) = \sum_{i=2}^{\infty} H^i(y), \tag{23}$$

where each $H^i(y)$ is a homogeneous polynomial of degree i in the symplectic coordinates $y_1, \ldots, y_{2n}$ and a first integral of the linear system (22), that is,

$$H^i(e^{tA}y) \equiv H^i(y) \quad \text{for all} \quad t \quad (i = 1, 2, \ldots).$$

The formal power series $H^*(y)$ is called a *normal form* for the Hamiltonian $H(x)$.

As a corollary, we obtain the well-known Birkhoff's normal form.

Corollary 6.1 *(Birkhoff's normal form). Suppose that the quadratic part of the Hamiltonian (21) has the form*

$$H_2(x) = \sum_{i=1}^{n} \frac{\omega_i}{2}(x_i^2 + x_{n+i}^2), \tag{24}$$

and that $\omega_1, \ldots, \omega_n$ are linearly independent over the rationals. Then, there exists a formal symplectic change of variables $x = \Phi(y) = y + \ldots$ which tranforms the Hamiltonian $H(x)$ in (21) to the Hamiltonian $H^*(y)$ in (23), where now $H^i(y)$ is a homogeneous polynomial of degree $i-1$ in the variables

$$I_1 = \frac{1}{2}(y_1^2 + y_{n+1}^2), \ldots, I_n = \frac{1}{2}(y_n^2 + y_{2n}^2).$$

Proof. Notice that the linearized system has a diagonalizable matrix since the eigenvalues are all distinct. Hence, there exists a symplectic transformation $x = \Phi(y)$ carrying $H(x)$ to $H^*(y)$ with the properties stated in

Theorem 6.1. Now, it is convenient to pass to complex variables by means of the complex transformation $\eta = Py$, where P is the symplectic matrix

$$P = \frac{1}{1+i} \begin{pmatrix} iI & I \\ -iI & I \end{pmatrix},$$

with I the $n \times n$ identity matrix. The new Hamiltonian $\mathcal{H}(\eta) = H^*(y)$ is real if η corresponds to a real vector y. Since $\eta_j = \frac{1}{1+i}(iy_j + y_{n+j})$, $\eta_{n+j} = \frac{1}{1+i}(-iy_j + y_{n+j})$ $(j = 1, \ldots, n)$, we have $i\eta_j\eta_{n+j} = \frac{1}{2}(y_j^2 + y_{n+j}^2)$, and therefore, the quadratic part of $\mathcal{H}(\eta)$ is

$$\mathcal{H}^0(\eta) = \sum_{i=1}^{n} i\omega_j \eta_j \eta_{n+j}.$$

The flow of the corresponding linear system is $\psi(t, \eta) = e^{tB}\eta$, where

$$B = \mathrm{diag}\ [i\omega, \ldots, i\omega_n; -i\omega_1, \ldots, -i\omega_n],$$

hence

$$\psi(t, \eta) = \mathrm{diag}\ [e^{i\omega t}\eta_1, \ldots, e^{i\omega_n t}\eta_n; e^{-i\omega_1 t}\eta_{n+1}, \ldots, e^{-i\omega_n t}\eta_{2n}].$$

Since $H^m(y)$ is a first integral of the flow $\phi(t, y)$ of the linear system defined by H^*, it follows that $\mathcal{H}^m(\eta) = H^m(y)$ is a first integral of the flow $\psi(t, \eta)$. Therefore, if

$$\mathcal{H}^m(\eta) = \sum h_{k_1 \cdots k_n l_1 \cdots l_n} \eta_1^{k_1} \eta_{n+1}^{l_1} \cdots \eta_1^{k_n} \eta_{2n}^{l_n}, \tag{25}$$

where the summation runs over all the $(k, l) = (k_1, \ldots, k_n; l_1, \ldots, l_n)$ such that $k_1 + \cdots + k_n + l_1 + \cdots + l_n = m$, then

$$\mathcal{H}^m(\psi(t, \eta)) = \sum h_{k,l}\ e^{i[(k_1-l_1)\omega_1 + \cdots + (k_n-l_n)\omega_n]t} \eta_1^{k_1} \eta_{n+1}^{l_1} \cdots \eta_n^{k_n} \eta_{2n}^{l_n} \tag{26}$$

is constant along the time, hence $h_{k_1 \cdots k_n l_1 \cdots l_n} = 0$ for all (k, l) such that $(k_1 - l_1)\omega_1 + \cdots + (k_n - l_n)\omega_n \neq 0$. Then, we are left only with those coefficients h_{kl} for which $(k_1 - l_1)\omega_1 + \cdots + (k_n - l_n)\omega_n = 0$, but since $\omega_1, \ldots, \omega_n$ are assumed to be linearly independent over the rationals it follows that the only coefficients that may appear in $\mathcal{H}$ are those for which $k_1 = l_1, \ldots, k_n = l_n$. Therefore, $m = 2p$ is even and

$$\mathcal{H}^m(\eta) = \sum h_{k_1 \cdots k_n k_1 \cdots k_n} (\eta_1\eta_{n+1})^{k_1} \cdots (\eta_n\eta_{2n})^{k_n}.$$

Going back to the variables $y_1, \ldots, y_{2n}$, we get

$$H^{2p}(y) = \sum h^{(k_1 \cdots k_n)} (y_1^2 + y_{n+1}^2)^{k_1} \cdots (y_n^2 + y_{2n}^2)^{k_n},$$

where $h^{(k_1 \cdots k_n)} = (2i)^{-p} h_{k_1 \cdots k_n k_1 \cdots k_n}$ is real. This completes the proof of the corollary. ∎

The Hamiltonian $H_0(x)$ in (24) is an algebraic sum of uncoupled linear oscillators with frequencies $|\omega_i|$.

We say that the frequencies satisfy a resonance relation of order k if there exists a linear combination

$$m_1 \omega_1 + \cdots + m_n \omega_n = 0$$

with the integers $m_1, \ldots, m_n$ satisfying the equality $|m_1| + \cdots + |m_n| = k$.

If the normalization process giving the series that defines the transformation $x = \Phi(y) = y + \cdots$ guaranteed in Theorem 6.1 is performed up to terms of order $2N$ we get a convergent symplectic transformation $x = \Phi_N(y)$ that normalizes the Hamiltonian $H(x)$ up to order $2N$, that is,

$$H^*(y) = \sum_{i=2}^{N} H^i(y) + O(|y|^{2N+1}),$$

where each homogeneous polynomial $H^i(y)$ is a first integral of the flow defined by the quadratic Hamiltonian H_0.

In view of the above considerations, we see that the proof of the corollary on the Birkhoff's normal form works equally well in the context of the following theorem.

Theorem 6.2 *(Finite Birkhoff's normal form). Suppose that the frequencies do not satisfy any resonance relation up to order $2N$. Then, there exists a convergent symplectic transformation $x = \Phi_N(y) = y + \cdots$ such that the transformed Hamiltonian has the form*

$$H^*(y) = H^2(y) + H^4(y) + \cdots + H^{2N}(y) + H_{2N+1}(y),$$

where each $H^{2j}(y)$ is a homogeneous polynomial of degree j in the quantities $\frac{1}{2}(y_1^2 + y_{n+1}^2), \ldots, \frac{1}{2}(y_n^2 + y_{2n}^2)$,

$$H^2(y) = \sum_{k=1}^{n} \frac{1}{2}\omega_k(y_k^2 + y_{n+k}^2),$$

and the remainder term $H_{2N+1}(y)$ is a convergent series that begins with terms of degree at least $2N + 1$ in $y_1, \ldots, y_{2n}$.

We also see from the proof of the corollary that if there exists a resonance relation $m_1\omega_1 + \cdots + m_n\omega_n = 0$ of some order, then the coefficients h_{kl} in (25) for which $k_j - l_j = m_j$ $(j = 1, \ldots, n)$ are not necessarily zero. This allows us to slightly extend Birkhoff's normal form. More precisely, we consider the **Z**-module

$$M_\omega = \{(m_1, \ldots, m_n) \in \mathbf{Z}^n; \ m_1\omega_1 + \cdots + m_n\omega_n = 0\}$$

and say that the Hamiltonian $H^*(y)$ is in normal form if, in passing to $\mathcal{H}(\eta) = H^*(y)$ in the manner we have done before, the Hamiltonian

$$\mathcal{H}(\eta) = \mathcal{H}^2(\eta) + \mathcal{H}^3(\eta) + \cdots + \mathcal{H}^m(\eta) + \cdots$$

has the property that the coefficients h_{kl} of $\mathcal{H}^m(\eta)$ in (25) are zero for all (k, l) such that $k - l = (k_1 - l_1, \ldots, k_n - l_n) \notin M_\omega$.

This means that besides the Birkhoff's terms, that is, those for which $k_j = l_j$ $(j = 1, \ldots, n)$, only the "resonant" terms, namely, those for which $k - l \in M_\omega$, $k \neq l$, are present in the series for $\mathcal{H}(\eta)$. This is sometimes referred to as Gustavson's normal form and was first considered by the astronomer F. Gustavson [13]; for another proof see Moser [23].

Given an analytic Hamiltonian system in a neighborhood of the origin of $\mathbb{R}^{2n}$, which is also an equilibrium point, the normal form and the series that lead to it are formal power series that are generically divergent. This was proved by Siegel [25] in 1954 for the case of the Birkhoff's normal form. However, the process is convergent for the finite normal forms, that is, when the normalization of only finitely many terms is considered. Frequently this suffices for getting information about the stability of the equilibrium solution.

7   The Linear Normalization

In the process of Birkhoff's normalization we require that the quadratic part H_2 of the Hamiltonian (21) be already in the normal form (24). We have seen in section 5 that the characteristic polynomial of $JD^2H(0)$ is even, so the eigenvalues of this matrix come in pairs, say, $\lambda_1, \ldots, \lambda_n; -\lambda_1, \ldots, -\lambda_n$. In section 15 of Siegel and Moser [26], the normalization of H_2 is carried out when $\lambda_1, \ldots, \lambda_n$ are arbitrary distinct complex numbers. Since we are mainly interested in the question of stability of the equilibrium when the quadratic part is indefinite, we will describe a constructive linear normalization process due to Markeev [20], in which we assume that the eigenvalues, besides being distinct, are all purely imaginary complex numbers, $\pm i\omega_1, \ldots, \pm i\omega_n$, with $\omega_1, \ldots, \omega_n$ real numbers.

Now, $H_2(x)$ is given by $H_2(x) = \frac{1}{2}x^T S x$, where $S = S^T$ is the Hessian matrix of H_2 at 0. We write $\lambda_k = i\omega_k$, $\lambda_{n+k} = \overline{\lambda}_k$. Since $\omega_1, \ldots, \omega_n$ are distinct, the matrix JS is diagonalizable.

Theorem 7.1 *There exists a real linear symplectic transformation $x = Ay$ that takes the Hamiltonian $H_2(x)$ into a new Hamiltonian $H^2(y) = H_2(Ay)$, which is the sum of uncoupled linear oscillators*

$$H^2(y) = \sum \frac{\omega_j}{2}(y_j^2 + y_{n+j}^2) \tag{27}$$

Proof. The condition that the transformation $x = Ay$ carries the system $\dot{x} = JSx$ into the system $\dot{y} = JRy$ is translated into the matrix equation

$$AJR = JSA, \tag{28}$$

while the condition for A to be symplectic is

$$A^T JA = J. \tag{29}$$

We take A in the form of a product $A = BC$, where

$$C = \begin{pmatrix} iI & I \\ -iI & I \end{pmatrix},$$

and seek to determine B to reach our goal. The condition (28) gives us

$$B^{-1}JSB = CJRC^{-1}, \tag{30}$$

and since $JR = \begin{pmatrix} O & \Omega \\ -\Omega & O \end{pmatrix}$, where $\Omega = \text{diag}\,[\omega_1,\ldots,\omega_n]$, we compute the right-hand side of (30) and find it to be the diagonal form of JS, that is, $\begin{pmatrix} i\Omega & O \\ O & -i\Omega \end{pmatrix}$. Therefore, the column vectors of the matrix B are the eigenvectors of JS, that is,

$$B = \text{col}\,[\mathbf{e}_1,\ldots,\mathbf{e}_n,\mathbf{e}_{n+1},\ldots,\mathbf{e}_{2n}],$$

where for $k = 1,\ldots,n$,

$$JS\mathbf{e}_k = \lambda_k\mathbf{e}_k, \quad \text{and} \quad \mathbf{e}_{n+k} = \bar{\mathbf{e}}_k, \quad \text{the complex conjugate of} \quad \mathbf{e}_k.$$

The condition (29) regarding the symplecticity of A can be written in the form

$$C^T F C = J \qquad \text{with} \qquad F = B^T J B. \tag{31}$$

Since J is skew-symmetric, so is F, and since J and C are invertible, F is invertible too. Now, from the second equation (31), we see that the elements of F are given by

$$f_{kl} = (\mathbf{e}_k, J\mathbf{e}_l),$$

where $(\ ,\ )$ stands for the scalar product in $\mathcal{C}^{2n}$, namely, $(z,w) = z_1 w_1 + \cdots + z_{2n} w_{2n}$.

Therefore, we have

$$\lambda_l(\mathbf{e}_k, J\mathbf{e}_l) = (\mathbf{e}_k, J(JS\mathbf{e}_l)) = (\quad S\mathbf{e}_k, \mathbf{e}_l)$$
$$= (J(JS\mathbf{e}_k), \mathbf{e}_l) = \lambda_k(J\mathbf{e}_k, \mathbf{e}_l) = -\lambda_k f_{kl},$$

from which we conclude that $f_{kl} = 0$ for $|k - l| \neq n$. Thus,

$$F = \begin{pmatrix} O & G \\ -G & O \end{pmatrix},$$

where $G = \text{diag}\,[g_{11},\ldots,g_{nn}]$ with $g_{kk} = (\mathbf{e}_k, J\bar{\mathbf{e}}_k) \neq 0$.

The first equation in (31) shows that $G = \frac{1}{2i}I$, and so

$$2i(\mathbf{e}_k, J\bar{\mathbf{e}}_k) = 1 \qquad (k = 1,\ldots,n). \tag{32}$$

Setting $\mathbf{e}_k = \mathbf{r}_k + i\mathbf{s}_k$, where $\mathbf{r}_k$ and $\mathbf{s}_k$ are real vectors, the condition (32) on the symplecticity of A becomes

$$4(\mathbf{r}_k, J\mathbf{s}_k) = 1 \qquad (k = 1,\ldots,n). \tag{33}$$

On the other hand, the condition that $\mathbf{e}_k$ is an eigenvector of JS associated with the eigenvalue $\lambda_k = i\omega_k$ is translated into the system of equations

$$JS\mathbf{r}_k = -\omega_k\mathbf{s}_k, \qquad JS\mathbf{s}_k = \omega_k\mathbf{r}_k. \tag{34}$$

Now, $(\mathbf{r}_k, J\mathbf{s}_k) \neq 0$ because $g_{kk} \neq 0$. Therefore, replacing $\mathbf{e}_k$ by a multiple of it, we can get an eigenvector associated with $i\omega_k$ that satisfies equation (33) up to the sign. But, if the left-hand side of (33) is negative, then by changing the signs of $\mathbf{r}_k$ and ω_k we can get a solution of (34) for which $(\mathbf{r}_k, J\mathbf{s}_k)$ is positive. This corresponds to interchanging the positions of $i\omega_k$ and $-i\omega_k$ in the sequence originally taken $i\omega_1, \ldots, i\omega_n, -i\omega_1, \ldots, -i\omega_n$.

In this way, we can construct the matrix B so that the symplecticity condition (33) is satisfied. The symplectic transformation $x = Ay$ takes the system $\dot{x} = JSx$ into the system $\dot{y} = JRy$ corresponding to the Hamiltonian (27). It remains to show that the matrix A is real. Denoting by $X^{(l)}$ the lth-column of the matrix X, for $l = 1, \ldots, n$ we have

$$A^{(l)} = BC^{(l)} = i(\mathbf{e}_l - \overline{\mathbf{e}}_l) = -2\mathbf{s}_l, \quad A^{(n+l)} = BC^{(n+l)} = \mathbf{e}_l + \overline{\mathbf{e}}_l = 2\mathbf{r}_l,$$

from which we see that A is a real matrix and, in fact, finishes the proof of the theorem with an explicit construction of A, namely,

$$A = \mathrm{col}\,[-2\mathbf{s}_1, \ldots, -2\mathbf{s}_n, 2\mathbf{r}_1, \ldots, 2\mathbf{r}_n].$$

■

8   Some Stability Results

In this section we prove a stability lemma for time-periodic Hamiltonian systems with one degree of freedom and apply it to the study of the stability of an equilibrium of a Hamiltonian system with two degrees of freedom. The application includes Arnold's stability theorem and also the case when resonances are present among the frequencies of the linearized system. In our presentation we will follow the work of Cabral and Meyer [6].

We will need Moser's invariant curve theorem, which we state below.

Let r, θ be polar coordinates in the plane and consider the mapping $M_0 : (r, \theta) \mapsto (r_1, \theta_1)$ defined on the annulus $0 \leq a \leq r \leq b$, $\theta \in \mathbb{R}$ by the equations

$$r_1 = r, \qquad \theta_1 = \theta + \omega(r),$$

where $\omega(r)$ is an analytic function with $\omega'(r) \neq 0$ for all $r \in [a, b]$. For each fixed r, the mapping M_0 leaves invariant the circle with center at the origin and radius r, rotating it by the angle $\omega(r)$. We refer to M_0 as a *twist mapping*, the function $\omega(r)$ being the twist term.

Now, if we perturb M_0 to a mapping M given by

$$M : \quad r_1 = r + f(r, \theta), \quad \theta_1 = \theta + \omega(r) + g(r, \theta), \qquad (35)$$

where f and g are real analytic functions of r, θ and 2π-periodic in θ, does M have closed invariant curves near the invariant circles of M_0? The answer is clearly negative if we require only that f and g are small: by taking $f(r, \theta) \equiv \epsilon$, $g(r, \theta) \equiv 0$, then, however small the constant ϵ may be, the radius r always increases under the application of the mapping, so this M has no closed invariant curves $r = \varphi(\theta) \equiv \varphi(\theta + 2\pi)$. However, if in addition to a smallness condition on f and g, we also require a certain topological property of the mapping, it turns out that closed invariant curves do exist. What we require from M is the following *intersection property*:

Any closed curve $\gamma : \quad r = \varphi(\theta) \equiv \varphi(\theta + 2\pi)$ *intersects its image curve* $M\gamma$.

It certainly holds if M is area-preserving and leaves invariant the inner boundary of the annulus, for example, if M is defined on a disk, has the center of the disk as a fixed point, and is area-preserving.

The invariant circles of M_0 that will survive as invariant curves of the perturbed mapping M are those whose rotation numbers $\omega(r)$ are badly approximated by rationals, in the sense of the following result.

Lemma 8.1 *Let Δ be a closed interval of length l. For $\epsilon > 0$, consider the set $\Delta(\epsilon)$ of the numbers ω in Δ such that*

$$\left| \omega - \frac{p}{q} \right| > \frac{\epsilon}{q^3}$$

for all integers p, q with $q \geq 1$. Then, if ϵ is sufficiently small, $\Delta(\epsilon)$ has positive Lebesgue measure, $m(\Delta(\epsilon)) > 0$. Moreover, $m(\Delta(\epsilon)) \to l$ as $\epsilon \to 0$.

Proof. Let $\Sigma(\epsilon) = \Delta - \Delta(\epsilon)$ be the complement of $\Delta(\epsilon)$ in Δ, that is, $\Sigma(\epsilon)$ is the set of $\alpha \in \Delta$ for which there exists a pair of integers p, q with $q \geq 1$ such that

$$\left| \omega - \frac{p}{q} \right| \leq \frac{\epsilon}{q^3}. \tag{36}$$

For a fixed integer $q \geq 1$, there exists at most $([l] + 1)q$ integers p such that the centers $\frac{p}{q}$ of the intervals (36) belong to Δ; $[l]$ denotes here the largest integer contained in l. It follows that

$$m(\Sigma(\epsilon)) \leq \sum_{q=1}^{\infty} ([l] + 1)q \cdot \frac{2\epsilon}{q^3} \leq 2\epsilon(l+1) \sum_{q=1}^{\infty} \frac{1}{q^2}.$$

Since the series $\sum_{q=1}^{\infty} \frac{1}{q^2}$ converges, $m(\Sigma(\epsilon))$ can be made smaller than l by taking ϵ small. Therefore, $m(\Delta(\epsilon)) = l - m(\Sigma(\epsilon)) > 0$. Clearly, $m(\Delta(\epsilon)) \to l$, as $\epsilon \to 0$.

We now state Moser's theorem; see Meyer and Hall [21] and Siegel and Moser [23].

Theorem 8.1 *(The Invariant curve theorem). Consider the mapping M_ϵ : $(I, \phi) \mapsto (I_1, \phi_1)$ defined for I, ϕ, ϵ in the region $0 \leq a \leq I \leq b$, $\phi \in \mathbb{R}$, $0 \leq \epsilon \leq \epsilon_0$, by*

$$I_1 = I + \epsilon^{r+s} f(I, \phi, \epsilon),$$
$$\phi_1 = \phi + \omega + \epsilon^r h(I) + \epsilon^{r+s} g(I, \phi, \epsilon),$$

where h is a real analytic function of I, and f, g are real analytic functions of I, ϕ, ϵ and 2π-periodic in ϕ. The numbers r and s are integers with $r \geq 0$ and $s \geq 1$. Suppose that M_ϵ has the intersection property and assume that the twist condition $h'(I) \neq 0$ takes place for all I, $a \leq I \leq b$. Choose the number ω as in the lemma.

Then, for ϵ_0 sufficiently small, the mapping M_ϵ has a closed invariant curve C_ϵ parametrized by

$$\phi = \xi + u(\xi), \qquad I = v(\xi),$$

where u and v are real analytic 2π-periodic functions of $\xi \in \mathbb{R}$. If $\gamma = |h(b) - h(a)|$, the functions u and v satisfy the requirement that

$$|u| \quad and \quad |v - \gamma^{-1}\omega|$$

*remain small for all ξ. Moreover, under the parametrization $\Phi(\xi) = (I(\xi),$
$\phi(\xi))$ of the invariant curve C_ϵ, we have*

$$M(\Phi(\xi)) = \Phi(\xi + \omega) \qquad \textit{for all} \quad \xi.$$

This theorem was originally proved by Moser [24] under the smoothness hypothesis that the functions defining the mapping M_ϵ are of class C^{333}, and then one has the continuity of the invariant curve. Under the analyticity assumption of the mapping, the invariant curve is also analytic. The proof in the analytic case is somewhat simpler as we can use complex function theory, which helps to simplify some estimates and arguments; this proof, however, is still quite delicate and long, involving a lot of subtle details (see Siegel and Moser [26]). For further comments on the invariant curve theorem, see [21].

Another result that we need in the proof of the stability lemma is an instability theorem due to Chetaev [8].

Consider the ordinary differential equation (15), where f is locally Lipschitzian in a region $\Omega \subset \mathbb{R}^n$. Given a differentiable function V on Ω, denote by $\dot{V}(x)$ the function defined on Ω by

$$\dot{V}(x) = \langle \nabla V(x), f(x) \rangle,$$

where $\langle \ , \ \rangle$ denotes the Euclidean inner product in $\mathbb{R}^n$. By the chain rule, if $x(t)$ is a solution of the differential equation (15), then

$$\frac{d}{dt}V(x(t)) = \dot{V}(x(t)). \tag{37}$$

Theorem 8.2 *(Chetaev). Let x^* be an equilibrium of the differential equation (15). Suppose that there exists a differentiable function V on Ω and an open set $A \subset \Omega$ such that the following conditions are satisfied:*

(1) *$V(x)$ and $\dot{V}(x)$ are positive on A,*

(2) *$V(x) = 0$ for all $x \in \partial A \cap \Omega$,*

(3) *$x^* \in \partial A$, where ∂A denotes the boundary of A.*

Then the equilibrium x^* is unstable.

Proof. Fix $\epsilon > 0$ such that the closed ball $\overline{B_\epsilon(x^*)}$ lies inside Ω (a region is a connected open set). By condition (3) of the theorem, for any $\delta > 0$, there exists a point $x \in B_\delta(x^*) \cap A$. Let $x(t)$ be the solution of (15) that starts at x at $t = 0$. We will show that either $x(t)$ is not defined for all $t > 0$ or $\|x(t)\| \geq \epsilon$ for some $t > 0$. This will prove the instability of the equilibrium x^*.

From (37) and condition (1), it follows that as long as the solution $x(t)$ remains in A, we have

$$V(x(t)) = V(x) + \int_0^t \dot{V}(x(s)) \, ds \geq V(x) + \alpha t, \tag{38}$$

where $\alpha = \inf \left\{ \dot{V}(x(t)); \text{ for } x(t) \in A \right\}$.

It follows from (38) and condition (2) of the theorem that, if $x(t)$ does not remain in A for all time t in its domain of definition, then $x(t)$ cannot leave A through an interior point of Ω, so it must do it through the boundary of Ω. Hence, $\|x(t)\| > \epsilon$ for some $t > 0$.

Now, suppose that $x(t)$ remains inside A for all t in its domain of definition. Let $A_x = \{y \in A; \ V(y) \geq V(x)\}$. From condition (2) it is seen that the set $K = \overline{B_\epsilon(x^*)} \cap A_x$ contains the limits of their convergent sequences, hence it is compact. Since $x(t)$ remains in A for all t, we have $\dot{V}(x(t)) > 0$ for all t, so $V(x(t)) > V(x)$, hence $x(t) \in A_x$ for all t. If $x(t)$ remained inside the compact set $\overline{B_\epsilon(x^*)}$ for all t in its domain of definition, this domain would contain the half-line $t > 0$ and $x(t)$ would also lie in the compact set K for all $t > 0$. Since $\dot{V}(x)$ is positive on K, we would have $\alpha > 0$. But then (38) would imply that $V(x(t)) \to \infty$ as $t \to \infty$, a contradiction because V is bounded on K. This contradiction resulted from the assumption that $x(t)$ remains in $\overline{B_\epsilon(x^*)}$, hence, $\|x(t)\| > \epsilon$ for some $t > 0$. This concludes the proof of Chetaev's theorem. ∎

We are now ready to prove the following result (Cabral and Meyer [6]).

Stability Lemma 1 *Let* $K(r, \phi, t) = \Psi(\phi)r^n + O(r^{n+\frac{1}{2}})$, *where* $n = m/2$ *with* $m \geq 3$, *an integer. Suppose that* K *is an analytic function of* $\sqrt{r}, \phi, t$, τ-*periodic in* ϕ *and* T-*periodic in* t. *If* $\Psi(\phi) \neq 0$ *for all* ϕ, *then the origin*

$r = 0$ *is a stable equilibrium for the Hamiltonian system*

$$\dot{r} = \frac{\partial K}{\partial \phi}, \quad \dot{\phi} = -\frac{\partial K}{\partial r},$$

in the sense that given $\epsilon > 0$, there exists $\delta > 0$ such that if $r(0) < \delta$, then the solution is defined for all t and $r(t) < \epsilon$. If $\Psi(\phi)$ has a simple zero, that is, if there exists ϕ^* such that $\Psi(\phi*) = 0$ and $\Psi'(\phi^*) \neq 0$, then the equilibrium $r = 0$ is unstable.

Proof. Suppose that $\Psi(\phi) \neq 0$ for all ϕ, say, $\Psi(\phi) > 0$. Consider the truncated Hamiltonian

$$k = \Psi(\phi)r^n,$$

and for each $k > 0$ define the variable $I = I(k)$ by

$$I = \frac{1}{2\pi} \int_0^\tau r(k, \phi) \, d\phi,$$

where

$$r(k, \phi) = \frac{k^{1/n}}{\Psi(\phi)^{1/n}}.$$

Consider a generating function $S(I, \phi)$ defined by

$$S(I, \phi) = \int_0^\phi r(k, \theta) \, d\theta.$$

By inserting in $S(I, \phi)$ the expression for the factor $k^{1/n}$ obtained from I, we get

$$S(I, \phi) = \beta I G(\phi),$$

where

$$\beta = 2\pi \left\{ \int_0^\tau \frac{d\theta}{\Psi(\theta)^{1/n}} \right\}^{-1}, \quad G(\phi) = \int_0^\phi \frac{d\theta}{\Psi(\theta)^{1/n}}.$$

Now, S defines a symplectic transformation $(r, \phi) \to (I, W)$ by the relations

$$W = \frac{\partial S}{\partial I} = \beta G(\phi), \quad r = \frac{\partial S}{\partial \phi} = \beta I G'(\phi),$$

and the original Hamiltonian

$$K(r, \phi, t) = \Psi(\phi)r^n + O(r^{n+\frac{1}{2}})$$

is transformed into the new Hamiltonian (analytic in $\sqrt{I}, W, t$)

$$K(I, W, t) = \beta^n I^n + O(I^{n+\frac{1}{2}}),$$

since $G'(\phi) = \Psi(\phi)^{-1/n}$, and hence $\Psi(\phi)G'(\phi)^n = 1$.

Notice that, since $\Psi(\phi)$ is τ-periodic,

$$W(r, \phi + \tau) = \beta G(\phi + \tau) = \beta \int_0^{\phi+\tau} \frac{d\theta}{\Psi(\theta)^{1/n}}$$

$$= \beta \int_0^\tau \frac{d\theta}{\Psi(\theta)^{1/n}} + \beta \int_\tau^{\phi+\tau} \frac{d\theta}{\Psi(\theta)^{1/n}} = 2\pi + W(r, \phi),$$

so W is a true angular variable; therefore, $K(I, W, t)$ is 2π-periodic in W, and of course, T-periodic in t.

Consider the change of variables $(I, W) \to (J, \psi)$ defined by

$$I = \sigma \gamma J, \qquad W = \psi,$$

where $\sigma > 0$ is a small parameter, $1 \leq J \leq 2$ and γ is chosen so that $\beta^n \gamma^{n-1} = 1/n$. The mapping $(I, W) \mapsto (J, \psi)$ is a symplectic change of variables with multiplier $\frac{1}{\sigma\gamma}$, so the new Hamiltonian is given by

$$\mathcal{K}(J, \psi, t) = \frac{1}{n}\sigma^{n-1}J^n + O\left(\sigma^{n-\frac{1}{2}}\right)$$

and the corresponding Hamiltonian equations are

$$\frac{dJ}{dt} = O\left(\sigma^{n-\frac{1}{2}}\right), \quad \frac{d\psi}{dt} = -\sigma^{n-1}J^{n-1} + O\left(\sigma^{n-\frac{1}{2}}\right),$$

with the right-hand side analytic in J, ψ, t, 2π-periodic in ψ and T-periodic in t, with $1 \leq J \leq 2$.

Integrating between $t = 0$ and $t = T$, and denoting by J, ψ the initial values and by J_1, ψ_1 the final values, we obtain the mapping (with $\epsilon = \sqrt{\sigma}$)

$$J_1 = J + \epsilon^{2n-1}F_1(J, \psi, \epsilon),$$
$$\psi_1 = \psi - \epsilon^{2n-2}TJ^{n-1} + \epsilon^{2n-1}F_2(J, \psi, \epsilon),$$

defined and analytic in the region $1 \leq J \leq 2$, $\psi \in \mathbb{R}$, $|\epsilon| < \epsilon_0$, with F_1, F_2 periodic in ψ. This map is area-preserving by virtue of the Hamiltonian character of the differential equations. Therefore, by Moser's invariant curve theorem, for small σ there exist invariant curves $J = J(\psi) =$

$J(\psi + 2\pi)$ close to the circles centered at the origin. Since $I = \sigma\gamma J$, the corresponding curve $I = I(\psi)$ can be taken inside small neighborhoods of the origin (by taking σ sufficiently small). In the three-dimensional space (I, ψ, t), identifying the sections $t = 0$ and $t = T$, we get a torus formed by solutions curves that begin on the closed curve $I = I(\psi)$. By uniqueness of solutions, any solution $(I(t), \psi(t))$ that starts at a point inside the region bounded by the curve $I = I(\psi)$ cannot cross the torus, and therefore $I(t)$ remains small. Thus $I(t)$ remains small for all t. Since the solutions stay inside a compact set, they are defined for all time. This proves the stability, in the sense defined in the statement of the theorem.

Now we will give the proof of the instability statement. Assume that $\Psi(\phi^*) = 0$ and $\Psi'(\phi^*) > 0$. Choose $\delta > 0$ so small that

$$\Psi(\phi) \neq 0, \quad \text{and} \quad \Psi'(\phi) > 0 \quad \text{for} \quad 0 < |\phi - \phi^*| \leq \delta.$$

Consider the function

$$V = r^n \sin \Phi,$$

where $\Phi = (\pi/2\delta)(\phi - \phi^* + \delta)$.

Define a region Ω as the set of points (r, ϕ, t) such that

$$\phi^* - \delta < \phi < \psi^* + \delta.$$

Then, $V > 0$ in Ω and $V = 0$ on $\partial\Omega$, the boundary of Ω. The derivative of V along the solutions of the system of equations

$$\dot{r} = \frac{\partial K}{\partial \phi} = r^n \Psi'(\phi) + O(r^{n+\frac{1}{2}}),$$

$$\dot{\phi} = -\frac{\partial K}{\partial r} = -nr^{n-1}\Psi(\phi) + O(r^{n-\frac{1}{2}})$$

is given by

$$\frac{dV}{dt} = \frac{\partial V}{\partial r}\dot{r} + \frac{\partial V}{\partial \phi}\dot{\phi} + \frac{\partial V}{\partial t}$$

$$= nr^{2n-1}\left[\Psi'(\phi)\sin \Phi - \frac{\pi}{2\delta}\Psi(\phi)\cos \Phi\right] + O(r^{2n-\frac{1}{2}}).$$

For $0 < \phi - \phi^* < \delta$, we have $\pi/2 < (\pi/2\delta)(\phi - \phi^* + \delta) < \pi$ so that $\cos \Phi < 0$. Also for $-\delta < \phi - \phi^* < 0$, we have $\cos \Phi > 0$ and therefore

$$\Psi(\phi) \cos \Phi < 0 \quad \text{on} \quad 0 < |\phi - \phi^*| < \delta.$$

Since $\Psi'(\phi) > 0$ and $\sin\Phi > 0$ on $|\phi - \phi^*| < \delta$, we have $\Psi'(\phi)\sin\Phi > 0$ in this interval. Since for the function inside the brackets the two summands do not vanish simultaneously on the compact interval $|\phi - \phi^*| \leq \delta$, it follows that it has a positive minimum, and therefore we conclude that $dV/dt > 0$ on Ω if r is sufficiently small. It then follows from Chetaev's theorem that the equilibrium is unstable. ∎

Stability Lemma 2 *Let $K(r, \phi, t, \epsilon) = \epsilon^m \Psi(\phi) r^n + \epsilon^{m+1} \tilde{K}(r, \phi, t, \epsilon)$, where m and $2n \leq m$ are positive integers. Suppose that K is an analytic function of $\sqrt{r}, \phi, t, \epsilon$, τ-periodic in ϕ, T-periodic in t for all $0 \leq r \leq 3$ and all $0 \leq \epsilon \leq \epsilon_0$ and that $\tilde{K} = O(r^{n+\frac{1}{2}})$. If $\Psi(\phi) \neq 0$ for all ϕ and if ϵ_0 is sufficiently small, then any solution of*

$$\dot{r} = \frac{\partial K}{\partial \phi}, \quad \dot{\phi} = -\frac{\partial K}{\partial r}$$

which starts with $|r(0)| \leq 1$ for $0 \leq \epsilon \leq \epsilon_0$ satisfies $|r(t)| \leq 2$ for all t.

Proof. As in the proof of the stability part of the previous lemma we show that there are invariant curves for the section map that separate $r = 1$ from $r = 2$. ∎

As an application of the stability lemmas we consider the stability and the instability of an equilibrium point for an analytic autonomous Hamiltonian system with two degrees of freedom $H = H(x_1, x_2, y_1, y_2)$. The classical Liapunov theory shows that the origin is unstable unless the eigenvalues of the linearized system are purely imaginary numbers, and so we consider a system whose linear part has eigenvalues $\pm i\omega_1$, $\pm i\omega_2$. If the quadratic part of the Hamiltonian is sign definite, Dirichlet's theorem asserts that the equilibrium is stable. Therefore, we shall consider the case when the frequencies ω_1, ω_2 have opposite sign, that is, when the Hamiltonian has an indefinite quadratic part

$$H = \frac{\omega_1}{2}(x_1^2 + y_1^2) - \frac{\omega_2}{2}(x_2^2 + y_2^2),$$

with ω_1 and ω_2 positive constants. Furthermore, we assume that the frequencies satisfy the resonance relation

$$p\omega_1 - q\omega_2 = 0,$$

where p and q are relatively prime positive integers. If $p = q = 1$, we assume also that the matrix of the linearized system is diagonalizable. We write the Hamiltonian in action-angle variables $(I, \phi) = (I_1, I_2, \psi_1, \psi_2)$ defined by

$$x_j = \sqrt{2I_j}\cos\phi_j, \quad y_j = \sqrt{2I_j}\sin\phi_j \quad (j = 1, 2)$$

and assume that the Hamiltonian H is in normal form through terms of order m, where $m = 2l - 1$ or $m = 2l$. This means that

$$H(I, \phi_1, \phi_2) = H_2(I) + \cdots + H_{2l-2}(I) + H_m(I, p\phi_1 + q\phi_2) + \cdots \quad (39)$$

where

- $H_2 = \omega_1 I_1 - \omega_2 I_2$,

- H_{2j} is a homogeneous polynomial of degree j in I_1, I_2,

- $H_m(I, p\phi_1 + q\phi_2)$ is a homogeneous polynomial of degree m in $\sqrt{I_1}$, $\sqrt{I_2}$ with coefficients that are finite Fourier series in the single angle $p\phi_1 + q\phi_2$,

- the ellipses denote terms of order greater than m in the variables $\sqrt{I_1}$, $\sqrt{I_2}$.

- H is an analytic function of the variables $\sqrt{I_1}$, $\sqrt{I_2}$, ϕ_1, ϕ_2 and 2π periodic in ϕ_1 and ϕ_2.

That H_m is a function of the single angle $p\phi_1 + q\phi_2$ is equivalent to the fact that H_m is constant along the solutions of the linear equations whose Hamiltonian is H_2, that is, H_m is constant along the solutions of

$$\dot\phi_1 = -\omega_1, \quad \dot\phi_2 = \omega_2, \quad \dot I_1 = 0, \quad \dot I_2 = 0.$$

Let

$$\Psi(\phi) = H_m(\omega_2, \omega_1, p\phi),$$

where

$$\phi = \phi_1 + \frac{q}{p}\phi_2.$$

Let $D_{2j} = H_{2j}(\omega_2, \omega_1)$. If we have $D_{2j} \neq 0$ for some $j = 2, \ldots, l - 1$, then Arnold's stability theorem guarantees the stability of the equilibrium solution $x_i = y_i = 0$. Therefore, we assume in addition that

$$D_{2j} = 0 \quad \text{for} \quad j = 2, \ldots, l - 1,$$

and so H_m is the term that will decide the stability or instability of the equilibrium.

Theorem 8.3 *If $\Psi(\phi) \neq 0$ for all ϕ, then the equilibrium solution $q_i = p_i = 0$ is stable. If Ψ has a simple zero, that is, if there exists ϕ^* such that $\Psi(\phi^*) = 0$ and $\Psi'(\phi^*) \neq 0$, then the equilibrium solution is unstable.*

Remark For the stability statement we do not need the resonance condition, and so H_m could be independent of an angle. Thus, this theorem includes Arnold's Theorem.

Proof.   We follow the ideas in the proof of Arnold's stability theorem as given in [21], [22]. Since $D_{2j} = 0$, the homogeneous polynomial H_{2j} has H_2 as a factor, that is, $H_{2j} = H_2 F_{2j-2}$, where the second factor is a homogeneous polynomial of degree $j - 1$ in I_1, I_2. We have

$$H = H_2 F + H_m(I, \phi) + \cdots,$$

with $F = 1 + F_2 + \cdots + F_{2l-4}$. Near the origin the values of F_{2j} are small and we can take the reciprocal of the function F,

$$F^{-1} = 1 + \cdots,$$

where the ellipses represent terms of degree at least 1 in I_1, I_2. Therefore, $\tilde{H} = F^{-1}H$ can be written as

$$\tilde{H} = H_2 + H_m(I_1, I_2, p\phi_1 + q\phi_2) + \cdots,$$

where the ellipses represent terms of degree at least $m + 1$ in $\sqrt{I_1}, \sqrt{I_2}$.

Since $H = F\tilde{H}$ the equations of motion are of the form $\dot{z} = J\nabla H = F J \nabla \tilde{H} + \tilde{H} J \nabla F$ where $z = (I_1, I_2, \phi_1, \phi_2)$ and J is the usual 4×4 skew-symmetric matrix of mechanics. If we change time by $d\tau = F dt$ and let $' = d/d\tau$, the equations of motion on the set $H = 0$ (or $\tilde{H} = 0$) are

$$z' = J\nabla \tilde{H}.$$

Thus, near the equilibrium, the flow defined by $\tilde{H}$ on $\tilde{H} = 0$ is a reparametrization of the flow defined by H on $H = 0$.

It suffices to prove instability on the surface $H = 0$ or, what is the same, on $\tilde{H} = 0$. Solving the equation

$$0 = \tilde{H} = \omega_1 I_1 - \omega_2 I_2 + H_m(I_1, I_2, p\phi_1 + q\phi_2) + \cdots \tag{40}$$

for I_2, we get

$$I_2 = \frac{\omega_1}{\omega_2} I_1 + \frac{1}{\omega_2} H_m\left(I_1, \frac{\omega_1}{\omega_2} I_1, p\phi_1 + q\phi_2\right) + O(I_1^{\frac{m+1}{2}})$$

or

$$I_2 = \frac{q}{p} I_1 + \frac{1}{\omega_2^{\frac{m+2}{2}}} H_m(\omega_2, \omega_1, p\phi_1 + q\phi_2) I_1^{\frac{m}{2}} + O(I_1^{\frac{m+1}{2}}).$$

The right-hand sides of these equations are analytic functions of $\sqrt{I_1}, \phi_1, \phi_2$.

Let $H^\dagger(I_1, \phi_1, \phi_2)$ be the negative of the right-hand side of this expression for I_2. From the equations of motion we see that ϕ_2 is an increasing function of τ and so we can take it as the new independent variable (time). The function $H^\dagger$ then defines a time-dependent Hamiltonian with one degree of freedom, 2π-periodic in ϕ_1 and ϕ_2.

We now make the symplectic change of variables

$$\phi = \phi_1 + \frac{q}{p} \phi_2, \qquad r = I_1,$$

which is generated by the function

$$S(r, \phi_1, \phi_2) = \left(\phi_1 + \frac{q}{p} \phi_2\right) r.$$

Since the derivative of S with respect to the time ϕ_2 is

$$\frac{\partial S}{\partial \phi_2} = r \frac{q}{p} = \frac{q}{p} I_1,$$

the new Hamiltonian function is given by

$$K(r, \phi, \phi_2) = \Psi(\phi) r^n + O(r^{n+\frac{1}{2}}),$$

where $n = m/2$ and

$$\Psi(\phi) = -\frac{1}{\omega_2^{n+1}} H_m(\omega_2, \omega_1, p\phi).$$

We notice that K is 2π-periodic in ϕ and $2p\pi$-periodic in ϕ_2. By hypothesis, $\Psi(\phi)$ has a simple zero. Therefore, the Stability Lemma 1 implies that $r = 0$ is an unstable equilibrium for the Hamiltonian system defined by K. Consequently, the equilibrium $x_i = y_i = 0$ is unstable.

If we just want to prove the stability of the equilibrium point on the level set $H = 0$, we can simply apply the Stability Lemma 1, but with a little extra effort we can get the full stability statement. First we scale the action variables $I_i = \epsilon^2 J_i$, where ϵ is a small scaling variable. This is a symplectic change of coordinates with multiplier ϵ^{-2}; so, in the variables J_1, J_2, ϕ, the Hamiltonian (39) becomes

$$H = H_2 F + \epsilon^{m-2} H_m + O(\epsilon^{m-1}),$$

where, now,

$$F = 1 + \epsilon^2 F_2 + \cdots + \epsilon^{2l-4} F_{2l-4}.$$

We fix a bounded neighborhood of the origin, say, $|J_i| \leq 4$, so that the remainder term is uniformly $O(\epsilon^{m-1})$ in it and henceforth restrict our attention to this neighborhood. Let h be a new parameter in the interval $[-1, 1]$. Since $F = 1 + \cdots$, we have

$$H - \epsilon^{m-1} h = KF,$$

where

$$K = H_2 + \epsilon^m H_{m-2} + O(\epsilon^{m-1}). \tag{41}$$

For sufficiently small ϵ, the function F is positive in the neighborhood under consideration and so the level set $H = \epsilon^{m-1} h$ is the same as the level set $K = 0$. Let $z = (J_1, J_2, \phi_1, \phi_2)$ and let ∇ be the gradient operator with respect to these variables. The equations of motion are

$$\dot{z} = J\nabla H = F(J\nabla K) + K(J\nabla F).$$

On the level set $K = 0$, the equations become

$$\dot{z} = F(J\nabla K).$$

As we noticed, for small ϵ, F is positive. So the reparametrization $d\tau = F dt$ transforms this equation to

$$z' = J\nabla K(z), \tag{42}$$

where the prime denotes the derivative with respect to τ.

We have thus shown that in the considered neighborhood, and for small c, the flow defined by H on each level set $H = \epsilon^{m-1}h$ is a reparametrization of the flow defined by K on the level set $K = 0$. Now, by varying the parameter h, the stability of the equilibrium on each level set $H = \epsilon^{m-1}h$ guarantees the stability of the equilibrium. Thus, it suffices to prove the stability of the origin for the system (42) on the level set $K = 0$.

Now, from (41), we have

$$K = \omega_1 J_1 - \omega_2 J_2 + \epsilon^{m-2} H_m(J_1, J_2, p\phi) + O(\epsilon^{m-1}).$$

From now on we proceed to compute the Hamiltonian in the $K = 0$ set just as in the instability case to get (with $n = m/2$)

$$K(r, \phi, \phi_2) = \epsilon^{m-2}\Psi(\phi)r^n + O(\epsilon^{m-1}).$$

The difference is that K is analytic for $\frac{1}{2} \leq r \leq 3$ for all small ϵ, and so by the Stability Lemma 2 there exist invariant tori that separate the $r = 1$ torus from the $r = 2$ torus for all small ϵ, say $0 \leq \epsilon \leq \epsilon_0$. For all $0 \leq \epsilon \leq \epsilon_0$ all solutions that start with $r \leq 1$ must have $r \leq 2$ for all τ. Since on the set $K = 0$ we have $J_2 = (\omega_1/\omega_2)J_1 + \cdots$, it follows that a bound on $r - J_1$ implies a bound on J_2. Thus there are constants c and k such that if $J_1(\tau), J_2(\tau)$ satisfy the system for $0 \leq \epsilon \leq \epsilon_0$, start on $K = 0$, and satisfy $|J_1(0)|, |J_2(0)| \leq c$, then $|J_1(\tau)|, |J_2(\tau)| \leq k$ for all τ and $0 \leq \epsilon \leq \epsilon_0$.

Returning to the original unscaled variables with the original Hamiltonian H, this means that for $0 \leq c \leq \epsilon_0$, all solutions that start on $H = \epsilon^m h$ and satisfy $|I_1(0)|, |I_2(0)| \leq \epsilon^2 c$ must satisfy $|I_1(\tau)|, |I_2(\tau)| \leq \epsilon^2 k$ for all t and all $-1 \leq h \leq 1$, $\epsilon \leq \epsilon_0$. Thus the equilibrium is stable. ∎

As applications, consider the classical counterexample of Cherry and theorems of Markeev [19, 20] and Alfriend [1, 2].

Cherry's counterexample: In the second edition (1917) of Whittaker's book on dynamics, the equations of motion about the Lagrange triangular libration point L_4 of the circular three-body problem are linearized and the assertion is made that the libration point is stable for $0 < \mu < \mu_1$, where μ is the mass ratio parameter and $\mu_1 = \frac{1}{2}(1 - \sqrt{69}/9)$ is the critical mass ratio parameter of Routh. In the third edition of Whittaker [28], this assertion was dropped and an example due to Cherry [7] was included. Cherry's example is a polynomial Hamiltonian system of two degrees of freedom and the linearized equations are two harmonic oscillators with frequencies in a ratio of 2:1. The author explicitly gives the solution and shows that the higher order terms can destabilize the system. However, a closer look reveals that the Hamiltonian is in Birkhoff's normal form.

Cherry's counterexample in action-angle variables is

$$H = 2I_1 - I_2 + I_1^{1/2} I_2 \cos(\phi_1 + 2\phi_2),$$

and by the above theorem the equilibrium is unstable.

1:2 resonance: Consider the case where the linear system is in 1:2 resonance, that is, when the linearized system has exponents $\pm i\omega_1$ and $\pm i\omega_2$ with $\omega_1 = 2\omega_2$. Let $\omega = \omega_2$. The normal form for the Hamiltonian is a function of I_1, I_2 and the single angle $\theta_1 + 2\theta_2$. Assume the system has been normalized through terms of degree 3; that is, assume the Hamiltonian is of the form

$$H = 2\omega I_1 - \omega I_2 + \delta I_1^{1/2} I_2 \cos \psi + H^\dagger, \tag{43}$$

where $\psi = \theta_1 + 2\theta_2$, $H^\dagger(I_1, I_2, \theta_1, \theta_2) = O((I_1 + I_2)^2)$.

Corollary 8.1 *(Alfriend–Markeev theorem). If in the presence of 1:2 resonance, the Hamiltonian system is in the normal form (43) with $\delta \neq 0$, then the equilibrium is unstable.*

1:3 resonance: Now consider the case when the linear system is in 1:3 resonance, that is, $\omega_1 = 3\omega_2$. Let $\omega = \omega_2$. The normal form for the Hamiltonian is a function of I_1, I_2 and the angle $\theta_1 + 3\theta_2$. Assume the system has been normalized through terms of degree 4; that is, assume the Hamiltonian is of the form

$$\begin{aligned} H = 3\omega I_1 - \omega I_2 + \delta I_1^{1/2} I_2^{3/2} \cos \psi \\ + \frac{1}{2} \{ A I_1^2 + 2B I_1 I_2 + C I_2^2 \} + H^\dagger, \end{aligned} \tag{44}$$

where $\psi = \theta_1 + 3\theta_2$, $H^\dagger = O((I_1 + I_2)^{5/2})$. Let

$$D = A + 6B + 9C. \tag{45}$$

Corollary 8.2 *(Alfriend–Markeev theorem). If in the presence of 1:3 resonance, the Hamiltonian system is in the normal form (44) and if $6\sqrt{3}|\delta| > |D|$, then the equilibrium is unstable, whereas if $6\sqrt{3}|\delta| < |D|$, the equilibrium is stable.*

9 The Restricted Three-Body Problem

We have seen in section 2 that the Hamiltonian of the circular restricted three-body problem in rotating coordinates is given by

$$H = \frac{1}{2}(y_1^2 + y_2^2) + (x_2 y_1 - x_1 y_2) - V(x_1, x_2),$$

where the potential is

$$V = \frac{1-\mu}{\rho_1} + \frac{\mu}{\rho_2},$$

with

$$\rho_1^2 = (x_1 + \mu)^2 + x_2^2, \quad \rho_2^2 = (x_2 - 1 + \mu)^2 + x_2^2.$$

Annihilating the gradient $\nabla H = (-y_2 - V_{x_1}, y_1 - V_{x_2}, y_1 + x_2, y_2 - x_1)$, we find the equilibria to be at the points (x_1, x_2, y_1, y_2), where $y_1 = -x_2$, $y_2 = x_1$, and (x_1, x_2) are the solutions of the equations

$$V_{x_1} + x_1 = 0, \qquad V_{x_2} + x_2 = 0. \tag{46}$$

Since

$$V_{x_1} = -\left(\frac{1-\mu}{\rho_1^3}(x_1 + \mu) + \frac{\mu}{\rho_2^3}(x_1 - 1 + \mu)\right), \quad V_{x_2} = -\left(\frac{1-\mu}{\rho_1^3} + \frac{\mu}{\rho_2^3}\right) x_2,$$

then, for $x_2 \neq 0$, the second equation (46) implies that $\frac{1-\mu}{\rho_1^3} + \frac{\mu}{\rho_2^3} = 1$. Taking this into the first equation (46), we obtain $-(x_1 + \mu) + \frac{\mu}{\rho_2^3} + x_1 = 0$. Thus, $\rho_2 = 1$, hence $\frac{1-\mu}{\rho_1^3} = 1 - \mu$, so that $\rho_1 = 1$. Therefore, the infinitesimal particle at (x_1, x_2) together with the primaries form an equilateral triangle.

If $x_2 = 0$, the second equation (46) is automatically satisfied while the first reduces to

$$f(x_1) = x_1 - \frac{1-\mu}{\rho_1^2}\frac{(x_1 + \mu)}{\rho_1} - \frac{\mu}{\rho_2^2}\frac{(x_1 - 1 + \mu)}{\rho_2} = 0, \qquad (47)$$

with $\rho_1 = |x_1 + \mu|$ and $\rho_2 = |x_1 - 1 + \mu|$. Since $\frac{\partial \rho_1}{\partial x_1} = \frac{|x_1 + \mu|}{\rho_1}$ and $\frac{\partial \rho_1}{\partial x_2} = \frac{|x_1 - 1 + \mu|}{\rho_2}$, we compute

$$f'(x_1) = 1 + \frac{2(1-\mu)}{\rho_1^3} + \frac{2\mu}{\rho_2^3} > 0 \qquad \text{for all} \qquad x_1 \neq -\mu, 1 - \mu.$$

Notice that $f(x_1) \to +\infty$, as $x_1 \to +\infty$ or as x_1 goes to $-\mu$ or $1 - \mu$, from the left. Also, $f(x_1) \to -\infty$, as $x_1 \to -\infty$ or as x_1 goes to $-\mu$ or to $1 - \mu$, from the right. Since $f(x_1)$ is increasing on each of the intervals $(-\infty, -\mu)$, $(-\mu, 1-\mu)$, and $(1-\mu, +\infty)$, it follows that in each one of these intervals $f(x_1)$ has exactly one zero. Consequently, there exist exactly three equilibria for which $x_2 = 0$; they are the collinear equilibria found by Euler in 1767 [12]. The one in the interval $(-\infty, -\mu)$ is denoted by L_1, that in the middle interval by L_2, and the one in $(1 - \mu, +\infty)$ by L_3.

The equilateral equilibria found above were discovered by Lagrange [16] (together with the three Euler equilibria) in 1772. There are two such equilateral equilibria and they are located at the points $(\frac{1}{2} - \mu, \pm\frac{\sqrt{3}}{2})$, the one corresponding to the plus sign being denoted by L_4 and the other by L_5. They are known as the Lagrangean equilibria.

Let us now compute

$$V_{x_1 x_1} = -\left(\frac{1-\mu}{\rho_1^3} + \frac{\mu}{\rho_2^3}\right) + 3\left(\frac{1-\mu}{\rho_1^5}(x_1 + \mu)^2 + \frac{\mu}{\rho_2^5}(x_1 - 1 + \mu)^2\right),$$

$$V_{x_1 x_2} = 3\left(\frac{1-\mu}{\rho_1^5}(x_1 + \mu) + \frac{\mu}{\rho_2^5}(x_1 - 1 + \mu)\right)x_2,$$

$$V_{x_2 x_2} = -\left(\frac{1-\mu}{\rho_1^3} + \frac{\mu}{\rho_2^3}\right) + 3\left(\frac{1-\mu}{\rho_1^5} + \frac{\mu}{\rho_2^5}\right)x_2^2.$$

Setting $a_k = \frac{1-\mu}{\rho_1^3} + \frac{\mu}{\rho_2^3}$, at L_k $(k = 1, 2, 3)$, we get $V_{x_1 x_1} = 2a_k$, $V_{x_1 x_2} = 0$, and $V_{x_2 x_2} = -a_k$ for L_k, $(k = 1, 2, 3)$, while for L_k $(k = 4, 5)$, we have $\rho_1 = \rho_2 = 1$, $x_1 = \frac{1}{2} - \mu$, $x_2 = \pm\frac{\sqrt{3}}{2}$, the one with the plus sign corresponding to L_4.

The matrix of the linearized system at the equilibrium L_k is

$$\begin{pmatrix} J & I \\ V_{xx} & J \end{pmatrix} = \begin{pmatrix} 0 & 1 & 1 & 0 \\ -1 & 0 & 0 & 1 \\ V_{x_1x_1} & V_{x_1x_2} & 0 & 1 \\ V_{x_2x_1} & V_{x_2x_2} & -1 & 0 \end{pmatrix}, \tag{48}$$

where the partial derivatives are computed at L_k. The characteristic polynomial is then

$$\lambda^4 + (2 - a_k)\lambda^2 + (1 - a_k)(1 + 2a_k) = 0 \qquad (k = 1, 2, 3)$$

for the collinear equilibria and

$$\lambda^4 + \lambda^2 + \frac{27}{4}\mu(1 - \mu) = 0 \tag{49}$$

for the equilateral equilibria.

If λ and $\mu \neq \pm\lambda$ are roots of the first equation, we have $\lambda^2\mu^2 = (1 - a_k)(1 + 2a_k)$. Therefore, if $a_k > 1$, the roots cannot be all purely imaginary numbers.

Now, let $x_1^{(k)}$ be the abscissa of L_k ($k = 1, 2, 3$). Then, for $x_1^{(1)}$, we have $\rho_2 = 1 + \rho_1$ and $\rho_1 = -(x_1^{(1)} + \mu)$, hence

$$f(x_1^{(1)}) = x_1^{(1)} + \frac{(1 - \mu)}{\rho_1^2} + \frac{\mu}{\rho_2^2} = 0.$$

Adding and subtracting μ to this equation, we get

$$-\rho_1 + \frac{(1 - \mu)}{\rho_1^2} + \frac{\mu(1 - \rho_2^2)}{\rho_2^2} = 0,$$

and since $\rho_2 > 1$, we conclude that $\rho_1 < \frac{1-\mu}{\rho_1^2}$. Hence

$$1 < \frac{1 - \mu}{\rho_1^3} < \frac{1 - \mu}{\rho_1^3} + \frac{\mu}{\rho_2^3} = a_1.$$

With similar arguments we can also show that $a_2 > 1$ and $a_3 > 1$.

The conclusion is that *for any value of the parameter μ, the collinear equilibria are unstable.*

Let us now turn to the equilateral equilibria. The equation (49) is quadratic in λ^2. Solving for it, we have

$$\lambda^2 = \frac{1}{2}[-1 \pm \sqrt{1 - 27\mu(1 - \mu)}].$$

Let $\delta(\mu) = 27\mu(1 - \mu)$. If $\delta(\mu) < 1$, the linearized system has two pairs of distinct, purely imaginary eigenvalues. If $\delta(\mu) = 1$, the system has one pair of purely imaginary eigenvalues, each with multiplicity 2. And if $\delta(\mu) > 1$, there are two pairs of complex eigenvalues with nonzero real parts. The equation $\delta(\mu) = 1$ has two positive real roots, which are symmetric with respect to $\mu = \frac{1}{2}$. The smaller root, $\mu_1 = \frac{9 - \sqrt{69}}{18}$, is called the critical mass ratio of Routh.

We take $m_1 = \mu$ to be the smaller of the two finite masses, so $\mu \leq \frac{1}{2}$. From the above, we see at once that we have instability of the Lagrangean equilibria for $\mu_1 < \mu \leq \frac{1}{2}$ and linear stability for $0 < \mu < \mu_1$. For $\mu = \mu_1$, it turns out that the linearized system has a nondiagonalizable matrix, so the equilibrium is linearly unstable; see [21]. To study the stability of the nonlinear system, we begin by normalizing the quadratic part of the Hamiltonian at L_k ($k = 4, 5$).

A vector $x_\lambda = (u, v)$ is an eigenvector of the matrix (48) associated to the eigenvalue λ if and only if

$$v = (\lambda I - J)u \qquad \text{and} \qquad u \in \ker D(\lambda),$$

where

$$D(\lambda) = \lambda^2 I - 2\lambda J - (V_{xx} + I) = \begin{pmatrix} \lambda^2 - a - 1 & -(2\lambda + b) \\ 2\lambda - b & \lambda^2 - c - 1 \end{pmatrix},$$

with $\quad a = V_{x_1 x_1} = -\frac{1}{4}, \quad b = V_{x_1 x_2} = \pm \frac{3\sqrt{3}}{4}(1 - 2\mu), \quad c = V_{x_2 x_2} = \frac{5}{4}.$

We find that

$$x_\lambda = (2\lambda + b, \lambda^2 - a - 1, \lambda^2 + b\lambda + a + 1, \lambda^3 + (1 - a)\lambda + b).$$

For $\lambda = i\eta$, the real and imaginary parts of x_λ are

$$\mathbf{r}_\lambda = (b, -\eta^2 - a - 1, -\eta^2 + a + 1, b) \quad \text{and} \quad \mathbf{s}_\lambda = (2\eta, 0, b\eta, -\eta^3 + (1 - a)\eta).$$

We compute that

$$\frac{1}{\eta}(\mathbf{r}_\lambda, J\mathbf{s}_\lambda) = b^2 + [(\eta^2 + a^2) - 1] + 2(\eta^2 - a - 1),$$

and since η is a root of the equation $\eta^4 - \eta^2 + \frac{27}{4}\mu(1 - \mu) = 0$, we have

$$\eta^2 = \frac{1}{2}(1 \pm \sqrt{1 - \delta}), \qquad \text{where} \qquad \delta = 27\mu(1 - \mu).$$

Using the values of a, b, and η^2, we find that

$$\frac{1}{\eta}(\mathbf{r}_\lambda, J\mathbf{s}_\lambda) = \frac{1}{2}\left[(1 - \delta) \pm \frac{5}{2}\sqrt{1 - \delta}\right]; \tag{50}$$

Taking as values of η the positive numbers ω_1 and ω_2 given by

$$\omega_1 = \sqrt{\frac{1}{2}(1 + \sqrt{1 - \delta})} \quad \text{and} \quad \omega_2 = \sqrt{\frac{1}{2}(1 - \sqrt{1 - \delta})}, \tag{51}$$

the eigenvalues of the linearized system are $i\omega_1$, $i\omega_2$, $-i\omega_1$, and $-i\omega_2$.

From (50) we see that the value of $(\mathbf{r}_\lambda, J\mathbf{s}_\lambda)$ is positive for $\lambda = i\omega_1$ and negative for $\lambda = i\omega_2$. By Markeev's linear normalization process described in section 7, we must interchange the places of $i\omega_2$ and $-i\omega_2$. Therefore, the quadratic part of the Hamiltonian is indefinite and its normalized form is

$$H_2 = \frac{\omega_1}{2}(x_1^2 + y_1^2) - \frac{\omega_2}{2}(x_2^2 + y_2^2).$$

We now look for resonances among the frequencies $p\omega_1 = q\omega_2$, with p, q relatively prime positive integers. Since $\omega_1 \geq \omega_2$, we must have $q \geq p$. Using (51), we get $\sqrt{1 - \delta} = \frac{q^2 - p^2}{q^2 + p^2}$, hence $\mu(1 - \mu) = \frac{m}{n}$, where $\sqrt{m} = 2pq$ and $n = 27(q^2 + p^2)^2$. Since $\sqrt{m}$ is an even integer, the first four values of m are 4, 16, 36, and 64, which give the values 1, 2, 3, and 4, respectively, for the product qp. The corresponding pairs of integers q, p give the resonances $\omega_1 = \omega_2$, $\omega_1 = 2\omega_2$, $\omega_1 = 3\omega_2$, and $\omega_1 = 4\omega_2$, of orders 2, 3, 4, and 5, respectively. The first resonance $\omega_1 = \omega_2$ corresponds to the value μ_1 of Routh's critical ratio, while the value of μ for the next two can be computed from the equation $\mu(1 - \mu) = \frac{m}{n}$. We find

$$\mu_1 = \frac{9 - \sqrt{69}}{18} > \mu_2 = \frac{45 - \sqrt{1833}}{90} > \mu_3 = \frac{15 - \sqrt{213}}{30}.$$

If μ belongs to the interval $(0, \mu_1)$ of linear stability and $\mu \neq \mu_2, \mu_3$, then there is no resonance up to order 4 and we can write the Hamiltonian in Birkhoff's normal form

$$H = \omega_1 I_1 - \omega_2 I_2 + \frac{1}{2}(A I_1^2 + 2B I_1 I_2 + C I_2^2) + H_5.$$

The coefficients A, B, C were computed by Deprit and Deprit and Bartholomé [9, 10], and they found for $D_4 = H_4(\omega_2, \omega_1)$ the expression

$$D_4(\mu) = -\frac{36 - 541\omega_1^2\omega_2^2 + 644\omega_1^4\omega_4^4}{8(1 - 4\omega_1^2\omega_2^2)(4 - 25\omega_1^2\omega_2^2)} \qquad \text{for} \quad \mu \in (0, \mu_1), \quad \mu \neq \mu_1, \mu_2.$$

Notice that $D_4(\mu) \neq 0$, except for one value μ_c belonging to the interval $(0, \mu_1)$; this is not a resonance value, nor does it seem to have any mathematical or astronomical significance. Carrying the normalization up to order 6, Meyer and Schmidt [22] have shown that $D_6 \neq 0$ at $\mu = \mu_c$. Therefore, by Arnold's stability theorem, L_4 and L_5 are stable equilibria for all μ in the interval $0 < \mu < \mu_1$, with $\mu \neq \mu_2, \mu_3$. In the 1960s, Markeev [19, 20] and, shortly after, J. Alfriend [1, 2] showed that the Lagrangean equilibria are unstable for $\mu = \mu_2$ and $\mu = \mu_3$. Indeed, for the normal form (43) of the Hamiltonian of the restricted three-body problem at L_4, with $\mu = \mu_2$, we have $\delta = 11\sqrt{11}/18\sqrt[4]{5}$ and for the normal form (44) at L_4, with $\mu = \mu_3$, we have $\delta \approx 4.48074$ and $|D| \approx 8.34107$, where D is the quantity given by the expression (45) (see [6]). Since $\delta \neq 0$ in the first case and $6\sqrt{3}|\delta| > |D|$ in the second, the results follow from the corollaries at the end of the previous section.

Finally, Sokol'skii [27] proved that the equilibrium L_4 is stable when $\mu = \mu_1$. In this case the additional difficulty lies in the fact that the linearized system has a nondiagonalizable matrix, so the normalization process requires another approach (see chap. VII of Meyer and Hall [21]).

10   Deprit–Hori's Normalization Scheme. Proof of Theorem 6.1

To prove Theorem 6.1 we use the normalization scheme of Deprit [9] and Hori [15], which we now describe. We start with a Hamiltonian function that depends on a small parameter ϵ, $H = H(x, \epsilon)$. Then we use a function $W(x, \epsilon)$ as a Hamiltonian with the parameter ϵ as time,

$$\frac{dx}{d\epsilon} = J\nabla_x W(x, c), \tag{52}$$

and consider the solution $\phi(y, \epsilon)$ of this equation such that $\phi(y, 0) = y$. The mapping $x = \phi(y, \epsilon)$ defines a symplectic transformation that takes the Hamiltonian $H = H(x, \epsilon)$ to the new Hamiltonian $H^*(y, \epsilon) = H(\phi(y, \epsilon), \epsilon)$. The general idea is to choose $W(x, \epsilon)$ so as to get a simpler Hamiltonian H^*.

Suppose that the series expansions for H, W, and H^* in powers of ϵ are

$$H(x, \epsilon) = \sum_{i=0}^{\infty} \left(\frac{\epsilon^i}{i!}\right) H_i^0(x),$$

$$W(x, \epsilon) = \sum_{i=0}^{\infty} \left(\frac{\epsilon^i}{i!}\right) W_{i+1}(x),$$

$$H^*(y, \epsilon) = \sum_{i=0}^{\infty} \left(\frac{\epsilon^i}{i!}\right) H_0^i(y).$$

The basis of the process is the following theorem.

Theorem 10.1 *The coefficients H_0^i are the terms along the diagonal of the following scheme (the Lie triangle):*

$$H_0^0$$
$$|$$
$$H_1^0 - H_0^1$$
$$| \qquad |$$
$$H_2^0 - H_1^1 - H_0^2$$
$$| \qquad | \qquad |$$
$$H_3^0 - H_2^1 - H_1^2 - H_0^3$$
$$| \qquad | \qquad | \qquad |$$

with H_j^i $(i, j = 0, 1, \ldots)$ satisfying the relations

$$H_j^i = H_{j+1}^{i-1} + \sum_{k=0}^{j} C_j^k \left\{ H_{j-k}^{i-1}, W_{k+1} \right\}$$

$$(i = 1, 2, \ldots; j = 0, 1, \ldots), \tag{53}$$

where $\{\ ,\ \}$ is the Poisson bracket and $C_j^k = \frac{j!}{k!(j-k)!}$.

Proof. We define an operator $\mathcal{D}$ on the functions $F(x, \epsilon)$ by

$$\mathcal{D}F(x, \epsilon) = \{F, W\}(x, \epsilon) + \frac{\partial F}{\partial \epsilon}(x, \epsilon).$$

Notice that

$$\mathcal{D}F(x, \epsilon)\big|_{x=\phi(y,\epsilon)} = \frac{\partial}{\partial \epsilon} F(\phi(y, \epsilon), \epsilon).$$

Now consider the sequence of functions defined inductively by

$$H^0 = H, \quad H^{(i)} = \mathcal{D}H^{(i-1)}, \quad i \geq 1.$$

Let H_j^i be the coefficients in the series expansion of $H^{(i)}$, that is,

$$H^{(i)} = \sum_{j=0}^{\infty} \left(\frac{\epsilon^j}{j!}\right) H_j^i(x). \tag{54}$$

Since the result of applying $\mathcal{D}$ to the jth term of the series expansion of $H^{(i-1)}$ is

$$\left(\frac{\epsilon^j}{j!}\right) \{H_j^{i-1}, W\}(x, \epsilon) + \frac{\epsilon^{j-1}}{(j-1)!} H_j^{i-1}(x),$$

we have

$$H^{(i)} = \mathcal{D}H^{(i-1)} = \sum_{j=0}^{\infty} \left(\frac{\epsilon^j}{j!}\right) [\{H_j^{i-1}, W\} + H_{j+1}^{i-1}]. \tag{55}$$

The sum that involves the Poisson bracket can be worked out in the following way, where we make $l = j + k$ and, subsequently, write j instead of l:

$$\sum_{j=0}^{\infty} \left(\frac{\epsilon^j}{j!}\right) \{H_j^{i-1}, W\} = \sum_{j=0}^{\infty} \sum_{k=0}^{\infty} \left(\frac{\epsilon^{j+k}}{j!k!}\right) \{H_j^{i-1}, W_{k+1}\}$$

$$= \sum_{l=0}^{\infty} \sum_{k=0}^{l} \frac{l!}{(l-k)!k!} \left(\frac{\epsilon^l}{l!}\right) \{H_{l-k}^{i-1}, W_{k+1}\}$$

$$= \sum_{j=0}^{\infty} \left(\frac{\epsilon^j}{j!}\right) \sum_{k=0}^{j} C_j^k \left\{H_{j-k}^{i-1}, W_{k+1}\right\}.$$

By comparing the terms in the original development (54) of $H^{(i)}$ with those in the expansion (55) and by taking into consideration this last computation, we conclude that the coefficients H_j^i are related by the formula (53).

It remains to show that the coefficients H_0^i in the expansion of the transformed Hamiltonian H^* coincide with the coefficients H_j^i, $j = 0$, in the expansion (54) of $H^{(i)}$. For the sake of clarity, we denote by $\tilde{H}_0^i$ the coefficients in the expansion of $H^*(y, \epsilon)$. Then, by Taylor's formula, we have

$$\tilde{H}_0^i(y) = \frac{\partial^i H^*}{\partial \epsilon^i}(y, 0).$$

Notice now that

$$\frac{\partial H^*}{\partial \epsilon}(y, \epsilon) = \frac{\partial}{\partial \epsilon} H(\phi(y, \epsilon), \epsilon) = DH(x, \epsilon)_{x = \phi(y, \epsilon)} = H^{(1)}(\phi(y, \epsilon), \epsilon).$$

Inductively, we get

$$\frac{\partial^i H^*}{\partial \epsilon^i}(y, \epsilon) = H^i(\phi(y, \epsilon), \epsilon).$$

Setting $\epsilon = 0$, we conclude that $\tilde{H}_0^i(y) = H^{(i)}(y, 0) = H_0^i(y)$. The theorem is proved. ∎

This result says nothing about the simplification of the Hamiltonian, which we now describe.

Let $\{\mathcal{P}_i\}_{i=0}^{\infty}$ be a sequence of linear spaces of smooth functions defined in a neighborhood of the origin in $\mathbb{R}^{2n}$. Suppose that $\{H_0^0, f\} \in \mathcal{P}_i$ for every $f \in \mathcal{P}_i$ $(i = 0, 1, \ldots)$ and consider the linear operator

$$L_i : \mathcal{P}_i \to \mathcal{P}_i, \qquad f \mapsto \{H_0^0, f\} \qquad (i = 1, 2, \ldots). \tag{56}$$

We say that L_i is *simple* if $\mathcal{P}_i = \mathcal{Q}_i \oplus \mathcal{R}_i$, where $\mathcal{Q}_i = \ker L_i$ and $\mathcal{R}_i = \operatorname{Im} L_i$. For the next theorem, which guarantees the existence of normal forms, we need the following lemma.

Lemma 10.1 *If L_i is simple, then for all $D \in \mathcal{P}_i$ there exist $B \in \mathcal{Q}_i$ and $C \in \mathcal{R}_i$ such that*

$$B = D + \{H_0^0, C\}. \qquad \text{(Lie equation)}$$

Moreover, B and C are unique.

Proof. The vector D decomposes as $D = B + D' \in \mathcal{Q}_i \oplus \mathcal{R}_i$. Take $\tilde{C} \in \mathcal{P}_i$ such that $L_i(\tilde{C}) = -D'$. Then $\tilde{C} = C' + C \in \mathcal{Q}_i \oplus \mathcal{R}_i$ and $L_i(C) = -D'$. It follows that $D + L_i(C) = B$, and this proves the existence of B and C. As to the uniqueness, let $B_1 \in \mathcal{Q}_i$ and $C_1 \in \mathcal{R}_i$ be such that $B_1 = D + \{H_0^0, C_1\}$. Then $B_1 - B = L_i(C_1 - C) \in \mathcal{Q}_i \cap \mathcal{R}_i$, hence $B_1 = B$. It follows that $C_1 - C \in \mathcal{Q}_i \cap \mathcal{R}_i$, so $C_1 = C$.   ∎

We can now prove the following theorem.

Theorem 10.2 *Suppose that the operator (56) is simple and that the following properties hold:*

(a) $H_i^0 \in \mathcal{P}_i$ $(i = 0, 1, \ldots)$,

(b) $\{\mathcal{P}_i, \mathcal{R}_j\} \subset \mathcal{P}_{i+j}$ $(i = 0, 1, \ldots; j = 1, 2, \ldots)$,

It is then possible to choose the functions W_i and H_0^i so that $W_i \in \mathcal{R}_i$ and $H_0^i \in \mathcal{Q}_i$ $(i = 1, 2, \cdots)$.

Proof. Let us make the following inductive hypothesis on the functions H_j^i, W_i, and H_0^i appearing in the Lie triangle:

(I_n) $H_j^i \in \mathcal{P}_{i+j}$, $0 \le i + j \le n$, and $W_i \in \mathcal{R}_i$, $H_0^i \in \mathcal{Q}_i$, $1 \le i \le n$.

We observe that (I_0) is true, since $H_0^0 \in \mathcal{P}_0$ and $\mathcal{Q}_0$ and $\mathcal{R}_0$ are empty sets. Let us now show that (I_n) is true under the assumption that (I_{n-1}) is true.

By Theorem 10.1, we have

$$H_{n-1}^1 = H_n^0 + \sum_{k=0}^{n-2} C_{n-1}^k \left\{ H_{n-1-k}^0, W_{k+1} \right\} + \left\{ H_0^0, W_n \right\},$$

hence

$$H_{n-1}^1 = L^1 + \left\{ H_0^0, W_n \right\},$$

where, by the hypothesis (I_{n-1}) and the hypotheses in the theorem, we have $L^1 \in \mathcal{P}_n$.

Again by Theorem 10.1, we have

$$H_{n-2}^2 = H_{n-1}^1 + \sum_{k=0}^{n-2} C_{n-2}^k \left\{ H_{n-2-k}^1, W_{k+1} \right\}$$

$$= L^1 + \sum_{k=0}^{n-2} C_{n-2}^k \left\{ H_{n-2-k}^1, W_{k+1} \right\} + \left\{ H_0^0, W_n \right\},$$

and therefore $H_{n-2}^2 = L^2 + \left\{ H_0^0, W_n \right\}$, with $L^2 \in \mathcal{P}_n$. Analogously we have

$$H_{n-s}^s = L^s + \left\{ H_0^0, W_n \right\}, \quad \text{with} \quad L^s \in \mathcal{P}_n, \quad (s = 1, 2, \ldots, n). \tag{57}$$

In particular, $H_0^n = L^n + \left\{ H_0^0, W_n \right\}$ with $L^n \in \mathcal{P}_n$. By the above lemma, we can choose a solution H_0^n, W_n of this equation with $H_0^n \in \mathcal{Q}_n$ and $W_n \in \mathcal{R}_n$. It now follows from (57) and hypothesis (b) of the theorem that $H_{n-s}^s \in \mathcal{P}_n$ for $s = 1, 2, \ldots, n$. This shows that the hypothesis (I_n) is true and concludes the proof of the theorem. ∎

Since $\mathcal{Q}_i \subset \mathcal{P}_i$, the subspace $\mathcal{Q}_i$ is, in principle, simpler than $\mathcal{P}_i$, and so the theorem says that we are simplifying the Hamiltonian as the coefficients H_0^i are simpler than the coefficients H_i^0. We say that $H^*(y, \epsilon)$ is a *normal form* of $H(x, \epsilon)$

We are now ready to prove Theorem 6.1.

Proof of Theorem 6.1. We observe that the equalities $H^i(e^{tA}y) = H^i(y)$ in the theorem are equivalent to the equalities $\left\{ H^i, H_0 \right\} = 0$, $i = 1, 2, \ldots$.

For $\epsilon \neq 0$, $x = \epsilon u$ defines a symplectic transformation with multiplier ϵ^2. The new Hamiltonian is

$$\tilde{H}(u, \epsilon) = \frac{1}{\epsilon^2} H(\epsilon u) = \sum_{i=0}^{\infty} \left(\frac{\epsilon^i}{i!} \right) H_i^0(u),$$

where $H_i^0(u) = i! H_{i+2}^0(u)$ is a homogeneous polynomial of degree $i + 2$.

Let $\mathcal{P}_i$ be the linear space of all the homogeneous polynomials of degree $i+2$ and let $L_i : \mathcal{P}_i \to \mathcal{P}_i$ be the linear operator defined by $L_i(G) = \{G, H_0\}$. Then, $L_i(G)(u) = DG(u) \cdot Au$.

Now, by hypothesis, the matrix A of the linear part is diagonalizable, so let $\mathbf{s}_1, \ldots, \mathbf{s}_{2n}$ be a basis of eigenvectors for A corresponding to the eigenvalues $\lambda_1, \ldots, \lambda_{2n}$ (we can order them so that $\lambda_{n+i} = \overline{\lambda}_i$, $\mathbf{s}_{n+i} = \overline{\mathbf{s}}_i$, $i = 1, \ldots, r$, the remaining eigenvalues and eigenvectors being all real).

Any $K \in \mathcal{P}_i$ can be written in the form

$$K(u) = \sum_{|m|} k_{m_1 \cdots m_{2n}} \xi_1^{m_1} \cdots \xi_{2n}^{m_{2n}},$$

where $u = \xi_1 \mathbf{s}_1 + \cdots + \xi_{2n} \mathbf{s}_{2n}$ and $m = (m_1, \ldots, m_{2n})$ are the $2n$-tuples of nonnegative integers such that $|m| = m_1 + \cdots + m_{2n} = i + 2$. This means that the set of monomials

$$\mathcal{B} = \{\xi_1^{m_1} \cdots \xi_{2n}^{m_{2n}}; \; m_1 + \cdots + m_{2n} = i + 2\}$$

forms a basis for the space $\mathcal{P}_i$. Since $Au = \lambda_1 \mathbf{s}_1 + \cdots + \lambda_{2n} \mathbf{s}_{2n}$, we compute

$$L_i(\xi_1^{m_1} \cdots \xi_{2n}^{m_{2n}}) = (m_1 \lambda_1 + \cdots + m_{2n} \lambda_{2n}) \xi_1^{m_1} \cdots \xi_{2n}^{m_{2n}}.$$

This says that the elements of $\mathcal{B}$ are eigenvectors of L_i with eigenvalues $m_1 \lambda_1 + \cdots + m_{2n} \lambda_{2n}$, $m_1 + \cdots + m_{2n} = i + 2$. We can now define two subspaces of $\mathcal{P}_i$, invariant under L_i,

$$\mathcal{K}_i = \langle \xi_1^{m_1} \cdots \xi_{2n}^{m_{2n}}; \; m_1 + \cdots + m_{2n} = i + 2, \; m_1 \lambda_1 + \cdots + m_{2n} \lambda_{2n} = 0 \rangle,$$

$$\mathcal{R}_i = \langle \xi_1^{m_1} \cdots \xi_{2n}^{m_{2n}}; \; m_1 + \cdots + m_{2n} = i + 2, \; m_1 \lambda_1 + \cdots + m_{2n} \lambda_{2n} \neq 0 \rangle.$$

Obviously, $\mathcal{K}_i = \ker L_i$, $\mathcal{R}_i = \operatorname{Im} L_i$, and $\mathcal{P}_i = \mathcal{K}_i \oplus \mathcal{R}_i$. Therefore, the previous theorem applies and guarantees the existence of a generating function $W(u, \epsilon)$ with $W_i \in \mathcal{R}_i$ and such that the transformed Hamiltonian $\tilde{H}(v, \epsilon) = \sum_{i=0}^{\infty} \left(\frac{\epsilon^i}{i!}\right) H_0^i(v)$ has the terms H_0^i in $\mathcal{K}_i$, that is, $L_i(H_0^i) = \{H_0^i, H_0\} = 0$.

The mapping $y = \epsilon v$ is symplectic with multiplier ϵ^2, so the new Hamiltonian is given by

$$H^*(y, \epsilon) = \epsilon^2 \sum_{i=0}^{\infty} \left(\frac{\epsilon^i}{i!}\right) \epsilon^{-i-2} H_0^i(y) = \sum_{i=0}^{\infty} \frac{1}{i!} H_0^i(y).$$

Since the composition $x \mapsto u = \epsilon^{-1} x \mapsto v \mapsto y = \epsilon v$ is symplectic, its inverse $x = \Phi(y)$ defines a symplectic transformation of the form $x = y + \cdots$ and is such that the transformed Hamiltonian $H^*(y) = H^*(y, \epsilon)$ has the expansion

$$H^*(y) = \sum_{i=2}^{\infty} H^i(y),$$

where $H^i = \frac{1}{(i-2)!}H_0^{i-2}$ satisfies the equation $\{H^i, H_0\} = 0$. The proof of the theorem is complete. ∎

References

[1] Alfriend, J., The stability of the triangular Lagrangian points for commensurability of order 2, *Celest. Mech.* **1**, 351–359 (1970).

[2] Alfriend, J., Stability of and motion about L_4 at three-to-one commensurability, *Celest. Mech.* **4**, 60–77 (1971).

[3] Arnold, V. I., Proof of A. N. Kolmogorov's theorem on the preservation of quasiperiodic motions under small perturbations of the Hamiltonian, *Russian Math. Surveys* **18**, 9–36 (1963).

[4] Arnold, V. I., Small divisors problems in classical and celestial mechanics, *Russ. Math. Surv.* **18**, 85–192 (1963).

[5] Birkhoff, G. D., *Dynamical Systems*, Colloq. #9, Amer. Math. Soc., Providence, Providence, RI, 1927.

[6] Cabral, H. E., and K. R. Meyer, Stability of equilibria and fixed points of conservative systems. *Nonlinearity* **12**, 1351–1362 (1999).

[7] Cherry, T. M., On periodic solutions of Hamiltonian systems of differential equations, *Phil. Trans. Roy. Soc.* **A 227**, 137–221 (1928).

[8] Chetaev, N., *The Stability of Motion*, Pergamon Press, 1961.

[9] Deprit, A., Canonical transformations depending on a small parameter, *Celestial Mech.* **72**, 173–179 (1969).

[10] Deprit, A. and Deprit-Bartholomé, A., Stability of the triangular Lagrangean points, *Astron. J.* **72**, 173–179 (1967).

[11] Dirichlet, G. L., Über die Stabilität des Gleichgewichts, *J. Reine Angew. Math.* **32**, 85–88 (1846).

[12] Euler, L., De motu rectilineo trium corporum se mutuo attrahentium. *Novi Comm. Acad. Sci. Imp. Petrop.* **11**, 144–151 (1767).

[13] Gustavson, F., On constructing formal integrals of a Hamiltonian system near an equilibrium point, *Astron. Journ.* **71**, 670–686 (1966).

[14] Hirsch, M. W. and S. Smale, *Differential Equations, Dynamical Systems and Linear Algebra*, Academic Press, New York, 1974.

[15] Hori, G., Theory of general perturbations with unspecified canonical variables, *Publ. Astron. Soc. Japan* **18**, 287–296 (1966).

[16] Lagrange, J. L., Essai sur le problème des trois corps, *Oeuvres VI*, 229–324 (1772).

[17] Leontovich, A. M., On the stability of the Lagrange periodic solutions of the reduced problem of three bodies, *Soviet Math. Doklady Akad. Nauk.* **3**, 425–429 (1963).

[18] Lyapunov, A., *Problème général de la stabilité du mouvement*, Ann. of Math. Studies 17, Princeton Univ. Press, Princeton, NJ, 1982.

[19] Markeev, A. P., On the stability of the triangular libration points in the circular bounded three body problem, *Appl. Math. Mech.* **33**, 105–110 (1966).

[20] Markeev, A. P., *Libration Points in Celestial Mechanics and Space Dynamics*, Nauka, Moscow, 1978, (in Russian).

[21] Meyer, K. R., and G. R. Hall, *Introduction to Hamiltonian Dynamical Systems and the N-Body Problem*, Springer-Verlag, New York, 1995.

[22] Meyer, K. R., and D. S. Schmidt, The stability of the Lagrange triangular point and a theorem of Arnold, *Journ. Diff. Equations* **62**, 222–236 (1986).

[23] Moser, J. K., *Lectures on Hamiltonian Systems*, Mem. Amer. Math. Soc., 81, AMS, Providence, RI, 1968.

[24] Moser, J. K., On invariant curves of area-preserving mappings of an annulus. *Nachr. Akad. Wiss. Göttingen Math.-Phys. Kl.* 1962, 1–10.

[25] Siegel, C. L., Über die Existenz einer Normalform analytischer Hamiltonscher Differentialgleichungen in der Nähe einer Gleichgewichtslösung. *Math. Ann.* **128**, 144–170 (1954).

[26] Siegel, C. L., and J. K. Moser, *Lectures on Celestial Mechanics*, Springer-Verlag, New York 1971.

[27] Sokol'skii, A. G., Stability of the Lagrange solutions of the restricted three body problem for the critical ratio of the masses. *Appl. Math. Mech.* **39**, 342–345 (1975).

[28] Whittaker, E. T., *A Treatise on the Analytical Dynamics of Particles and Rigid Bodies*, Cambridge University Press, Cambridge, UK, 1937.

Photograph: Ivan Feitosa

Poincaré's Compactification and

Applications to Celestial Mechanics

Ernesto Pérez-Chavela

UAM – IZTAPALAPA
DEPARTAMENTO DE MATEMÁTICAS
APARTADO POSTAL 55-534
MÉXICO, D.F. 09340, MÉXICO
e-mail: epc@xanum.uam.mx

1   Introduction

The kind of compactification we will discuss here was introduced by Henri Poincaré at the beginning of the 20th century to compactify certain polynomial vector fields. Since then, many authors used it to study, among other things, the behavior at infinity and the escapes for particle systems with laws of motion given by polynomial vector fields [1], [2], [3]. In celestial mechanics the Newtonian n-body problem is given by a vector field defined on a noncompact manifold. The boundary of this manifold contains all the singularities due to collisions and to escapes or captures at infinity. The vector field of the n-body problem can be written in polynomial form if using the ideas of Heggie [5] and other authors. Essentially, this process consists of the regularization of all binary collisions. To obtain this polynomial vector field for three or more bodies, it is necessary to introduce redundant variables. Since any polynomial vector field in $\mathbb{R}^m$ can be extended analytically to the m-dimensional sphere, we get an extension of the n-body problem to a compact manifold.

In these lecture notes, we study the Poincaré compactification for a general polynomial vector field and then give the global expressions for the Poincaré compactification of a polynomial Hamiltonian vector field. After that, we study generic properties for arbitrary Hamiltonian polynomial vector fields, especially at infinity, where the invariant subset at infinity is a smooth manifold. We then consider the generic properties of the invariant flow. In some n-body problems deriving from celestial mechanics, we find that the critical points that appear are degenerate. In this sense we study the case where the homogeneous part of the highest degree monomial in the Hamiltonian polynomial has a single term. Though not generic, this case is frequently found in applications.

In section 6 we apply the Poincaré compactification to study the Kepler problem on the line and in the plane. We first regularize the singularity due to collision in order to obtain a polynomial Hamiltonian vector field. Since this is the simplest problem in celestial mechanics, we get some insight into the advantages and disadvantages of this technique. We obtain the same results as if applying blow-up transformations, except that here we do not have to fix an energy level.

In the second part of these notes we generalize the Poincaré compactification to the vector fields defined by the sum of homogeneous functions. This generalization is useful when we apply it to problems in celestial mechanics, for example, when studying escape. The advantage of this generalization is that escape can be tackled without regularizing the singularities of the equations of motion. Regularization usually requires the introduction of redundant variables (see, for example, [5]), which we want to avoid.

Let us notice that when applying our technique to problems in celestial mechanics, we can encounter escapes in velocities, in positions, or in both of them. In most cases we will be interested only in escapes in positions. We exploit the fact that S^n is a differentiable manifold, so we will perform the computations in local coordinates [4], where we can easily analyze the equilibrium points and linearize the flow around the equilibria. In sections 8 and 9 we apply this technique to study the behavior of escapes in the Kepler and Hill problems. The novelty is that we achieve our goals without regularizing collisions. For the Kepler problem we obtain, in a different way, classical results regarding escape orbits for a fixed energy level, $h \geq 0$, for hyperbolic motions ($h > 0$), and for parabolic ones ($h = 0$).

The last example that we present is the Hill problem, for which the study of escapes is a bit more complicated. If we regularize the problem, we need to introduce some redundant variables with the natural consequence that we have to work in a higher dimensional space. Our generalization is relevant since we can finalize it without regularizing the singularities. We present the problem by introducing new variables, which lead to a Hamiltonian system. After that we apply the Poincaré compactification to the corresponding vector field, which is given by the sum of homogeneous functions. Passing to the local charts we find that on each level of energy the behavior of the escape to infinity is obtained from the analysis of the two unique equilibrium points. Coming back to the original coordinates and using the energy relation, we get the final evolution of escape orbits. These notes are based on a couple of papers written by this author together with J. Delgado, E. A. Lacomba, and J. Llibre [2], [3] and a forthcoming paper with A. Susín [8].

2 Poincaré Compactification for Polynomial Vector Fields

Let $X = (P^1, \ldots, P^n)$ be a polynomial vector field in $\mathbb{R}^n$. We identify $\mathbb{R}^n$ with the hyperplane $\pi = \{y = (y_1, \ldots, y_{n+1}) \in \mathbb{R}^{n+1} | y_{n+1} = 1\}$ tangent to the sphere $S^n = \{y \in \mathbb{R}^{n+1} | \Sigma_{i=1}^{n+1} y_i^2 = 1\}$ at the north pole; after that we take the central projection from the sphere S^n to the hyperplane π; that is, for each point in π we draw the straight line through this point and the origin of $\mathbb{R}^{n+1}$. In this way we obtain two antipodal points in S^n, one in the open northern hemisphere H^+ of S^n and the other in the open southern hemisphere H^- of S^n (see Figure 1).

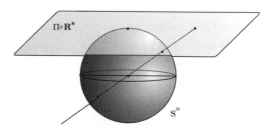

Figure 1. The central projection.

This construction defines the following two diffeomorphisms

$$\phi^+ : \mathbb{R}^n \longrightarrow H^+ \quad \text{and} \quad \phi^- : \mathbb{R}^n \longrightarrow H^-,$$

given by

$$\phi^+(x) = \tfrac{1}{\Delta(x)}(x_1, \ldots, x_n, 1),$$

$$\phi^-(x) = -\tfrac{1}{\Delta(x)}(x_1, \ldots, x_n, 1),$$

where

$$\Delta(x) = \left(1 + \sum_{i=1}^n x_i^2\right)^{1/2}.$$

In this form X induces a vector field $\hat{X}$ on $H^+ \cup H^-$ defined by

$$\hat{X} = \begin{cases} (D\phi^+)_x X(x) & \text{if} \quad y = \phi^+(x), \\ (D\phi^-)_x X(x) & \text{if} \quad y = \phi^-(x). \end{cases}$$

The expression for $\hat{X}(y)$ in $H^+ \cup H^-$ is

$$
\hat{X} = y_{n+1}
\begin{pmatrix}
1 - y_1^2 & -y_1 y_2 & \cdots & -y_1 y_n \\
-y_2 y_1 & 1 - y_2^2 & \cdots & -y_2 y_n \\
\vdots & & & \\
-y_n y_1 & -y_n y_2 & \cdots & 1 - y_n^2 \\
-y_{n+1} y_1 & -y_{n+1} y_2 & \cdots & -y_{n+1} y_n
\end{pmatrix}
\begin{pmatrix}
\hat{P}^1 \\
\hat{P}^2 \\
\vdots \\
\hat{P}^{n-1} \\
\hat{P}^n
\end{pmatrix},
$$

where $\hat{P}^i(y_1, \ldots, y_{n+1}) = P^i(y_1/y_{n+1}, \ldots, y_n/y_{n+1})$.

The equator $S^{n-1} = \{y \in S^n | y_{n+1} = 0\}$ of the Poincaré sphere corresponds to the infinity of $\mathbb{R}^n$. The key point of the Poincaré compactification is to extend the flow given by $\hat{X}$ on $S^n \backslash S^{n-1}$ to S^n. In this way we will be able to study the orbits of X going to, or coming from, infinity in $\mathbb{R}^n$. This extension is possible due to the polynomial character of X. Thus, the vector field

$$
\tilde{X}(y) - y_{n+1}^{m-1} \hat{X}(y) =
\begin{pmatrix}
1 - y_1^2 & -y_1 y_2 & \cdots & -y_1 y_n \\
-y_2 y_1 & 1 - y_2^2 & \cdots & -y_2 y_n \\
\vdots & & & \\
-y_n y_1 & y_n y_2 & \cdots & 1 - y_n^2 \\
-y_{n+1} y_1 & y_{n+1} y_2 & \cdots & -y_{n+1} y_n
\end{pmatrix}
\begin{pmatrix}
\tilde{P}^1 \\
\tilde{P}^2 \\
\vdots \\
\tilde{P}^{n\ 1} \\
\tilde{P}^n
\end{pmatrix}
$$

is analytic on the whole S^n. Here $m = \max\{\text{degree}(P^1), \ldots, \text{degree}(P^n)\}$ is the degree of X and

$$
\tilde{P}^k(y_1, \ldots, y_{n+1}) = y_{n+1}^m P^k(y_1/y_{n+1}, \ldots, y_n/y_{n+1}).
$$

Notice that $\tilde{P}^k$ are homogeneous polynomials of degree m. The vector field $\ddot{X}$ is called "The Poincaré compactification of X."

In many cases it is better to do the computations in local charts. In particular, this is true for obtaining the equilibrium points of the flow and their respective linearization. Therefore, we will consider S^n as a differentiable manifold and we will cover it by $2(n + 1)$ local charts given by $F_i : U_i \longrightarrow \mathbb{R}^n$ and $G_i : V_i \longrightarrow \mathbb{R}^n$, where

$$
U_i = \{y \in S^n | y_i > 0\} \quad \text{and} \quad V_i = \{y \in S^n | y_i < 0\}
$$

for $i = 1, 2, \ldots, n+1$. F_i is defined by

$$F_i(y) = \left(\frac{y_1}{y_i}, \ldots, \frac{y_{i-1}}{y_i}, \frac{y_{i+1}}{y_i}, \ldots, \frac{y_{n+1}}{y_i} \right) = (z_1, z_2, \ldots, z_n)$$

and G_i is given by the same formula. Let us observe that the local co-ordinates are denoted by $(z_1, \ldots, z_n)$, although these variables have a different meaning in each chart. Due to the symmetry that the vector field has on the sphere, its expression in the local chart (V_i, G_i) is the same as for (U_i, F_i) multiplied by the factor $(-1)^{m-1}$. That is, in V_i we have $X_{n+1}^{m-1} = [-z_n/\Delta(z)]^{m-1} = (-1)^{m-1}[z_n/\Delta(z)]^{m-1}$. In any local chart the points at infinity have the coordinate z_n equal to zero. Notice that all the local charts contain points at infinity except for (U_{n+1}, F_{n+1}) and (V_{n+1}, G_{n+1}).

We will now get the analytical expression of $\tilde{X}$ in each local chart. We start with doing the computations for U_1. Let $y \in U_1 \cap H^+$. Since the differential $(DF_1)y$ goes from $T_y U_1$ to $T_{F_1(y)} \mathbb{R}^n$, we get

$$(DF_1)_y(\tilde{X}(y)) = (DF_1)_y(y_{n+1}^{m-1} \hat{X}(y))$$

$$= y_{n+1}^{m-1}(DF_1)_y(D\phi^+)_x X(x) = y_{n+1}^{m-1} D(F_1 \circ \phi^+)_x X(x),$$

where $y = \phi^+(x)$. Then

$$D(F_1 \circ \phi^+)_x X(x) = \frac{1}{x_1^2}(-x_2 P^1 + x_1 P^2, \ -x_3 P^1 + x_1 P^3, \ldots, -x_n P^1 + x_1 P^n, \ -P^1),$$

where $P^i \equiv P^i(x_1, \ldots, x_n)$.

In the local chart U_1 we have

$$(z_1, z_2, \ldots, z_n) = F_1(y_1, \ldots, y_{n+1}) = \left(\frac{y_2}{y_1}, \ldots, \frac{y_{n+1}}{y_1} \right),$$

and we get

$$D(F_1 \circ \phi^+)_x X(x) = z_n(-z_1 P^1 + P^2, -z_2 P^1 + P^3, \ldots, -z_{n-1} P^1 + P^n, -z_n P^1),$$

where $P^i \equiv P^i(1/z_n, z_1/z_n, \ldots, z_{n-1}/z_n)$. Since $y_{n+1}^{m-1} = [z_n/\Delta(z)]^{m-1}$, the vector field $\tilde{X}$ becomes

$$\frac{z_n^m}{\Delta(z)^{m-1}}(-z_1 P^1 + P^2, \ -z_2 P^1 + P^3, \ldots, -z_{n-1} P^1 + P^n, \ -z_n P^1).$$

If $y \in U_1 \cap H^-$, we obtain the same expression as above. Scaling the expression of $\tilde{X}$ by $\Delta(z)^{m-1} > 0$ in each local chart and doing similar computations for the other local charts, we obtain the following expressions for $\tilde{X}$.

In $U_1 : (-z_1\tilde{P}^1 + \tilde{P}^2, \ -z_2\tilde{P}^1 + \tilde{P}^3, \ldots, -z_{n-1}\tilde{P}^1 + \tilde{P}^n, \ -z_n\tilde{P}^1)$

where $\tilde{P}^i \equiv \tilde{P}^i(1, z_1, \ldots, z_{n-1})$;

In $U_2 : (-z_1\tilde{P}^2 + \tilde{P}^1, \ -z_2\tilde{P}^2 + \tilde{P}^3, \ldots, -z_{n-1}\tilde{P}^2 + \tilde{P}^n, \ -z_n\tilde{P}^2)$

where $\tilde{P}^i \equiv \tilde{P}^i(z_1, 1, \ldots, z_{n-1})$;

$$\vdots \qquad\qquad\qquad\qquad\qquad\qquad\qquad \vdots$$

In $U_n : (-z_1\tilde{P}^n + \tilde{P}^1, \ -z_2\tilde{P}^n + \tilde{P}^2, \ldots, -z_{n-1}\tilde{P}^n + \tilde{P}^{n-1}, \ -z_n\tilde{P}^n)$

where $\tilde{P}^i \equiv \tilde{P}^i(z_1, \ldots, z_{n-1}, 1)$;

In $U_{n+1} : (P^1, P^2, \ldots, P^n)$ where $P^i \equiv \tilde{P}^i(z_1, z_2, \ldots, z_n)$.

From the expressions of $\tilde{X}$ in the local charts, we deduce that the infinity is invariant by the flow of $\tilde{X}$.

We will call "finite" (respectively, "infinite") critical points of X or $\tilde{X}$, the critical points of $\tilde{X}$ that lie on $S^n \backslash S^{n-1}$ (respectively, S^{n-1}). Notice that the orbits in S^n are always symmetric with respect to the origin of $\mathbb{R}^{n+1}$, but the vector field $\tilde{X}$ is only symmetric when m is odd. Furthermore, due to this symmetry, if $y \in S^{n-1}$ is an infinite critical point, then $-y$ is another one.

3 Poincaré Compactification for Polynomial Hamiltonian Vector Fields

A polynomial function $H : \mathbb{R}^{2d} \longrightarrow \mathbb{R}$ will be called a polynomial Hamiltonian and $X_H = \left(\frac{\partial H}{\partial y_{d+1}}, \ldots, \frac{\partial H}{\partial y_{2d}}, -\frac{\partial H}{\partial y_1}, \ldots, -\frac{\partial H}{\partial y_d} \right)$ will be called the polynomial Hamiltonian vector field associated with H. As we would expect in the case of a polynomial Hamiltonian vector field, the Poincaré compactification can also be described in terms of the derivatives of H. By a straightforward computation we get that the Poincaré compactification of X_H is given by the equations

$$y'_k = \frac{\partial \tilde{H}}{\partial y_{k+d}} + \lambda y_k,$$

$$y'_{k+d} = -\frac{\partial \tilde{H}}{\partial y_k} + \lambda y_{k+d}$$

$$y'_{n+1} = \lambda y_{n+1}$$

for $k = 1, \ldots, d$, where

$$\lambda = \sum_{k=1}^{d} \left(y_{k+d} \frac{\partial \tilde{H}}{\partial y_k} - y_k \frac{\partial \tilde{H}}{\partial y_{k+d}} \right),$$

and $\tilde{H}$ is obtained from the original Hamiltonian by the rule

$$\tilde{H}(y_1, y_2, \ldots, y_{n+1}) = y_{n+1}^{m+1} H(y_1/y_{n+1}, \ldots, y_n/y_{n+1}).$$

We will call $\tilde{H}$ the generating function of the vector field $\tilde{X}_H$.

The next step is to study how the energy levels $E_h = H^{-1}(h)$ are transformed into the Poincaré sphere S^n. For doing this we start with the expression of the compactified vector field restricted to infinity in terms of the highest degree homogeneous part of H.

Proposition 3.1 *Consider a polynomial Hamiltonian H in the $n = 2d$ variables $x_1, \ldots, x_d,\ x_{d+1}, \ldots, x_{2d}$ and write it as*

$$H = H_0 + H_1 + \cdots + H_{m+1},$$

where each term H_k is a homogeneous polynomial of degree k. Then the invariant flow at infinity $y_{n+1} = 0$ is given by the system of equations

$$y'_k = \frac{\partial H_{m+1}}{\partial y_{k+d}} + \lambda y_k,$$

$$y'_{k+d} = -\frac{\partial H_{m+1}}{\partial y_k} + \lambda y_{k+d},$$

for $k = 1, \ldots, d$, with

$$\lambda = \sum_{k=1}^{d} \left(y_{k+d} \frac{\partial H_{m+1}}{\partial y_k} - y_k \frac{\partial H_{m+1}}{\partial y_{k+d}} \right).$$

Proof. From the equations of the Poincaré compactification of X_H, it follows that $y_{n+1} = 0$ is invariant. By definition

$$\tilde{H} - y_{n+1}^{m+1} H(y_1/y_{n+1}, \ldots, y_n/y_{n+1})$$

$$= y_{n+1}^{m+1}(H_0(y_1/y_{n+1}, \ldots, y_n/y_{n+1}) + \cdots + H_{m+1}(y_1/y_{n+1}, \ldots, y_n/y_{n+1}))$$

but each term H_k is homogeneous of degree k. Therefore, by letting $y_{n+1} = 0$ the corresponding terms with $k \leq m$ vanish, and for $k = m + 1$, we get

$$\tilde{H}(y_1, \ldots, y_{n+1}) = H_{m+1}(y_1, \ldots, y_n).$$

This complete the proof.   ∎

4   Generic Properties

To study generic properties of polynomial Hamiltonian vector fields, it is better to write the Hamiltonian as

$$H = H_0 + H_1 + \cdots + H_{m+1},$$

where H_k is the homogeneous part of H of degree k, $k = 0, 1, \ldots, m + 1$. Notice that with this formulation it is sufficient to study the energy level $H = 0$.

We have shown in the first part of these lectures that the behavior at infinity for the respective compactified vector field $X_{\tilde{H}}$ is given by H_{m+1}. Assuming that the Poincaré sphere is S^n, we define

$$E^\infty = \{y \in S^{n-1} | H_{m+1}(y_1, \ldots, y_n) = 0\}.$$

Let us observe that $E^\infty = (H_{m+1}|S^{n-1})^{-1}(0)$. From the preimage theorem [4], we know that if 0 is a regular value of $H_{m+1}|S^{n-1}$, then E^∞ is an $(n - 2)$-dimensional smooth manifold. Now we will show first that generically E^∞ is a smooth manifold. Afterwards we will prove also that generically all finite and infinite critical points of $\tilde{X}_H$ are hyperbolic.

Let $\mathbf{H}^{m+1}$ be the set of all polynomial Hamiltonians of degree exactly $m + 1$ in the variables $(y_1, \ldots, y_n)$ with $n = 2d$. Let $\Phi : \mathbf{H}^{m+1} \longrightarrow \mathbb{R}^N$ be a map associating with each polynomial Hamiltonian H of degree $m+1$ its vector of coefficients in a given fixed order; then H is identified with $\Phi(H) \in \mathbb{R}^N$. Let us observe that $\mathbf{H}^{m+1}$ cannot be identified with the whole $\mathbb{R}^N$ because the polynomials of $\mathbf{H}^{m+1}$ have at least one of the coefficients of degree $m+1$ different from zero. We endowed $\mathbf{H}^{m+1}$ with the Euclidean topology of $\mathbb{R}^N$, usually called the "coefficient topology." Let

$$\mathbf{Y}^{m+1} = \{H \in \mathbf{H}^{m+1} | E^\infty \text{ is a smooth manifold}\}.$$

We will prove that $\mathbf{Y}^{m+1}$ is an open and dense subset of $\mathbf{H}^{m+1}$ with the coefficient topology. We will show that its complement is contained in an algebraic hypersurface of $\mathbb{R}^N$. To achieve this we need to introduce the notion of resultant of two polynomials.

Let P and Q be two polynomials in the variable x with coefficients in $\mathbb{R}$, with degrees n and m, respectively, that is,

$$P(x) = a_0\, x^n + a_1\, x^{n-1} + \cdots + a_{n-1}\, x + a_n,$$

$$Q(x) = b_0\, x^m + b_1\, x^{m-1} + \cdots + b_{m-1}\, x + b_m.$$

Multiplying the first equation successively by $x^{m-1}, x^{m-2}, \ldots, x^2, x$ and the second by $x, x^2, \ldots, x^{n-1}$, we get the linear system

$$a_0 x^{n+m-1} + a_1 x^{n+m} + \cdots + a_{n-1}x^m + a_n x^{m-1} = 0,$$

$$\vdots \qquad\qquad\qquad\qquad \vdots$$

$$a_0 x^{n+1} + a_1 x^n + \cdots + a_{n-1}x^2 + a_n x = 0,$$

$$a_0 x^n + a_1 x^{n-1} + \cdots + a_{n-1}x + a_n = 0,$$

$$b_0 x^m + b_1 x^{m-1} + \cdots + b_{m-1}x + b_n = 0,$$

$$b_0 x^{m+1} + b_1 x^m + \cdots + b_{m-1}x^2 + b_m x = 0,$$

$$\vdots \qquad\qquad\qquad\qquad \vdots$$

$$b_0 x^{m+n-1} + b_1 x^{m+n} + \cdots + b_{m-1}x^n + b_m x^{n-1} = 0.$$

The determinant of this system is called the resultant of P and Q, and we will denote it by $Res_x(P, Q)$. Let us observe that if P and Q have a common real root x^*, then $Res_{x^*}(P, Q) = 0$.

Theorem 4.1 *Generically, E^∞ is a smooth manifold.*

Proof. It is enough to prove that the complement of the set $\mathbf{Y}^{m+1}$ is contained in an algebraic hypersurface of $\mathbb{R}^N$, because then $\mathbf{Y}^{m+1}$ contains an open and dense set in $\mathbf{H}^{m+1}$.

In this way we will prove that there exists a polynomial function $\chi : \mathbb{R}^N \longrightarrow \mathbb{R}$ such that if $H \in \mathbf{H}^{m+1}$, then $\chi(\Phi^{-1}(H)) = 0$.

We know that if $H \notin \mathbf{Y}^{m+1}$, then $(H_{m+1}|S^{n-1})^{-1}(0)$ has a critical point $(y_1^*, y_2^*, \ldots, y_n^*)$, otherwise, by the implicit function theorem, E^∞ would be a smooth manifold. Then the system of equations

$$\frac{\partial H_{m+1}}{\partial y_i}(y_1^*, y_2^*, \ldots, y_n^*) + \mu(y_i^*) = 0, \qquad i = 1, \ldots, n,$$

$$\sum_{i=1}^{n}(y_i^*)^2 - 1 = 0,$$

has a nontrivial solution, where μ is a Lagrange multiplier. Multiplying the ith equation by y_i^* and adding them together, we get

$$\sum_{i=1}^{n} y_i^* \frac{\partial H_{m+1}}{\partial y_i}(y_1^*, y_2^*, \ldots, y_n^*) + \mu \sum_{i=1}^{n}(y_i^*)^2 = 0.$$

Then, from the homogeneity of H_{m+1} and using the fact that the critical point is in $S^{n-1} \cap H_{m+1}^{-1}(0)$, we have that the left part of the above equation should be zero and therefore $\mu = 0$. We thus obtain a set of homogeneous polynomial equations,

$$\frac{\partial H_{m+1}}{\partial y_i}(y_1^*, y_2^*, \ldots, y_n^*) = 0,$$

which for $i = 1, \ldots, n$ have a nontrivial solution $(y_1^*, y_2^*, \ldots, y_n^*) \in S^{n-1}$. Let $L_k = \frac{\partial H_{m+1}}{\partial y_k}$. We inductively define $(k = 2, \ldots, n)$

$$R_k(y_3, \ldots, y_n) = Res_{y_2}(L_1(1, y_2, y_3, \ldots, y_n), \ L_k(1, y_2, y_3, \ldots, y_n)),$$

$$R_{2k}(y_4, \ldots, y_n) = Res_{y_3}(R_2(y_3, y_4, \ldots, y_n), \ R_k(y_3, y_4, \ldots, y_n)),$$

$$R_{23k}(y_5, \ldots, y_n) = Res_{y_4}(R_{23}(y_4, \ldots, y_n), \ R_{2k}(y_4, \ldots, y_n)),$$

$$\vdots$$

$$R_{23\ldots n} = Res_{y_n}(R_{23\ldots(n-1)}(y_n), \ R_{23\ldots n}(y_n)),$$

where Res_{y_i} is the resultant of polynomials with respect to the variable y_i. We want to show that there exists a polynomial function $\chi : \mathbb{R}^N \longrightarrow \mathbb{R}$ such that if the polynomials L_k have a common real root, then $\chi = 0$.

At the final step, $R_{23...n}$ is just a number that is a polynomial function of the coefficients of the polynomial H. We claim that if the solution $(y_1^*, \ldots, y_n^*)$ has $y_1^* \neq 0$, then $R_{23...n}$ must be zero. Since if $y_1^* \neq 0$, then, by the homogeneity of L_k, it follows that if $\overline{y}_k = \frac{y_k^*}{y_1^*}$ for $k = 2, \ldots, n$, then the systems $L_1(1, \xi, \overline{y}_3, \ldots, \overline{y}_n) = 0$ and $L_k(1, \xi, \overline{y}_3, \ldots, \overline{y}_n) = 0$ for $k = 2, \ldots, n$ have the common solution $\xi = \frac{y_2^*}{y_1^*}$. Therefore, $R_k(\overline{y}_3, \ldots, \overline{y}_n) = 0$ for $k = 2, \ldots, n$. Again, considering the polynomials $R_k(\xi, \overline{y}_4, \ldots, \overline{y}_n)$, they have the common root $\xi = \overline{y}_3$, therefore $R_{2k}(\overline{y}_4, \ldots, \overline{y}_n) = 0$. In this way we obtain that $R_{23...n} = 0$. In general we rename $R_{23...n}$ as $\Re_1$, which is not necessarily zero if $y_1^* = 0$.

Now if the solution has $y_j \neq 0$, we just replace $L_k(1, y_2, \ldots, y_n)$ by $L_k(y_1, \ldots, 1, \ldots, y_n)$, where 1 appears in the jth position in the definition of R_k; then we proceed as above to get $\Re_j$. Finally defining $\Re = \prod_{i=1}^n \Re_j$, for a nontrivial solution of the original system of homogeneous polynomial equations we must have $\Re = 0$. This polynomial depends only on the coefficients of the highest degree terms of H, but it can be considered as a polynomial in all the coefficients. This completes the proof.   ∎

We will now study the finite and infinite critical points of polynomial Hamiltonians. For this we need the following result.

Lemma 4.2 *Given the real polynomial $q(x)$, we define the real polynomials $q_1(x)$ and $q_2(x)$ by $q(ix) = q_1(x) + iq_2(x)$, where $i = \sqrt{-1}$. If $q(x)$ has a pure imaginary root ib, then the resultant $Res_x(q_1, q_2) = 0$.*

Proof. Since $q(ib) = 0$, then $q_1(b) = q_2(b) = 0$, and therefore the resultant of q_1 and q_2 must be zero.   ∎

Let Δ^{m+1} be the set $\{H \in \mathbf{H}^{m+1}|$ critical points of the compactified Hamiltonian flow associated to H are hyperbolic$\}$. We will show that $\mathbf{H}^{m-1} \backslash \Delta^{m+1}$ is contained in an algebraic hypersurface. Using this, we will prove the following result.

Theorem 4.3 *Generically, the compactifications of polynomial Hamiltonians have all their finite and infinite critical points hyperbolic.*

Proof. As in the above theorem, we will prove that if $H \in \mathbf{H}^{m+1}\backslash\Delta^{m+1}$, then there exists a polynomial function $\mathbf{P}:\mathbb{R}^N \longrightarrow \mathbb{R}$ such that $\mathbf{P}(\Phi^{-1}(H)) = 0$. In this way we consider the equations that define the critical points of the compactified Hamiltonian flow associated to H in the local chart (U_1, F_1) with coordinates $(z_1, z_2, \ldots, z_n)$. Let $q(\xi)$ be the characteristic polynomial at a critical point and $q_1(\xi), q_2(\xi)$ be as in the above lemma. Then if $H \in \mathbf{H}^{m+1}\backslash\Delta^{m+1}$ the set of equations

$$P_1' = z_1(z_1, z_2, \ldots, z_n, 0) = 0,$$
$$P_2' = z_2(z_1, z_2, \ldots, z_n, 0) = 0,$$
$$\vdots \qquad \vdots$$
$$P_n' = z_n(z_1, z_2, \ldots, z_n, 0) = 0,$$
$$q_1(z_1, z_2, \ldots, z_n, \xi) = 0,$$
$$q_2(z_1, z_2, \ldots, z_n, \xi) = 0$$

has a common real solution. The above equations involve polynomials in the coefficients of the original Hamiltonian H. Using a similar argument as in the first theorem of this section, it follows that there exists a polynomial function of the coefficients such that $\Re_1 = 0$. Proceeding similarly with the rest of the charts, we obtain the rest $\Re_j$ for $j = 1, 2, \ldots, n + 1$. Then defining $\Re = \prod_{j-1}^{n+1} \Re_j$, we have that if $H \in \mathbf{H}^{m+1}\backslash\Delta^{m+1}$, then $\Re = 0$. This completes the proof. ∎

5 Behavior at Infinity in the Monomial Case

In this section we will analyze the case in which the homogeneous part of highest degree in the polynomial Hamiltonian has a single term. In this case we can give a complete description of the behavior of the vector field near or at infinity. Let us observe that this case is not generic but appears frequently in applications to celestial mechanics.

Given the polynomial Hamiltonian H, suppose its highest homogeneous part has the form

$$H_{m+1}(y_1, \ldots, y_n) = y_1^{\alpha_1} y_2^{\alpha_2} \cdots y_n^{\alpha_n},$$

where $\alpha_i \geq 0$ for $j = 1, 2, \ldots, n$ and $\alpha_1 + \alpha_2 + \cdots + \alpha_n = m + 1$. Let

$$S_j^{n-2} = \{(y_1, \ldots, y_n) \in S^{n-1} | y_j = 0\}.$$

The following theorem summarizes all the results obtained about the behavior at infinity in this case.

Theorem 5.1 *Suppose H_{m+1} is a single monomial. Then*

(a) $E^\infty = \cup_{\alpha_i \geq 1} S_j^{n-2}$;

(b) *if $\alpha_i \geq 2$, then S_j^{n-2} consists entirely of critical points;*

(c) *if $\alpha_i, \alpha_k \geq 1$ for $i \neq k$ and $n \geq 4$, then $S_i^{n-2} \cap S_k^{n-2}$ is an $(n-3)$-sphere of critical points;*

(d) *if $\alpha_i = 1$, then on S_i^{n-2}, there are no critical points other than those described in (c). If $\alpha_{\iota+d} \geq 1$, then any solution starts in some sphere of critical points $S_\iota^{n-2} \cap S_k^{n-2}$ and ends in $S_j^{n-2} \cap S_{j+d}^{n-2}$. If $\alpha_{\iota+d} = 0$, then any solution starts and ends in some sphere of critical points $S_j^{n-2} \cap S_k^{n-2}$.*

Proof. The set E^∞ is defined by the equations $H_{m+1}(y_1, \ldots, y_n) = 0$ and $y_1^2 + \cdots + y_n^2 = 1$. Since H_{m+1} contains explicitly only factors y_i for $\alpha_i > 0$, part (a) follows easily. Suppose $\alpha_i \geq 2$. Then in the computation of $\partial H_{m+1}/\partial y_k$, there exists some positive power of y_i. Therefore, on setting $y_i = 0$ in the global equations

$$y_i' = \frac{\partial H_{m+1}}{\partial y_{i+d}} + \lambda y_i,$$

$$y_{i+d}' = -\frac{\partial H_{m+1}}{\partial y_i} + \lambda y_{i+d}$$

for $i = 1, 2, \ldots, d$, all right-hand sides vanish. This means that S_i^{n-2} consists of critical points, and consequently (b) is proved. For proving (c), a similar argument applies. Now suppose $\alpha_\iota = 1$ and, without lost of generality, that $1 \le \iota \le d$; then it is readily seen that S_ι^{n-2} is invariant, and the global equations restricted to S_j^{n-2} reduce to

$$y'_{\iota+d} = (-1 + y_{\iota+d}^2)\frac{\partial H_{m+1}}{\partial y_\iota},$$

$$y'_k = y_{\iota+d}y_k\frac{\partial H_{m+1}}{\partial y_\iota} \quad \text{for } k \ne \iota + d.$$

For any initial condition not contained in the sphere of critical points $S_\iota^{n-2} \cap S_k^{n-2}$, we can reparametrize the solution by the common factor $(\partial H_{m+1}/\partial y_j)^{-1}$. Then the system becomes

$$\dot{y}_{\iota+d} = -1 + y_{\iota+d}^2, \quad \dot{y}_k = y_{\iota+d}y_k \quad \text{for} \quad k \ne \iota + d,$$

where the $(\cdot)$ means derivative with respect to the new time variable. From the first equation above, we have that y_{j+c} decreases monotonically from $y_{\iota+d} = 1$ to $y_{\iota+d} = -1$, passing through $y_{\iota+d} = 0$. In the interval of values $-1 < y_{\iota+d} < 0$, no other coordinate y_k vanishes, therefore the solution dies in $y_{\iota+d} = 0$ if the exponent $\alpha_{\iota+d} \ge 1$, or passes through and dies in $y_n = 0$ if $\alpha_{\iota+d} = 0$. This proves (d). ∎

6 The Kepler Problem

Now we will apply the Poincaré compactification to the Kepler problem. Since it is the simplest problem in celestial mechanics, it will be a parameter to measure the advantages and disadvantages of this technique. We will start with the Kepler problem on the line.

6.1 The Kepler Problem on the Line

The Hamiltonian function in this case is

$$H = \frac{p^2}{2} - \frac{1}{q}$$

with $q > 0$ and $p \in \mathbb{R}$. To obtain a polynomial Hamiltonian, we regularize the singularity by a Levi-Civita transformation together with an appropriate scaling of time. On a fixed energy level $H = h$, through the change of variables

$$q = Q^2, \qquad p = \frac{P}{2Q}, \qquad \frac{dt}{d\zeta} = 4Q^2,$$

the new Hamiltonian becomes

$$\overline{H} = \frac{P^2}{2} - 4hQ^2 - 4.$$

We are interested in its flow on the energy level $\overline{H} = 0$. Here, the Poincaré sphere is $S^2 = \{(y_1, y_2, y_3) \in \mathbb{R}^3 | y_1^2 + y_2^2 + y_3^2 = 1\}$, the degree of the associated polynomial Hamiltonian vector field is $m = 1$, and $d = 1$ is the number of degrees of freedom, $n = 2d = 2$.

In order to study the vector field on S^2, we change the variables

$$Q = \frac{y_1}{y_3}, \qquad P = \frac{y_2}{y_3}$$

in $\overline{H}$. Multiplying by the factor y_3^2 we get the generating function $\tilde{H}$ of the vector field $\tilde{X}$ in the Poincaré sphere S^2:

$$\tilde{H} = \frac{y_2^2}{2} - 4hy_1^2 - 4y_3^2.$$

The vector field on S^2 is given by

$$y_1' = y_2(1 - y_1^2(8h + 1)),$$

$$y_2' = y_1(8h - y_2^2(8h + 1)),$$

$$y_3' = -(8h + 1)y_1 y_2 y_3.$$

The only critical points $y \in S^2$ that also satisfy the energy relation $\tilde{H} = 0$ are located at infinity and are given by

$$y_1 = \pm\frac{1}{\sqrt{8h + 1}}, \qquad y_2 = \pm\sqrt{\frac{8h}{8h + 1}}, \qquad y_3 = 0 \quad \text{with} \quad h \geq 0.$$

If $h > 0$, these are four points on the equator (only two points if $h = 0$). The energy levels then extended to the infinity $y_3 = 0$ as critical points.

Let us provide an interpretation of the critical points. For $h > 0$, if $(y_1(t), y_2(t), y_3(t))$ is a solution (with energy h) going to a critical point, then

$$Q = \frac{y_1}{y_3} \longrightarrow \infty \quad \text{and} \quad P = \frac{y_2}{y_3} \longrightarrow \infty.$$

This implies that in the original coordinates, $q \to \infty$ and $p = \frac{P}{2Q} = \frac{y_2}{2y_1} \to \pm\sqrt{2h}$, which corresponds to hyperbolic motion. For $h = 0$, we get

$$q \to \infty \quad \text{and} \quad p \to 0,$$

corresponding to parabolic motion. On each positive energy level the critical points are nondegenerate. We will now study this problem in $\mathbb{R}^2$.

6.2 The Kepler Problem in the Plane

The Hamiltonian in this case is given by

$$H = \frac{|p|^2}{2} - \frac{1}{|q|},$$

with $q, p \in \mathbb{R}^2$ and $q \neq 0$. As usual, we regularize the singularity with a Levi-Civita transformation by regarding (q, p) and (Q, P) as complex variables,

$$q = Q^2, \qquad p = \frac{P}{2Q}, \qquad \frac{dt}{d\zeta} = 4|Q|^2.$$

The regularized Hamiltonian is

$$\overline{H} = \frac{|P|^2}{2} - 4h|Q|^2 - 4.$$

In this case we have a problem with two degrees of freedom, that is, $d = 2$ and $n = 2d = 4$ is the number of variables, therefore the Poincaré compactification gives us a vector field on S^4. The degree of the associated polynomial Hamiltonian vector field is $m = 1$.

With the change of variables

$$Q_1 = y_1/y_5, \qquad\qquad P_1 = y_3/y_5,$$

$$Q_2 = y_2/y_5, \qquad\qquad P_2 = y_4/y_5,$$

and multiplying by y_5^2, we get the generating function for the vector field on S^4,

$$\tilde{H} = \frac{1}{2}(y_3^2 + y_4^2) - 4h(y_1^2 + y_2^2) - 4y_5^2.$$

The equations of the compactified vector field are

$$y_1^1 = y_3 + \lambda y_1, \qquad y_3' = 8hy_1 + \lambda y_3,$$

$$y_2^1 = y_4 + \lambda y_2, \qquad y_4' = 8hy_2 + \lambda y_4,$$

$$y_5' = \lambda y_5,$$

where $\lambda = -(8h + 1)(y_1 y_3 + y_2 y_4)$.

Let us observe that the infinity $y_5 = 0$ is invariant, and as in the Kepler problem on the line we have that if $h \geq 0$ we extend the energy relation $\tilde{H} = 0$ to the sphere at infinity

$$S_\infty^3 = \{(y_1, y_2, y_3, y_4, y_5) \in S^4 |\quad y_5 = 0,\ y_1^2 + y_2^2 + y_3^2 + y_4^2 = 1\}.$$

We will consider first the case $h > 0$. We have to glue to the finite part of the level energy E_h the set

$$E_h^\infty = \{(y_1, \ldots, y_5) \in S_\infty^3 |\quad y_3^2 + y_4^2 - 8h(y_1^2 + y_2^2) = 0\},$$

which in view of the relation $y_1^2 + \cdots + y_4^2 = 1$ reduces to

$$E_h^\infty = \left\{(y_1, \ldots, y_5) \in S_\infty^3 |\quad y_1^2 + y_2^2 = \frac{1}{8h + 1},\ y_3^2 + y_4^2 = \frac{8h}{8h + 1}\right\};$$

that is, E_h^∞ is a 2-torus.

The only critical points satisfying the energy relation $\tilde{H} = 0$ are at infinity and are given by the following equations:

$$y_3 = -\lambda y_1, \qquad y_1 = -(\lambda/8h)y_3,$$

$$y_4 = -\lambda y_2, \qquad y_2 = -(\lambda/8h)y_4.$$

Doing a simple calculation we obtain $\lambda^2 = 8h$, and therefore

$$y_3 = \pm\sqrt{8h}\,y_1,$$

$$y_4 = \pm\sqrt{8h}\,y_2.$$

From these equations we easily obtain that the critical points are two circles $C_h^\pm$ in the 2-torus E_h^∞, which can be parametrized by an angle θ:

$$(y_1, y_2, y_3, y_4) = \frac{1}{\sqrt{8h+1}}(\cos\theta, \sin\theta,\ \varepsilon\sqrt{8h}\cos\theta,\ \varepsilon\sqrt{8h}\sin\theta),$$

where the choice of $\varepsilon = \pm 1$ gives the two circles of equilibria.

Let us take local coordinates to compute the linearization of the vector field at the set of critical points. In terms of the local chart (U_1, F_1), the vector field is

$$z_1' = -z_1 z_2 + z_3, \qquad z_3' = -z_2 z_3 + 8hz_1,$$

$$z_2' = -z_2^2 + 8h, \qquad z_4' = -z_2 z_4.$$

The circles of equilibria C_h^ε are locally given by the equations

$$z_2 = a, \quad z_3 = az_1, \quad z_4 = 0,$$

where $a = \varepsilon\sqrt{8h}$ and the characteristic polynomial of the linearization of the vector field at an equilibrium point is

$$p(\overline{\lambda}) = \overline{\lambda}(2a + \overline{\lambda})^2(a + \lambda).$$

Since there is only one zero eigenvalue, the circles C_h^ε are normally hyperbolic: C_h^+ is an attractor and C_h^- is a repeller.

In fact, the flow on the invariant manifold $E_h^\infty \equiv S^1 \times S^1$, where

$$y_1 = r_1 \cos\theta, \qquad y_3 = r_2 \cos\phi,$$

$$y_2 = r_1 \sin\theta, \qquad y_4 = r_2 \sin\phi,$$

$$r_1 = \frac{1}{\sqrt{8h+1}}, \qquad r_2 = \sqrt{\frac{8h}{8h+1}},$$

is given in terms of angular variables by

$$\theta' = \sqrt{8h}\sin(\phi - \theta),$$

$$\phi' = -\sqrt{8h}\sin(\phi - \theta).$$

Therefore, the trajectories are straight lines of slope -1 in the covering plane (θ, ϕ) and the critical points correspond to the lines $\phi = \theta$ and $\phi = \theta + \pi$.

It is interesting to note that in this case the same information is obtained if we blow up the infinity [6]. The treatment of the case $h = 0$ is similar, except that E_h^∞ is a circle of degenerate critical points. In [2] we have completely analyzed the escapes in the collinear three-body problem. In these lectures, however, we employ a novel generalization of the Poincaré compactification and present new examples.

7   The Poincaré Compactification for Homogeneous Functions

Let $X = (f^1, \ldots, f^n)$ be a vector field in $\mathbb{R}^n$, where the f^i are sums of homogeneous functions, that is, $f^i(x_1, \ldots, x_n) = \sum_{j=1}^{k_i} f_j^i(x_1, \ldots, x_n)$, where f_j^i is a homogeneous function with degree of homogeneity m_{ij}. We identify $\mathbb{R}^n$ with the hyperplane $\pi = \{x \in \mathbb{R}^{n+1} | x_{n+1} = 1\}$, tangent to the Poincaré sphere $S^n = \{y \in \mathbb{R}^{n+1} | \sum_{i=1}^{n+1} y_i^2 = 1\}$ at the north pole, and as in section 2, we take the central projection from the sphere S^n to the hyperplane π. This construction defines the following two diffeomorphisms:

$$\phi^+ : \pi \longrightarrow E^+ \quad \text{and} \quad \phi^- : \pi \longrightarrow E^-$$

given by

$$\phi^+(x) = \frac{1}{\Delta(x)}(x_1, \ldots, x_n, 1), \tag{1}$$

$$\phi^-(x) = -\frac{1}{\Delta(x)}(x_1, \ldots, x_n, 1),$$

where

$$\Delta(x) = \left(1 + \sum_{i=1}^{n} x_i^2\right)^{1/2}.$$

In this form X induces a vector field $\hat{X}$ on $E^+ \cup E^-$ defined by

$$\hat{X}(y) = \begin{cases} (D\phi^+)_x X(x) & \text{if} \quad y = \phi^+(x), \\ (D\phi^-)_x X(x) & \text{if} \quad y = \phi^-(x). \end{cases}$$

The expression for $\hat{X}$ in $E^+ \cup E^-$ is

$$\hat{X}(y) = y_{n+1} \begin{pmatrix} 1-y_1^2 & -y_1 y_2 & \cdots & -y_1 y_n \\ -y_2 y_1 & 1-y_2^2 & \cdots & -y_2 y_n \\ \vdots & & & \\ -y_n y_1 & -y_n y_2 & \cdots & 1-y_n^2 \\ -y_{n+1}y_1 & -y_{n+1}y_2 & \cdots & -y_{n+1}y_n \end{pmatrix} \begin{pmatrix} \hat{f}^1 \\ \hat{f}^2 \\ \vdots \\ \hat{f}^{n-1} \\ \hat{f}^n \end{pmatrix},$$

where

$$\hat{f}^i(y_1, \ldots, y_{n+1}) = \sum_{j=1}^{k_i} f_j^i \left(\frac{y_1}{y_{n+1}}, \ldots, \frac{y_n}{y_{n+1}} \right).$$

The equator $S^{n-1} = \{ y \in S^n | y_{n+1} = 0 \}$ on the Poincaré sphere corresponds to the infinity of $\mathbb{R}^n$. We point out that the vector field given by $\hat{X}$ is not defined on S^{n-1}. Until now we have done the same construction as the one of section 2. Now, in order to extend the flow to the whole S^n, we will use the fact that the functions f_j^i are homogeneous. Let $m = \max_{i,j}(m_{ij})$ and note that m_{ij} is the degree of homogeneity of f_j^i.

We define $\tilde{X}$ by

$$\tilde{X}(y) = y_{n+1}^{m-1} \hat{X}(y).$$

This vector field defined on S^n is known as the Poincaré compactification of the vector field X.

$$\tilde{X}(y) = \begin{pmatrix} 1-y_1^2 & -y_1 y_2 & \cdots & -y_1 y_n \\ -y_2 y_1 & 1-y_2^2 & \cdots & -y_2 y_n \\ \vdots & & & \\ -y_n y_1 & -y_n y_2 & \cdots & 1-y_n^2 \\ -y_{n+1}y_1 & -y_{n+1}y_2 & \cdots & -y_{n+1}y_n \end{pmatrix} \begin{pmatrix} y_{n+1}^m \hat{f}^1 \\ y_{n+1}^m \hat{f}^2 \\ \vdots \\ y_{n+1}^m \hat{f}^{n-1} \\ y_{n+1}^m \hat{f}^n \end{pmatrix}.$$

Let us observe that if the f_i are polynomials for $i = 1, 2, \ldots, n$, then our definition agrees with the classical one for polynomial vector fields. The last row on the above matrix shows that the infinity corresponding to $y_{n+1} = 0$ is invariant by the extended flow. With this construction we have proved the following result.

Theorem 7.1 *For any vector field X in $\mathbb{R}^n$ defined by the sum of homogeneous functions, the Poincaré compactification exists. That is, we can extend it to S^n in $\mathbb{R}^{n+1}$ and this extension preserves the same degree of differentiability as X.*

As in section 2, we will work in local charts. So we regard S^n as a differentiable manifold and cover it by $2(n+1)$ local charts given by the sets $U_i^{+(-)}$ and the functions $F_i^+ : U_i^+ \longrightarrow \mathbb{R}^n$ and $F_i^- : U_i^- \longrightarrow \mathbb{R}^n$, where

$$U_i^+ = \{y \in S^n | y_i > 0\} \quad \text{and} \quad U_i^- = \{y \in S^n | y_i < 0\}$$

for $i = 1, 2, \ldots, n+1$. F_i^+ is defined by

$$F_i^+(y) = \left(\frac{y_1}{y_i}, \ldots, \frac{y_{i-1}}{y_i}, \frac{y_{i+1}}{y_i}, \ldots, \frac{y_{n+1}}{y_i} \right) = (z_1, z_2, \ldots, z_n),$$

and F_i^- is defined by the same formula. Again, the local coordinates are denoted by $(z_1, \ldots, z_n)$, although these variables have a different meaning in each chart. Due to the symmetries, we will restrict our analysis to the local charts (U_i^+, F_i^+) because we thus cover the entire equator S^{n-1} except the subsets $\{(y_1, \ldots, y_n) | y_i < 0\}$ for $i = 1, 2, \ldots, n$. Since we have to identify the opposite points, we recover all the points. To simplify notation, from here on we rename (U_i^+, F_i^+) as (U_i, F_i). In any local chart the points of infinity have the coordinate z_n equal to zero and again all the local charts contain points of infinity except (U_{n+1}, F_{n+1}).

We will further obtain the analytical expression of $\tilde{X}$ in each local chart starting with U_1. Let $y \in U_1 \cap E^+$. Then, since the differential $(DF_1)_y$ goes from $T_y U_1$ to $T_{F_1(y)} \mathbb{R}^n$, we get

$$(DF_1)_y(\tilde{X}(y)) = (DF_1)_y(y_{n+1}^{m-1} \hat{X}(y))$$

$$= y_{n+1}^{m-1}(DF_1)_y(D\phi^+)_x X(x)$$

$$= y_{n+1}^{m-1} D(F_1 \circ \phi^+)_x X(x),$$

where $y = \phi^+(x)$. Then

$$D(F_1 \circ \phi^+)_x X(x) = \frac{1}{x_1^2}(-x_2 f^1 + x_1 f^2, -x_3 f^1 + x_1 f^3, \ldots, -x_n f^1 + x_1 f^n, -f^1),$$

where $f^i \equiv f^i(x_1, \ldots, x_n)$.

Thus, in the local chart (U_1, F_1), we have

$$(z_1, z_2, \ldots, z_n) = F_1(y_1, \ldots, y_{n+1}) = \left(\frac{y_2}{y_1}, \ldots, \frac{y_{n+1}}{y_1} \right),$$

and we get

$$D(F_1 \circ \phi^+)_x X(x) = z_n(-z_1 \tilde{f}^1 + \tilde{f}^2, \ -z_2 \tilde{f}^1 + \tilde{f}^3, \ldots, -z_{n-1} \tilde{f}^1 + \tilde{f}^n, \ -z_n \tilde{f}^1),$$

where $\tilde{f}^1 \equiv \hat{f}^1(1/z_n, z_1/z_n, \ldots, z_{n-1}/z_n)$. Doing similar computations in each local chart, using the homogeneity of the functions f_j^i we obtain the expressions for $\tilde{X}$:

in (U_1, F_1) :

$$z_n^m(-z_1 \tilde{f}^1 + \tilde{f}^2, \ -z_2 \tilde{f}^1 + \tilde{f}^3, \ldots, -z_{n-1} \tilde{f}^1 + \tilde{f}^n, \ -z_n \tilde{f}^1),$$

where $\tilde{f}^i = \sum_{j=1}^{k_i} z_n^{-m_{ij}} f_j^i(1, z_1, \ldots, z_{n-1})$;

in (U_2, F_2) :

$$z_n^m(-z_1 \tilde{f}^2 + \tilde{f}^1, \ -z_2 \tilde{f}^2 + \tilde{f}^3, \ldots, -z_{n-1} \tilde{f}^2 + \tilde{f}^n, \ -z_n \tilde{f}^2),$$

where $\tilde{f}^i \equiv \sum_{j=1}^{k_i} z_n^{-m_{ij}} f_j^i(z_1, 1, \ldots, z_{n-1})$;

$$\vdots$$

in (U_n, F_n) :

$$z_n^m(-z_1 \tilde{f}^n + \tilde{f}^1, \ -z_2 \tilde{f}^n + \tilde{f}^2, \ldots, -z_{n-1} \tilde{f}^n + \tilde{f}^{n-1}, \ -z_n \tilde{f}^n),$$

where $\tilde{f}^i \equiv \sum_{j=1}^{k_i} z_n^{-m_{ij}} f_j^i(z_1, \ldots, z_{n-1}, 1)$;

in (U_{n+1}, F_{n+1}) :

$$(f^1, f^2, \ldots, f^n),$$

where $f^i \equiv f^i(z_1, z_2, \ldots, z_n)$.

8 The Kepler Problem without Regularization

The Poincaré compactification for vector fields defined by homogeneous functions can be used to study the behavior at infinity of a large family of vector fields. In this paper, however, we are interested only in some applications to celestial mechanics. We start by analyzing the Kepler problem on the line in order to fix the geometrical ideas that we will use for the general Kepler problem.

8.1   The Kepler Problem on the Line

In this case the Hamiltonian is given by the function

$$H(q,p) = \frac{p^2}{2} - \frac{1}{q},$$

with $q > 0$ and $p \in \mathbb{R}$; that is, the configuration space is $\mathbb{R}^+$. The Hamilton equations are given by

$$\dot{q} = \frac{\partial H}{\partial p} = p,$$
$$\dot{p} = -\frac{\partial H}{\partial q} = -\frac{1}{q^2}. \tag{2}$$

In this example, $f_1(q,p) = p$ with $m_1 = 1$ and $f_2(q,p) = -\frac{1}{q^2}$ with $m_2 = -2$, therefore $m = \max(m_1, m_2) = 1$.

The system in the Poincaré sphere S^2 takes the form

$$y_1' = (1 - y_1^2)y_2 + \frac{y_2 y_3^3}{y_1},$$
$$y_2' = -y_1 y_2^2 - \frac{(1 - y_2^2)y_3^3}{y_1^2},$$
$$y_3' = -y_3 \left(y_1 y_2 - \frac{y_2 y_3^3}{y_1^2} \right).$$

Let us observe that the above system has singularities when $y_1 = 0$, which correspond to the collision singularity $q = 0$ in the original system. Since the configuration space of (2) is $\mathbb{R}^+$ we should preserve this restriction on the Poincaré sphere, which means $y_1 > 0$. On the other hand, the system (3) looks more complicated than the original system (2), however, when we take the local chart, this difficulty disappears.

In terms of the local chart (U_1, F_1), the vector field (3) is expressed as

$$z_1' = -(z_1^2 + z_2^3), \quad z_2' = -z_2 z_1.$$

The above system has an unique equilibrium point given by $(z_1, z_2) = (0,0)$; this equilibrium point is not hyperbolic.

In the local chart (U_2, F_2), we have

$$z_1' = \frac{z_2^3}{z_1} + 1, \quad z_2' = \frac{z_2^4}{z_1^2}.$$

Here the vector field does not have any equilibrium point. Strictly speaking, by the restriction $y_1 > 0$, the analysis in this local chart is not necessary, however we will do it in order to preserve the symmetries.

In the local chart (U_3, F_3) we have the original system given by (2), which does not have any equilibrium point. Therefore, the system has only one equilibrium point in the local chart (U_1, F_1); this equilibrium point corresponds to the point $(\pm 1, 0, 0)$ in the Poincaré sphere S^2. Fixing an energy level, we have

$$h = H = \frac{p^2}{2} - \frac{1}{q}.$$

Now we will study the possible escapes of the orbits as a function of the energy.

(i) If $h > 0$, from the relations in the local chart (U_1, F_1) we have

$$z_1 = \frac{y_2}{y_1} = \frac{p}{q} - 0, \qquad z_2 = \frac{y_3}{y_1} = \frac{1}{q} = 0.$$

Using the energy relation we have that $q \to \infty$ and $p \to \pm\sqrt{2h}$, which corresponds to hyperbolic motion. On the Poincaré sphere these orbits die at the equilibrium points $(\pm 1, 0, 0)$ with asymptotic slope $\pm\sqrt{2h}$.

For $h = 0$, we have the same conclusion, but here $q \to \infty$ and $p \to 0$, which correspond to parabolic motion.

(ii) If $h < 0$, then $h + \frac{1}{q} \geq 0$, which implies that $q \leq -\frac{1}{h}$. This means that the system is bounded, therefore the escape to infinity occurs only in the momentum variables. In the original variables we have that, when the particles go to collision, $q \to 0$ and $|p| \to \infty$. On the Poincaré sphere this means that the orbits end up at the points $(0, \pm 1, 0)$, which we have excluded previously.

8.2   The Kepler Problem in the Plane

Here the Hamiltonian is given by

$$H(q,p) = \frac{|p^2|}{2} - \frac{1}{|q|},$$

where $q, p \in \mathbb{R}^2$ and $q \neq (0,0)$. The Hamilton equations take the form

$$\dot{q}_1 = \frac{\partial H}{\partial p_1} = p_1, \qquad \dot{p}_1 = -\frac{\partial H}{\partial q_1} = -\frac{q_1}{\sqrt{(q_1^2 + q_2^2)^3}},$$

$$\dot{q}_2 = \frac{\partial H}{\partial p_2} = p_2, \qquad \dot{p}_2 = -\frac{\partial H}{\partial q_2} = -\frac{q_2}{\sqrt{(q_1^2 + q_2^2)^3}},$$

where $p = (p_1, p_2)$ and $q = (q_1, q_2)$. Here the homogeneous functions are given by $f_1 = p_1$, $f_2 = p_2$, $f_3 = -q_1/\sqrt{(q_1^2 + q_2^2)^3}$, and $f_4 = -q_2/\sqrt{(q_1^2 + q_2^2)^3}$. The degree of homogeneity of f_1 and f_2 is 1, and the degree of homogeneity f_3 and f_4 is -2. Here we have $m = 1$. The vector field on the Poincaré sphere S^4 is given by

$$y_1' = (1 - y_1^2)y_3 - y_1 y_2 y_4 + \frac{y_1 y_5^3}{\sqrt{(y_1^2 + y_2^2)^3}}(y_1 y_3 + y_2 y_4),$$

$$y_2' = -y_1 y_2 y_3 + (1 - y_2^2)y_4 + \frac{y_2 y_5^3}{\sqrt{(y_1^2 + y_2^2)^3}}(y_1 y_3 + y_2 y_4),$$

$$y_3' = -y_1 y_3^2 - y_2 y_3 y_4 - \frac{y_5^3}{\sqrt{(y_1^2 + y_2^2)^3}}(y_1(1 - y_3^2) - y_2 y_3 y_4),$$

$$y_4' = -y_1 y_3 y_4 - y_2 y_4^2 - \frac{y_5^3}{\sqrt{(y_1^2 + y_2^2)^3}}(-y_1 y_3 y_4 + y_2(1 - y_4^2)),$$

$$y_5' = y_5(-y_1 y_3 - y_2 y_4) + \frac{y_5^4}{\sqrt{(y_1^2 + y_2^2)^3}}(y_1 y_3 + y_2 y_4).$$

Let us observe that the system is not defined when $y_1 = y_2 = 0$, which for the Poincaré sphere S^4 corresponds to $S^2 = \{y \in S^4 | y_3^2 + y_4^2 + y_5^2 = 1\}$. This is because the original system is defined at the point $q_1 = q_2 = 0$, which corresponds to the singularity due to double collision. The equator of S^4 given by $y_5 = 0$ is invariant under the flow.

To find the critical points we will study the compactified vector field in the five local charts (U_i, F_i), $i = 1, \ldots, 5$.

In the local chart (U_1, F_1), the system takes the form:

$$z_1' = -z_1 z_2 + z_3,$$

$$z_2' = -z_2^2 - \frac{z_4^3}{\sqrt{(1+z_1^2)^3}},$$

$$z_3' = -z_3 z_2 - \frac{z_1 z_4^3}{\sqrt{(1+z_1^2)^3}},$$

$$z_4' = -z_2 z_4.$$

The critical points are given by $(z_1, 0, 0, 0)$, which on the Poincaré sphere S^4 corresponds to the set $\{y \in S^4 | y_1^2 + y_2^2 = 1\}$. That is, we have a circumference minus two points $(0, \pm 1)$ (which corresponds to $y_1 = 0$).

In the local chart (U_2, F_2), the system is given by

$$z_1' = -z_1 z_3 + z_2,$$

$$z_2' = -z_2 z_3 - \frac{z_1 z_4^3}{\sqrt{(1+z_1^2)^3}},$$

$$z_3' = -z_3^2 - \frac{z_4^3}{\sqrt{(1+z_1^2)^3}},$$

$$z_4' = -z_4 z_3.$$

We obtain that the critical points are given by the same set as that in the local chart (U_1, F_1). In the local chart (U_3, F_3), we have

$$z_1' = \frac{z_1^2 z_4^3}{\sqrt{(z_1^2 + z_2^2)^3}} + 1,$$

$$z_2' = \frac{z_1 z_2 z_4^3}{\sqrt{(z_1^2 + z_2^2)^3}} + z_3,$$

$$z_3' = \frac{z_1 z_3 z_4^3}{\sqrt{(z_1^2 + z_2^2)^3}} - \frac{z_2 z_4^3}{\sqrt{(z_1^2 + z_2^2)^3}},$$

$$z_4' = -\frac{z_1 z_4^4}{\sqrt{(z_1^2 + z_2^2)^3}}.$$

This system does not have any critical point. In the same way we check that the compactified vector field does not have critical points in the local charts (U_4, F_4) and (U_5, F_5).

Therefore, there is only one circumference $y_1^2 + y_2^2 = 1$ of critical points, all of which are at infinity; that is, they belong to the equator ($y_5 = 0$) of the Poincaré sphere S^4. If we analyze the linearization of the vector field in each local chart we get that the circumference of critical points is very degenerate in the sense that all the eigenvalues are zero. However, analyzing the set of critical points together with the energy relation we can give the physical interpretation of the critical points. For any fixed energy level $H = h$, we have:

(i) If $h > 0$, from the characterization of the set of critical points in the local chart (U_1, F_1),

$$z_2 = \frac{y_3}{y_1} = \frac{p_1}{q_1} = 0, \quad z_3 = \frac{y_4}{y_1} = \frac{p_2}{q_1} = 0, \quad z_4 = \frac{y_5}{y_1} = 0.$$

From the energy relation $\frac{|p|^2}{2} - \frac{1}{|q|} = h$, we have $(p_1, p_2) \neq (0,0)$, then $|q| \to \infty$ and $|p| \to \pm\sqrt{2h}$, which corresponds to hyperbolic motion. In the case $h = 0$, we have that $|q| \to \infty$ and $|p| \to 0$, which corresponds to parabolic motion.

(ii) If $h < 0$, from the energy relation

$$\frac{|p^2|}{2} - \frac{1}{|q|} = h$$

we have $|q| \leq -\frac{1}{h}$, which means that the configuration space is bounded. As in the Kepler problem on the line, we can still go to infinity in the momenta variables. Remember that in this case the configuration space is $\{q \in \mathbb{R}^2 | q \neq (0,0)\}$, which implies that on the Poincaré sphere we have to remove the set $\{(y_1, y_2, \ldots, y_5)|y_1 = y_2 = 0, y_3^2 + y_4^2 + y_5^2 = 1\}$. Then when $q \to 0$, $|p|^2 \to \infty$. In other words, the escapes to infinity in the momentum variables are asymptotic to the set $\{y_1 = y_2 = y_5 = 0, y_3^2 + y_4^2 = 1\}$.

The analysis of the critical points on the Poincaré sphere given by the local chart (U_2, F_2) is similar. Summarizing, when we directly apply the Poincaré compactification to the Kepler problem we recover the classical results about escapes to infinity. In other words, we have proved a classical result in a different way.

Theorem 8.1 *In the Kepler problem, if we fix the level of energy $H = h$ we have:*

(a) *For $h < 0$, the motion is bounded in the configuration space. We have escapes to infinity only in the momentum variables; this happens when the particles approach collisions.*

(b) *For $h \geq 0$, the escapes are in the configuration space with limiting velocity $\pm\sqrt{2h}$. For $h > 0$ this corresponds to hyperbolic motion; for $h = 0$ we get parabolic motions.*

9   Hill's Problem

The Hill problem was introduced in order to study the motion of the moon under the influence of the earth and the sun (see [10] for more details). It describes the motion of a massless particle in the neighborhood of one of the massive particles when the third one is far away. Since we are interested in escapes to infinity in the configuration space, we restrict our analysis to values of the Jacobi constant $C \gg 1$, which is defined from the Jacobian integral of the Hill problem

$$\dot{x}^2 + \dot{y}^2 - 3x^2 + \frac{2}{\sqrt{x^2 + y^2}} - C.$$ (3)

Here (x, y) are the coordinates of the massless particle in a synodic frame with the origin fixed at the nearest massive particle (see [10]).

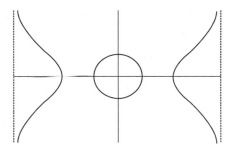

Figure 2. The configuration space in the Hill problem.

In Figure 2 we show the shape of the possible regions of motion in the xy plane for these values of the Jacobian constant. As we can see, there are three possible regions of motion bounded by the zero velocity curves. Using the symmetry of the problem, we can restrict our study of escape orbits to the positive unbounded region.

In order to apply the Poincaré compactification, we first write the equations in Hamiltonian form, introducing a new set of variables as given in [9]:

$$x_1 = x, \quad x_2 = y, \quad x_3 = \dot{x} - y, \quad x_4 = \dot{y} + x. \tag{4}$$

Using these variables, the Hamiltonian function for the Hill problem takes the form

$$H = \frac{1}{2}(x_3^2 + x_4^2) + x_2 x_3 - x_1 x_4 - x_1^2 + \frac{x_2^2}{2} - \frac{1}{\sqrt{x_1^2 + x_2^2}},$$

and Hamilton's equations become

$$x_1' = x_3 + x_2,$$
$$x_2' = x_4 - x_1,$$
$$x_3' = 2x_1 + x_4 - \frac{x_1}{\sqrt{(x_1^2 + x_2^2)^3}}, \tag{5}$$
$$x_4' = -x_2 - x_3 - \frac{x_2}{\sqrt{(x_1^2 + x_2^2)^3}},$$

where (x_1, x_2) are the coordinate variables and (x_3, x_4) their associated momenta. We will restrict our study to a fixed value of the energy $H = h$, where $h = -\mathcal{C}/2$.

In this section we will not write the equations (5) on the Poincaré sphere, since from the previous two sections we have seen that we can write (5) directly in local coordinates.

In (U_1, F_1), the system takes the form

$$z_1' = -z_1(z_1 + z_2) + z_3 - 1,$$
$$z_2' = -z_2(z_1 + z_2) + 2 + z_3 - \frac{z_4^3}{\sqrt{(1 + z_1^2)^3}},$$
$$z_3' = -z_3(z_1 + z_2) - z_1 - z_2 - \frac{z_1 z_4^3}{\sqrt{(1 + z_1^2)^3}},$$
$$z_4' = -z_4(z_1 + z_2).$$

In this chart there are no equilibrium points restricted at infinity.

In the local chart (U_2, F_2), the system takes the form

$$z_1' = -z_1(z_3 - z_1) + z_2 + 1,$$

$$z_2' = -z_2(z_3 - z_1) + 2z_1 + z_3 - \frac{z_1 z_4^3}{\sqrt{(1 + z_1^2)^3}},$$

$$z_3' = -z_3(z_3 - z_1) - 1 - z_2 - \frac{z_4^3}{\sqrt{(1 + z_1^2)^3}}, \qquad (6)$$

$$z_4' = -z_4(z_3 - z_1).$$

In this chart, we get just one equilibrium point at infinity $z_4 = z_3 = z_1 = 0$, and $z_2 = -1$. On the Poincaré sphere restricted to this chart ($y_2 > 0$), this point corresponds to the equilibrium point

$$A_2 - \left\{\left(0, \frac{1}{\sqrt{2}}, -\frac{1}{\sqrt{2}}, 0, 0\right)\right\}.$$

From (6) it is easy to check that this equilibrium point has two zero eigenvalues and two conjugate pure imaginary eigenvalues, so it is not hyperbolic.

In the local chart (U_3, F_3), the system takes the form

$$z_1' = -z_1\left(2z_1 + z_3 - \frac{z_1 z_4^3}{\sqrt{(z_1^2 + z_2^2)^3}}\right) + z_2 + 1,$$

$$z_2' = -z_2\left(2z_1 + z_3 - \frac{z_1 z_4^3}{\sqrt{(z_1^2 + z_2^2)^3}}\right) + z_3 - z_1,$$

$$z_3' = -z_3\left(2z_1 + z_3 - \frac{z_1 z_4^3}{\sqrt{(z_1^2 + z_2^2)^3}}\right) - z_2 - 1 - \frac{z_2 z_4^3}{\sqrt{(z_1^2 + z_2^2)^3}},$$

$$z_4' = -z_4\left(2z_1 + z_3 - \frac{z_1 z_4^3}{\sqrt{(z_1^2 + z_2^2)^3}}\right).$$

In this chart, we have again only one equilibrium point at infinity $z_4 = z_3 = z_1 = 0$, and $z_2 = -1$. On the Poincaré sphere this corresponds to the point $A_3 = -A_2$, which is not hyperbolic.

Finally, in the local chart (U_4, F_4) the system takes the form

$$z_1' = -z_1 \left(-z_2 - z_3 - \frac{z_2 z_4^3}{\sqrt{(z_1^2 + z_2^2)^3}} \right) + z_2 + z_3,$$

$$z_2' = -z_2 \left(-z_2 - z_3 - \frac{z_2 z_4^3}{\sqrt{(z_1^2 + z_2^2)^3}} \right) + 1 - z_1,$$

$$z_3' = -z_3 \left(-z_2 - z_3 - \frac{z_2 z_4^3}{\sqrt{(z_1^2 + z_2^2)^3}} \right) + 2z_1 + 1 - \frac{z_1 z_4^3}{\sqrt{(z_1^2 + z_2^2)^3}},$$

$$z_4' = -z_4 \left(-z_2 - z_3 - \frac{z_2 z_4^3}{\sqrt{(z_1^2 + z_2^2)^3}} \right).$$

In this chart there are no equilibrium points at infinity.

Now we will use the study on the local charts to describe the final evolution of the escape orbits in the positive unbounded region that we study. In this region we are restricted only to $x_1 > 0$, which corresponds to taking $y_1 > 0$ on the Poincaré sphere. Therefore, the only possible equilibrium points at infinity are A_2 and A_3.

Now we can give a physical interpretation of the possible escapes in the Hill problem. For this we return to the original coordinates and will use the Jacobian integral. If one orbit in the Poincaré sphere ends at A_2, then it has the point $(y_1, y_2, y_3, y_4, y_5) = (0, \frac{1}{\sqrt{2}}, -\frac{1}{\sqrt{2}}, 0, 0)$ as its ω-limit (for the other point the analysis is analogous by reversing the time and studying the α-limit).

From (1) we know that

$$y_1 = \frac{x_1}{\Delta}, \qquad y_2 = \frac{x_2}{\Delta}, \qquad y_3 = \frac{x_3}{\Delta}, \qquad y_4 = \frac{x_4}{\Delta}, \qquad y_5 = \frac{1}{\Delta},$$

where $\Delta = \sqrt{x_1^2 + x_2^2 + x_3^2 + x_4^2 + 1}$.

Now, from relation (4) between the original coordinates and the Hamiltonian ones, we get

$$y_1 = \frac{x}{\Delta}, \qquad y_2 = \frac{y}{\Delta}, \qquad y_3 = \frac{\dot{x} - y}{\Delta}, \qquad y_4 = \frac{\dot{y} + x}{\Delta}. \qquad (7)$$

Since we are analyzing the infinity, $y_5 \to 0$, which implies that $\Delta \to 0$, since $y_2 = \frac{y}{\Delta} \to \frac{1}{\sqrt{2}}$ we obtain that $y \to \infty$. Using (7) again, we have $y_1 = \frac{x}{\Delta} \to 0$. So the coordinate x can be bounded or unbounded.

(i) If x is bounded, from the Jacobian integral, (3), we obtain that both $\dot{x}$ and $\dot{y}$ must be bounded. In the particular case when $\dot{x} \to 0$ and $\dot{y} \to 0$, escapes are of parabolic type; if at least one of the above limits are different from zero, we have escapes of hyperbolic type. Finally using the fact that $y_4 = \frac{x_4}{\Delta} = \frac{\dot{y}+x}{\Delta} \to 0$ we have that, asymptotically, $\dot{y} = -x$, so for parabolic escapes, $x(t) \to 0$, in other words, the escapes are asymptotic to vertical lines in the $x - y$ plane. If the vertical lines agree with the y-axis, the escapes are of parabolic type.

(ii) The case "x unbounded" is not interesting from the celestial mechanics viewpoint; it represents the escapes to infinity (in both coordinates) with infinite limit velocity.

We can summarize our result as follows.

Theorem 9.1 *The escapes to infinity in the Hill problem restricted to the positive unbounded region are asymptotic to one of the two equilibrium points on the equator of the Poincaré sphere. The possible final evolution of the escape orbits is determined by the respective limit point.*

Acknowledgments I would like to thank Hildeberto Cabral for the invitation to lecture on this topic in the Department of Mathematics of UFPE, Recife, Brazil.

References

[1] A. Cima and J. Llibre, *Bounded polynomial vector fields*, Trans. Amer. Math. Soc. **318**, 557–579, 1990.

[2] J. Delgado, E. A. Lacomba, J. Llibre, and E. Pérez, *Poincaré compactification of the Kepler and collinear 3-body problems.* Proceedings of the Second Workshop on Dynamical Systems of the Euler Institute, Saint Petersburg, Russia, 1993.

[3] J. Delgado, E. A. Lacomba, J. Llibre, and E. Pérez, *Poincaré compactification of polynomial Hamiltonian vector fields*. Hamiltonian Dynamical Systems, History, Theory and Applications. The IMA Volumes in Mathematics and Its Applications. Volume **63**. Springer-Verlag, New York, 99–115, 1994.

[4] V. Guillemin and A. Pollack, *Differential Topology*. Prentice-Hall, Englewood Cliffs, NJ (1994).

[5] D. C. Heggie, *A global regularization of the gravitational n-body problem*, Cel. Mech. **10**, 217–241, 1974.

[6] R. McGehee, *Triple collision in the collinear three body problem*, Inventiones Math. **27**, 191–227 (1974).

[7] Ernesto A. Lacomba and Ernesto Pérez-Chavela, *A compact model for the planar rhomboidal four body problem*, Celestial Mechanics and Dynamical Astronomy **54**, 343–355 (1992).

[8] E. Pérez-Chavela and A. Susín, *Poincaré compactification for homogeneous functions with applications to celestial mechanics*, submitted for publication.

[9] C. Simó, and T. Stuchi, *Central stable/unstable manifolds and the destruction of KAM tori in the planar Hill problem*, Phys. D **140**, 1–32 (2000).

[10] V. Szebehely, *Theory of Orbits*. Academic Press, New York, 1967.

Photograph: Carlos Oliveira

The Motion of the Moon

Dieter Schmidt

DEPARTMENT OF COMPUTER SCIENCE
UNIVERSITY OF CINCINNATI
CINCINNATI, OH 45221-0008, USA
e-mail: dieter.schmidt@uc.edu

1 Remarks about the Accuracy of the Solution

The distance from the Earth to the Moon is on the average 384,400 km. The distance from the Earth to the Sun is about 1.49×10^8 km. The mass of the Earth is 5.98×10^{27} g, the Moon weighs 7.34×10^{25} g, and the Sun 1.993×10^{33} g. It is worthwhile to visualize the solar system in a model and actually to construct and lay out this model in order to appreciate these values. The scale factor we will use is 1:500,000,000, as then the Sun is 27.85 cm in diameter – roughly the size of a basketball. The Earth shrinks to a diameter of 2.5 mm, the size of the sphere found in some ball bearings. The Moon is then 0.7 mm, the size of the sphere found in a medium ballpoint pen. The distance from Earth to Sun is around 30 m, whereas the distance from Earth to Moon is just near 8 cm. The Earth will fit about 1.3 million times into the Sun. When we consider masses, the Sun is 300,000 times heavier than the Earth, whereas the Moon has only 1/80 of the Earth's mass.

When one sees this model it is not hard to be convinced that in a first approximation the three bodies can be treated like point masses and that any deviation caused by the oblateness of the Earth or the uneven mass distribution on the Moon can be considered later on with other perturbations.

The planets will also cause perturbations, which can be treated later on. For example, Venus, roughly the same size as the Earth in the above model, is about 20 m away from the Sun. Jupiter will be 150 m away from the Sun, and due to its mass will still cause some noticeable perturbations.

The magnitudes of the different gravitational attractions follow from Newton's law. Due to the enormous size of the Sun it turns out that the gravitational attraction of the Sun on the Moon is twice as large as that of the Earth on the Moon. More precisely, this value is $(1.993 \times 10^{33}/5.98 \times 10^{27})(3.84 \times 10^5/1.49 \times 10^8)^2 = 2.2$. Although we can assume that the Moon will stay close to Earth for all times, the above estimate shows that it is not a good choice to start out with the Moon on an elliptical orbit around the Earth and to treat the influence of the Sun as a perturbing factor. Delaunay and others before him had tried this approach, and the limited accuracy of these theories is due in part to the choice of an elliptic orbit as the starting point for the calculations.

It was Hill [7] who recognized how to account for the effect of the Sun. In essence he proposed to move the Sun infinitely far away, but he kept its gravitational attraction on the Earth–Moon system constant. It was Brown [1] who carried out Hill's idea and calculated the motion of the Moon to an accuracy that satisfied the needs of his time.

Today the measurement of the distance from Earth to Moon is done by laser ranging. From an observatory on Earth to reflectors left on the Moon by astronauts, this distance is measured to within a few centimeters. Also the position of the Moon in its orbit is found with unprecedented accuracy via occultations of stars. The Moon moves 1 second of arc in the sky for every 2 seconds of time. At a distance of 400,000 km this represents a velocity of 2 km/sec. Occultations can be measured within 10^{-4} sec or better, so that the Moon can be fixed in its orbital path to within 20 cm. An analytic theory for the motion of the Moon should therefore have an accuracy of 11 decimal digits. Since the solution for longitude, latitude, and sine parallax will be found in the form

$$\sum A_{k,l} r_1^{k_1} \cdots r_m^{k_m} \frac{\sin}{\cos} (l_1 \zeta_1 + \cdots + l_n \zeta_n), \tag{1}$$

it will be necessary to increase the accuracy with which each coefficient is found by several orders of magnitudes in order to offset the loss of accuracy through truncation and roundoff; see [4] and [9].

Assume that the coefficients in (1) are normalized so that the largest is 1, and there are N terms of (1) that have been computed to some accuracy. There is no way to give an estimate of the error that one makes by truncating the infinite series (1) to N terms. The only question we can answer is, How accurate is the sum of these N terms for a given set of values for the radial and angular variables? A uniform error estimate might be too pessimistic, as some combination of angles may cause errors in the individual terms to add up. Instead we will look at the mean square error. The radial variables are all less than 1, and we will assume that they are known to any desired accuracy. It means that it suffices to require that the evaluated term $A_{k,l} r_1^{k_1} \cdots r_m^{k_m}$ has to be $< \epsilon$ in absolute value. The angular variables will be linear functions of time and they too are given to any needed accuracy. If the mean error should be $< 10^{-11}$ then we need $\sqrt{N}\epsilon < 10^{-11}$. If the series has around 100,000 terms it is advisable to compute the evaluated terms to an accuracy of 10^{-14} or better.

Actually, there is a very pragmatic answer to the question of how accurately each coefficient should be found. All computers have floating point arithmetic with just two or three different precisions. An accuracy of slightly better than 14 decimal digits is given by floating point numbers represented in eight bytes. The extended precision would give 30 decimal digits, but the extra space and time required to do all of the calculations in extended precision makes this choice impractical.

For a complete solution to the main problem of lunar theory in floating point arithmetic we refer the reader to [3]. In the following sections we follow the presentation given there, but then restrict ourselves to solving the first-order terms in rational arithmetic with the help of a computer algebra program. We have used MACSYMA, and give the corresponding programs in the appendix. We hope that readers will be able to translate these programs into programs for their favored computer algebra system.

2   The Equations of Motion

In a fixed but otherwise unspecified coordinate system, a three-body problem is given by the Hamiltonian function

$$H = \sum_{j=1}^{3} \frac{|p_j|^2}{2m_j} + G \sum_{1 \le i < j \le 3} \frac{m_i m_j}{|q_i - q_j|},$$

where we will use the convention that index 1 refers to the Earth, 2 to the Moon, and 3 to the Sun. The gravitational constant G cannot be set to 1, as this is a real-life problem where distance, mass, and time are measured in the centimeter-gram-second system where the units are specified a priori. The position coordinates q_j and the momenta p_j are vectors in $\mathcal{R}^3$ so that the problem has nine degrees of freedom. Measuring the position of the Moon from an Earth-based coordinate system is more natural, and therefore we introduce the Jacobi coordinates, with Q_1 the vector from the Earth pointing to the Moon, Q_2 the vector from the center of mass of the Earth–Moon system to the Sun, and Q_3 fixes the center of mass of the whole system. Let P_1, P_2, and P_3 be the corresponding momenta. Since the center of mass Q_3 is an integral for the system, the transformed Hamiltonian will not depend on P_3 and the dependency on Q_3 will be ignored in the transformed Hamiltonian.

An algorithm that generates the transformation to Jacobi coordinates is given in Appendix A. The case here is studied well enough and we will list the symplectic transformation of $\mathcal{R}^{18} \to \mathcal{R}^{18}$ in terms of block matrices for the position and momenta vectors:

$$
\begin{pmatrix} Q_1 \\ Q_2 \\ Q_3 \end{pmatrix} = \begin{pmatrix} -1 & 1 & 0 \\ \frac{-m_1}{m_1+m_2} & \frac{-m_2}{m_1+m_2} & 1 \\ \frac{m_1}{m_1+m_2+m_3} & \frac{m_2}{m_1+m_2+m_3} & \frac{m_3}{m_1+m_2+m_3} \end{pmatrix} \begin{pmatrix} q_1 \\ q_2 \\ q_3 \end{pmatrix},
$$

$$
\begin{pmatrix} P_1 \\ P_2 \\ P_3 \end{pmatrix} = \begin{pmatrix} \frac{-m_2}{m_1+m_2} & \frac{m_1}{m_1+m_2} & 0 \\ \frac{-m_3}{m_1+m_2+m_3} & \frac{-m_3}{m_1+m_2+m_3} & \frac{m_1+m_2}{m_1+m_2+m_3} \\ 1 & 1 & 1 \end{pmatrix} \begin{pmatrix} p_1 \\ p_2 \\ p_3 \end{pmatrix}.
$$

The transformed Hamiltonian has thus six degrees of freedom and is

$$
H = \frac{m_1 + m_2 + m_3}{2m_3(m_1 + m_2)}|P_2|^2 + \frac{m_1 + m_2}{2m_1 m_2}|P_1|^2
$$
$$
- \frac{Gm_1 m_2}{|Q_1|} - \frac{Gm_1 m_3}{|Q_2 + \frac{m_2}{m_1+m_2}Q_1|} - \frac{Gm_2 m_3}{|Q_2 - \frac{m_1}{m_1+m_2}Q_1|}. \tag{2}
$$

Since $\lambda = |Q_1|/|Q_2|$ is approximately $1/400$, one is led to an expansion of the last two terms. This is accomplished with the help of the Legendre polynomials $P_j(\cos\gamma)$, whose generating function gives the following defining relation:

$$
(1 - 2\lambda \cos\gamma + \lambda^2)^{-1/2} = \sum \lambda^j P_j(\cos\gamma).
$$

Let γ be the angle between the two vectors Q_1 and Q_2, so that the expansion of the last two terms in (2) leads to

$$
H = \frac{m_1 + m_2 + m_3}{2m_3(m_1 + m_2)}|P_2|^2 - \frac{Gm_3(m_1 + m_2)}{|Q_2|} \tag{3}
$$
$$
+ \frac{m_1 + m_2}{2m_1 m_2}|P_1|^2 - \frac{Gm_1 m_2}{|Q_1|} - \frac{Gm_1 m_2 m_3}{m_1 + m_2} \sum_{j=2} c_j \frac{|Q_1|^j}{|Q_2|^{j+1}} P_j(\cos\gamma).
$$

The coefficients c_j are given by

$$c_j = \frac{m_1^{j-1} - (-m_2)^{j-1}}{(m_1 + m_2)^{j-1}}, \qquad j = 2, 3, \dots \; ,$$

and thus depend only on the mass ratio $\mu = m_2/m_1$. With $c_2 = 1$, $c_3 = 1 - \mu$, the dependency of the solution on this mass ratio is not very pronounced, and it is for this reason that μ is seldom treated as an independent parameter. Instead, its numerical value of $1/81.30068$ is used from the beginning.

The first line in (4) is the Hamiltonian for the two-body problem of the Earth–Moon system and the Sun. The second line describes the motion of the Moon around the Earth under the influence of the Sun. Looking at the corresponding differential equations one sees that they can be decoupled and that higher order dependencies can be dealt with later on among the perturbations.

The Hamiltonian function describing the motion of the Moon is therefore

$$H = \frac{m_1 + m_2}{2m_1 m_2}|P_1|^2 - \frac{Gm_1 m_2}{|Q_1|} - \frac{Gm_1 m_2 m_3}{m_1 + m_2} \sum_{j=2} c_j \frac{|Q_1|^j}{|Q_2|^{j+1}} P_j(\cos\gamma).$$

It has three degrees of freedom but it is time dependent, due to the elliptic motion of the center of mass of the Earth–Moon system around the Sun.

The motion of the Sun around the Earth–Moon system in polar coordinates is given by

$$r' = |Q_2| = a'(1 - e'\cos l' + \cdots),$$

$$\Phi' = g' + l' + 2e'\sin l' + \cdots \; ,$$

$$l' = n'(t - t_0) + l_0'.$$

It is customary in lunar theory to denote everything with a "$'$" that refers to the Sun, so that we have

$a' = $ semimajor axis,

$e' = $ eccentricity of Sun's orbit,

$l' = $ mean anomaly,

$n' = $ mean motion of Sun around Earth–Moon system,

$g' = $ argument of perihelion,

$l'_0 = $ mean anomaly at a given epoch t_0, that is, Jan. 1, 2000.

Kepler's law gives
$$n'^2 a'^3 = G(m_1 + m_2 + m_3).$$

In selecting the coordinate system that describes the Earth-based vector Q_1, we still have some freedom. The orbital plane of our two-body problem, the so-called ecliptic, is the natural choice. The first two components of Q_1 will be parallel to this plane, and the third one perpendicular to it.

If the Sun's motion could be circular, then it would be clear to choose a uniformly rotating coordinate system for Q_1 so that the x-component of this coordinate system always points to the Sun. In this case the Hamiltonian (4) in rotating coordinates would also be time independent. Since the Sun moves on an elliptic orbit, we do the next best thing. We use a uniformly rotating coordinate system, whose x-axis always points to the mean position of the Sun. With $\phi = g' + l'$ the mean longitude of the Sun, the transformation reads

$$Q_1 = b \begin{pmatrix} x \cos \phi - y \sin \phi \\ x \sin \phi + y \cos \phi \\ z \end{pmatrix},$$

and for the momenta

$$P_1 = b(n - n') \frac{m_1 m_2}{m_1 + m_2} \begin{pmatrix} X \cos \phi - Y \sin \phi \\ X \sin \phi + Y \cos \phi \\ Z \end{pmatrix}.$$

The factors have been selected so that the new Hamiltonian will depend only on essential parameters. The scale factor b will be determined shortly.

With n the mean motion of the Moon around the Earth we also introduce a new time by

$$\tau = (n - n')(t - t_0)$$

so that the basic motion is 2π–periodic. The above time-dependent transformation remains symplectic, provided that the Hamiltonian is multiplied with $(m_1 + m_2)/(m_1 m_2 (n - n') b^2)$. With all of these changes the Hamiltonian in the uniformly rotating coordinates becomes

$$H = \frac{1}{2}(X^2 + Y^2 + Z^2) - m(xY - yX) - \frac{G(m_1 + m_2)}{b^3(n - n')^2}\frac{1}{r}$$
$$- \frac{Gm_3}{a'^3(n - n')^2}\sum c_j \left(\frac{b}{a'}\right)^{j-2}\left(\frac{a'}{r'}\right)^{j+1} r^j P_j(\cos\gamma). \qquad (4)$$

We have set $r = |Q_1| = \sqrt{x^2 + y^2 + z^2}$, and the parameter m is the ratio of the mean motions and defined by

$$m = \frac{n'}{n - n'}.$$

In analogy to Kepler's law a quantity a is defined by

$$n^2 a^3 = G(m_1 + m_2).$$

It allows us to write the coefficient of the third term in (4) as

$$\frac{G(m_1 + m_2)}{b^3(n - n')^2} = \left(\frac{a}{b}\right)^3 (1 + m)^2,$$

whereas the coefficient in front of the sum becomes

$$\frac{Gm_3}{a'^3(n - n')^2} = \frac{n'^2}{(n - n')^2}\frac{m_3}{m_1 + m_2 + m_3} = m^2\left(1 - \frac{m_1 + m_2}{m_1 + m_2 + m_3}\right).$$

At this point our choice of parameters deviates from that of Brown, who sets $b = a$ and works with a parameter $\alpha = a/a'$. It leaves the ratio of masses in the above formula as another parameter for which Brown substitutes the numerical value from the beginning. Since these two parameters are related we find it more natural to select

$$\beta = \left(\frac{m_1 + m_2}{m_1 + m_2 + m_3}\right)^{\frac{1}{3}}$$

as the independent parameter instead of α. We also set

$$b = a(1+m)^{2/3},$$

which makes the coefficient of $1/r$ into 1. This in turn gives

$$\frac{Gm_3}{a'^3(n-n')^2} = m^2(1-\beta^3)$$

for the coefficient in front of the sum in (4), and inside this sum we find

$$\frac{b}{a'} = \beta m^{2/3}.$$

One advantage of this choice is that β and the other parameters will have numerically the same order of magnitude. A disadvantage could be that a completely analytical solution would have to be developed in powers of $m^{1/3}$. Since the solution found by the method of Brown is semianalytical, this is actually of no concern because m is replaced by its numerical value from the beginning. Another insignificant drawback is that with Brown's choice of the parameter α, one is able to determine if a given term belongs to the x-y coordinates or to the z coordinates simply by knowing if the power of α is even or odd. With the appearance of β^3 in (4), this rule, which follows from d'Alembert's characteristic, does not apply to powers of β beyond the second.

The Hamiltonian for the lunar problem in real coordinates is therefore

$$H = \frac{1}{2}(X^2 + Y^2 + Z^2) + m(yX - xY) - \Omega$$

with

$$\Omega = \Omega(\tau, x, y, z; m, \beta, e', \mu)$$
$$= m^2(1-\beta^3)\sum c_j m^{2(j-2)/3}\beta^{j-2}\left(\frac{a'}{r'}\right)^{j+1} P_j(\cos\gamma).$$

The Legendre polynomials can be constructed iteratively by

$$P_0(t) = 1$$
$$P_1(t) = t$$
$$P_{j+1}(t) = \frac{2j+1}{j+1}tP_j(t) - \frac{j}{j+1}P_{j-1}(t), \qquad j = 1, 2, \dots,$$

and their argument follows from the dot product of the vectors Q_1 and Q_2 and is given by

$$\cos \gamma = \frac{xx' + yy'}{rr'}.$$

It was Hill who recognized that not all terms of Ω should be considered as a perturbation. He devised a way to account for a significant portion of the Sun's force when he constructed the intermediate orbit. In essence he extracted from Ω the most significant term and wrote

$$\Omega = m^2 r^2 P_2 \left(\frac{x}{r}\right) + \tilde{\Omega}$$

and included the first term in the equation for the intermediate orbit. Physically this can be interpreted by saying that the Sun has been moved infinitely away, but its gravitational attraction on the Moon has been kept constant.

The differential equations are therefore

$$\ddot{x} - 2m\dot{y} - m^2 x - 2m^2 x + x/r^3 = \partial\tilde{\Omega}/\partial x,$$
$$\ddot{y} + 2m\dot{x} - m^2 y + m^2 y + y/r^3 = \partial\tilde{\Omega}/\partial y,$$
$$\ddot{z} \qquad\qquad + m^2 z + z/r^3 = \partial\tilde{\Omega}/\partial z.$$

They have been written so that each column can be identified with one of the forces acting on the Moon. They are, from left to right: inertial, Coriolis, centrifugal, Sun's gravitation reduced to a quadrupole, Earth's gravitation, and perturbations.

The motion of the Moon has to be viewed in a six-dimensional phase space. Its behavior is more easily understood for a time-independent system, that is, when $e' = 0$ and $r' = a'$, as then the trajectories lie on a five-dimensional manifold given by $H = \text{const}$. Since no additional integrals exist, one has to expect that the solution behaves ergodically on this manifold. This is not the case for the intermediate orbit of Hill, which is a periodic orbit. This intermediate orbit will be surrounded by three-dimensional tori, which are specified by the two radial variables e and k. The solution curve on these tori is parametrized by the three angular variables l, F, and τ of Delaunay.

For small e' these 3-tori are not destroyed but they change their size synchronously with the mean motion of the Sun. The tori change their size slowly when compared to the motion of the Moon. Since there is no loss of energy this change is called adiabatic. These heuristic arguments give the justification for the form of the solution as Poisson series:

$$x = \sum A_j \cos\left(j_1 l + j_2 F + j_3 l' + (j_4 + 1)\tau\right),$$

$$y = \sum B_j \sin\left(j_1 l + j_2 F + j_3 l' + (j_4 + 1)\tau\right),$$

$$z = \sum C_j \sin\left(j_1 l + j_2 F + j_3 l' + j_4 \tau\right),$$

where each coefficient is itself a series in the radial variables, that is,

$$A_j = \sum_i A_j^i e^{i_1} k^{i_2} e'^{i_3} \beta^{i_4}.$$

The coefficients A_j^i are functions of m and of μ. The characteristics of d'Alembert give some restriction on which indices can occur:

$$i_1 \geq |j_1|, \qquad i_2 \geq |j_2| \qquad \text{and} \qquad i_3 \geq |j_3|.$$

There is no restriction involving i_4. The angles are

$$l = c\tau + l_0,$$

$$F = g\tau + F_0,$$

$$l' = m\tau + l_0',$$

$$\tau - (n - n')(t - t_0),$$

where the mean motion of the perigee c and the mean motion of the node g have to be determined along with the rest of the solution in the form of the following two series:

$$c = c_0 + \sum c_i e^{i_1} k^{i_2} e'^{i_3} \beta^{i_4} \tag{5}$$

and

$$g = g_0 + \sum g_i e^{i_1} k^{i_2} e'^{i_3} \beta^{i_4}. \tag{6}$$

3   The Solution Method

Already Hill and Brown used complex variables in order to avoid the use of trigonometric identities when one has to manipulate Poisson series. For this reason they introduced the variable

$$\zeta = e^{i\tau}.$$

When we write

$$\zeta^c \quad \text{it will mean} \quad e^{i(c\tau + l_0)},$$
$$\zeta^g \quad \text{it will mean} \quad e^{i(g\tau + g_0)},$$
$$\zeta^m \quad \text{it will mean} \quad e^{i(m\tau + l'_0)}.$$

Therefore $\zeta^{j_1 c + j_2 g + j_3 m + j_4}$ really stands for $e^{i(j_1 l + j_2 F + j_3 l' + j_4 \tau)}$. Since we will perform only the basic arithmetic operations, like adding and multiplying of series, we can with the above convention simply add the exponents of ζ when we multiply the individual terms together. Also differentiation can be performed within the above convention, and for this it is beneficial to introduce the differential operator D by

$$D = \zeta \frac{d}{d\zeta} = -i\frac{d}{d\tau}$$

so that

$$D\zeta^n = n\zeta^n.$$

The solution in complex coordinates will be denoted by u and v. Simultaneously, we take into account that we are working in a rotating coordinate system and set

$$\zeta u = x + iy,$$
$$\zeta^{-1} v = x - iy.$$

The transformed differential equations then read

$$(D + m + 1)^2 u + \tfrac{1}{2}m^2 u + \tfrac{3}{2}m^2 \zeta^{-2} v - u/r^3 = -2\partial\tilde{\Omega}/\partial v,$$
$$(D - m - 1)^2 v + \tfrac{3}{2}m^2 \zeta^2 u + \tfrac{1}{2}m^2 v - v/r^3 = -2\partial\tilde{\Omega}/\partial u, \qquad (7)$$
$$D^2 z - m^2 z - z/r^3 = -\partial\tilde{\Omega}/\partial z.$$

Since $v = \bar{u}$, the second equation does not need to be considered any longer. Brown called $\lambda = e^{i_1} k^{i_2} e'^{i_3} \beta^{i_4}$ the characteristic of a term. Its degree is

$|\lambda| = i_1 + i_2 + i_3 + i_4$. Let u_0 stand for the terms of degree 0, and u_λ and z_λ be the terms with characteristic λ. Then the solution to the main problem of lunar theory will be found in the form

$$u = u_0 + \sum_{|\lambda| \geq 1} u_\lambda \quad \text{and} \quad z = \sum_{|\lambda| \geq 1} z_\lambda.$$

4 The Intermediate Orbit

Hill discovered that the Moon moves on a trajectory that stays close to a periodic orbit of (7) with $\tilde{\Omega} = 0$. This periodic orbit is also called the variational orbit, but this name has nothing to do with the variational principle of mechanics or the method of variation of constants. Instead, "variation" is one of the inequalities that have been observed in the motion of the Moon. Inequality is the traditional name given to the deviation from a true uniform motion, that is, a linear increase with time of the angle for longitude or latitude. Several inequalities have a very strong sinusoidal dependence on time, and this has been observed and measured for a very long time.

The largest inequality is caused by the motion of the Moon towards and away from the Earth. Its period is known as the anomalistic month and depends on the mean anomaly l. The inequality will show up as a term $\sin l$ in longitude. Its coefficient is called the principal term in longitude. Its value, as found from observation and agreed upon by the International Astronomical Union (IAU) for epoch 2000, is 22639.55″.

The second largest inequality is due to the motion perpendicular to the ecliptic. Its period is known as the draconitic month, which depends on the argument of latitude F. The inequality will show up as the term with $\sin F$ in the latitude. Its coefficient is called the principal term in latitude and its value as found from observations for epoch 2000 is given as 18461.40″. The two principal terms will be used to determine the values of the parameters e and k, which are used in the solution for Cartesian coordinates. The method consists of transforming the Cartesian coordinates into spherical coordinates and then comparing coefficients.

The third largest inequality was observed by Tycho Brahe and called "the variation" by him. It depends on the elongation τ and shows up as the term $\sin 2\tau$ in the longitude with a coefficient of about 2370″. Newton already pointed out how this inequality could be explained if one assumes that the Moon starts out on a circular orbit around the Earth. It was Hill's contribution to lunar theory to account for most of this inequality already at terms of order 0 and thus to achieve a more rapid convergence of the series solution in Cartesian coordinates. Because of this connection, Hill also called his intermediate orbit the variational orbit. It is a periodic solution of

$$(D + 1 + m)^2 u + \frac{1}{2}m^2 u + \frac{3}{2}m^2\zeta^{-2}v - \frac{u}{r^3} = 0. \tag{8}$$

There are different ways for finding solutions to this equation. Most of them keep m as a formal parameter and calculate u as a series in m. The speed of convergence for the series describing the intermediate orbit is acceptable, and it is even possible to give estimates for the radius of convergence of this series.

When the calculations are carried out by machine, it is sometimes easiest to compute the terms of the series directly by comparing coefficients. This is the approach to be used here. One checks easily that for $m = 0$ the solution to (8) is $u = 1$. Under the assumption that all terms with degree $< k$ in m have been computed, we write

$$u = 1 + \sum_{j=1}^{k} m^j u_j,$$

$$v = 1 + \sum_{j=1}^{k} m^j v_j,$$

and find for the terms of order k

$$(D + 1)^2 u_k - \frac{1}{2}u_k - \frac{3}{2}v_k = -2(D + 1)u_{k-1} - \frac{3}{2}u_{k-2}$$

$$-\frac{3}{2}\zeta^{-2}v_{k-2} + (u^{-1/2}v^{-3/2})_k. \tag{9}$$

Here $(u^{-1/2}v^{-3/2})_k$ stands for the terms at order k in $u^{-1/2}v^{-3/2}$ which depend on known terms, that is, on u_j and v_j with $j < k$.

The right-hand side of (9) is of the form

$$\sum_{l=-k/2}^{k/2} \alpha_{kl}\zeta^{2l},$$

and we therefore find

$$u_k = \sum_{l=-k/2}^{k/2} A_{kl}\zeta^{2l} \quad \text{and} \quad v_k = \sum_{l=-k/2}^{k/2} A_{kl}\zeta^{-2l}$$

from the following set of linear equations for $l > 0$:

$$\left((1+2l)^2 + \frac{1}{2}\right)A_{kl} + \frac{3}{2}A_{k,-l} = \alpha_{k,l},$$

$$\frac{3}{2}A_{kl} + \left((1-2l)^2 + \frac{1}{2}\right)A_{k,-l} = \alpha_{k,-l}.$$

For $l = 0$ there is only one equation that gives

$$A_{k0} = \frac{1}{3}\alpha_{k0},$$

and for $l > 0$ we obtain from the above equations

$$A_{kl} = \frac{((2l-1)^2 + 1/2)\alpha_{kl} - 3/2\alpha_{k,-l}}{4l^2(4l^2-1)}, \qquad -k/2 \leq l \leq k/2.$$

A MACSYMA program implementing these calculations is given in Appendix B.

5 The Terms of First Order in the Inclination

After computing the variational orbit, Hill [6] tried to find periodic planar orbits nearby. To accomplish this he developed the theory of infinite determinants. This tool was then also used by Cowell [2] to compute a three-dimensional orbit. Since this case is simpler than the planar one, we will treat it first. We will repeat Cowell's calculations but do not need to introduce any labor-saving shortcuts since our calculations are carried out by machine.

The parameter for the inclination is k. In terms of the notation for the characteristic $\lambda = k$ we want to compute kz_k. It has to satisfy the following second-order equation:

$$D^2 z_k - m^2 z_k - z_k/(u_0 v_0)^{3/2} = 0,$$

which is the linearized version of the third equation in (7). Let

$$M(\zeta) = \frac{1}{2}(m^2 + (u_0 v_0)^{-3/2}) = \sum_{j=-\infty}^{\infty} M_j \zeta^{2j} \tag{10}$$

be the series that is already available from computing the variational orbit. The coefficients satisfy $M_k = M_{-k}$ and the first few terms are given by

$$M_0 = \frac{1}{2} + m + \frac{5}{4}m^2 - \frac{9}{64}m^4 + 2m^5 + \frac{17}{3}m^6 + \cdots,$$

$$M_1 = \frac{3}{4}m^2 + \frac{19}{8}m^3 + \frac{10}{3}m^4 + \frac{43}{18}m^5 + \frac{18709}{27648}m^6 + \cdots,$$

$$M_2 = \frac{33}{32}m^4 + \frac{2937}{640}m^5 + \frac{23051}{2400}m^6 + \cdots,$$

$$M_3 = \frac{1393}{1024}m^6 + \cdots.$$

The following second order linear system therefore has to be solved:

$$D^2 z_k - 2M(\zeta)z_k = 0. \tag{11}$$

Due to the symmetry in the function $M(\zeta)$, the solution can be found as

$$z_k = \mathrm{i}(\zeta^g w - \zeta^{-g}\overline{w}),$$

where w is a series in ζ with real coefficients that have to be computed along with the term g_0 for the mean motion of the node. We will drop the subscript 0 for this section. Substituting the form of the solution into (11) leads to

$$(D + g)^2 w - 2M(\zeta)w = 0. \tag{12}$$

If w is assumed to be

$$w = \sum_{j=-\infty}^{\infty} \kappa_j \zeta^{2j},$$

then by comparing coefficients in (12) leads to the following infinite system of linear equations

$$\vartheta(g)\kappa = 0, \tag{13}$$

where the infinite symmetric matrix $\vartheta(g)$ is given by

$$\vartheta(g) = \begin{pmatrix} \ddots & \vdots & \vdots & \vdots & \\ \cdots & (-2+g)^2 - 2M_0 & -2M_1 & -2M_2 & \cdots \\ \cdots & -2M_{-1} & g^2 - 2M_0 & -2M_1 & \cdots \\ \cdots & -2M_{-2} & -2M_{-1} & (2+g)^2 - 2M_0 & \cdots \\ & \vdots & \vdots & \vdots & \ddots \end{pmatrix}$$

and the column vector of unknowns is

$$\kappa = (\cdots, \kappa_{-1}, \kappa_0, \kappa_1, \cdots)^T.$$

Since (13) is a homogeneous system of linear equations, a nontrivial solution exists only when g is determined such that $\vartheta(g)$ is singular.

Here we will not follow the standard method by approximating the infinite matrix with a finite one; instead, we will develop the solution directly in terms of powers of m. Due to the special form of the function $M(\zeta)$, we can write

$$M(\zeta) = \sum_{j=-\infty}^{\infty} \sum_{l \geq |j|} M_{jl} m^l \zeta^{2j}$$

$$= \sum_{l=0}^{\infty} \sum_{j=-l/2}^{l/2} M_{jl} \zeta^{2j} m^l$$

$$= \sum_{l=0}^{\infty} \tilde{M}_l m^l$$

and assume that the solution has the same form:

$$w = \sum_{l=0}^{\infty} w_l m^l = \sum_{l=0}^{\infty} \sum_{j=-l/2}^{l/2} \kappa_{jl} \zeta^{2j} m^l. \qquad (14)$$

Finally, set

$$g = \sum_{l=0}^{\infty} g_l m^l.$$

For terms at order 0 in m, we find from (12) that $g_0 = 1$ and $\kappa_{00} = 1$. For the coefficients of m^l in (12) we obtain

$$\sum_{j=0}^{l} \left(\sum_{i=0}^{j} \tilde{g}_j \tilde{g}_{j-i} - 2\tilde{M}_j \right) w_{l-j} = 0, \tag{15}$$

where $\tilde{g}_i = g_i$ for $i > 0$ but $\tilde{g}_0 = 1 + D$. The unknown terms w_l and g_l appear in (15) as

$$((D+1)^2 - 1)w_l + g_l + \alpha_l = 0, \tag{16}$$

and α_l depends on lower order terms. It is given by

$$\alpha_l = \sum_{j=1}^{l-1} \left(2g_j(1+D) + \sum_{i=1}^{j-1} g_i g_{j-i} - 2\tilde{M}_j \right) w_{l-j} - 2\tilde{M}_l + \sum_{j=1}^{l-1} g_j g_{l-j}$$

and has the form

$$\alpha_l = \sum_{i=-l}^{l} \alpha_{il} \zeta^{2i}.$$

With the form of the solution given in (14), we then get

$$\sum_{i=-l}^{l} (4i(i+1)\kappa_{il} + \alpha_{il})\zeta^{2i} + 2g_l = 0,$$

so that it follows easily that

$$\kappa_{il} = -\frac{\alpha_{il}}{4i(i+1)} \qquad \text{for } i \neq 0, -1,$$

and if we set $\kappa_{0l} = 0$, then

$$g_l = -\frac{\alpha_{0l}}{2}.$$

The coefficient that is not so easily found is that of ζ^{-2}, since it is in the kernel of (16). This also means that $\kappa_{-1,l-1}$ was not determined at the previous order but can now be used to eliminate the terms $\tilde{\alpha}_{-1,l}$ in $\alpha_{-1,l}$ which do not depend on it. One finds

$$\kappa_{-1,l-1} = \tilde{\alpha}_{-1,l}/(2 + 2g_1).$$

A MACSYMA program that implements the above formula is given in Appendix C and the results of a run are

$$w = 1 + \frac{3}{8}\zeta^{-2}m + \left(\frac{3}{16}\zeta^2 - \frac{29}{32}\zeta^{-2} \right) m^2$$

$$+ \left(\frac{1}{2}\zeta^2 - \frac{2029}{1536}\zeta^{-2} - \frac{9}{128}\zeta^{-4} \right) m^3$$

$$+ \left(\frac{25}{256}\zeta^4 + \frac{197}{384}\zeta^2 - \frac{18875}{18432}\zeta^{-2} - \frac{105}{512}\zeta^{-4} \right) m^4 \cdots$$

and

$$g = 1 + m + \frac{3}{4}m^2 - \frac{33}{32}m^3 - \frac{105}{128}m^4 + \frac{43}{2048}m^5 + \frac{2567}{24576}m^6$$

$$+ \frac{3\,47699}{5\,89824}m^7 + \frac{64\,42309}{70\,77888}m^8 + \frac{17118\,51619}{6794\,77248}m^9$$

$$+ \frac{30\,03648\,19183}{4\,07686\,34880}m^{10} + \frac{3355\,25486\,05553}{489\,22361\,85600}m^{11} + \cdots.$$

6   Terms at First Order in e

In order to find planar orbits in the vicinity of the variational orbit we set

$$u = u_0 + e u_e$$

so that the linearized system (7) reads

$$(D + 1 + m)^2 u_e + M(\zeta)u_e + N(\zeta)v_e = 0,$$

where

$$M(\zeta) = \frac{1}{2}(m^2 + (u_0 v_0)^{-3/2})$$

was already given in section 5 and

$$N(\zeta) = \frac{3}{2}(\zeta^{-2}m^2 + u_0^2/(u_0 v_0)^{5/2})$$

$$= \sum N_i \zeta^{2i}$$

through order 6 is given by

$$N_{-3} = \frac{1}{4}m^6 + \cdots ,$$

$$N_{-2} = \frac{123}{512}m^4 + \frac{823}{1280}m^5 + \frac{27899}{76800}m^6 + \cdots ,$$

$$N_{-1} = \frac{27}{16}m^2 - \frac{1}{4}m^3 - \frac{217}{96}m^4 - \frac{77}{18}m^5 - \frac{143911}{55296}m^6 + \cdots,$$

$$N_0 = \frac{3}{2} + 3m + \frac{9}{4}m^2 - \frac{417}{128}m^4 - \frac{551}{64}m^5 - \frac{4993}{256}m^6 + \cdots,$$

$$N_1 = \frac{69}{16}m^2 + \frac{29}{2}m^3 + \frac{2137}{96}m^4 + \frac{335}{18}m^5 - \frac{5737}{6912}m^6 + \cdots,$$

$$N_2 = \frac{4497}{512}m^4 + \frac{53121}{1280}m^5 + \frac{7201393}{76800}m^6 + \cdots,$$

$$N_3 = \frac{31549}{2048}m^6 + \cdots.$$

The form of the solution is

$$u_e = \zeta^c x + \zeta^{-c} y,$$

where c is the zero-order term in the motion of the perigee, which was denoted by c_0 in (5). The resulting equations for x and y are

$$(c + 1 + m + D)^2 x + M(\zeta)x + N(\zeta)\bar{y} = 0,$$
$$(c - 1 - m - D)^2 y + N(\zeta)\bar{x} + M(\zeta)y = 0.$$

Let

$$x = \sum \xi_j \zeta^{2j} \qquad \text{and} \qquad y = \sum \eta_j \zeta^{2j}$$

so that the coefficients ξ_j and η_j follow from the following infinite set of linear equations:

$$(c + 1 + m + 2j)^2 \xi_j + \sum_k M_{j-k}\xi_k + \sum_k N_{j-k}\bar{\eta}_{-k} = 0, \qquad (17)$$

$$(c - 1 - m + 2j)^2 \eta_{-j} + \sum_k N_{k-j}\bar{\xi}_k + \sum_k M_{k-j}\eta_{-k} = 0.$$

With the help of the 2 by 2 block matrices

$$\Delta(c) = \begin{pmatrix} (1 + 2j + c + m)^2 & 0 \\ 0 & (c - 1 - m + 2j)^2 \end{pmatrix}$$

and

$$L_j = \begin{pmatrix} M_j & N_{-j} \\ N_j & M_j \end{pmatrix},$$

the above system can be given in matrix form as

$$\Theta(c)\varepsilon = 0,$$

where

$$\Theta(c) = \begin{pmatrix} \ddots & \vdots & \vdots & \vdots & \\ \cdots \Delta_{-1}+L_0 & L_1 & L_2 & \cdots \\ \cdots & L_{-1} & \Delta_0+L_0 & L_1 & \cdots \\ \cdots & L_{-2} & L_{-1} & \Delta_1+L_0 & \cdots \\ & \vdots & \vdots & \vdots & \ddots \end{pmatrix}$$

and the vector of unknowns is

$$\varepsilon = (\cdots, \xi_{-1}, \eta_1, \xi_0, \eta_0, \xi_1, \eta_{-1}, \cdots)^T.$$

Since the above system of equations is homogeneous, it is at once clear that a nontrivial solution only exists when c is determined such that $\Theta(c)$ is singular.

Again we will not follow the standard approach of approximating this infinite determinant by a finite one. Instead we will compute the series for x and y directly order by order in m, and with it the series for c. Hill [5] had computed the series for c already, but since he had to do the calculations by hand his method was more sophisticated than what we need when we use the machine.

The form for the series $M(\zeta)$ was already given in section 5 and $N(\zeta)$ has the same form, that is,

$$N(\zeta) = \sum_{l=0}^{\infty} \sum_{j=-l/2}^{l/2} N_{jl}\zeta^{2j}m^l = \sum_{l=0}^{\infty} \tilde{N}_l m^l.$$

We can assume the same form for x and y except that we allow for a slightly larger range for the different powers of ζ at a given order in m:

$$x = \sum_{l=0}^{\infty} x_l m^l = \sum_{l=0}^{\infty} \sum_{j=-(l+1)/2}^{l/2} A_{jl}\zeta^{2j}m^l,$$

$$y = \sum_{l=0}^{\infty} y_l m^l = \sum_{l=0}^{\infty} \sum_{j=-l/2}^{(l+1)/2} B_{jl}\zeta^{2j}m^l,$$

and

$$c = \sum_{l=0}^{\infty} c_l m^l.$$

The reason for the different limits for the summation will become apparent later in (18), when it will be seen that terms quadratic in m determine the coefficient of ζ^{-2} in x_1 and the coefficient of ζ^2 in y_1.

For terms at order 0 in m the two equations (18) are

$$\left((1 + c_0)^2 + \frac{1}{2}\right) x_0 + \frac{3}{2} y_0 = 0,$$

$$\frac{3}{2} x_0 + \frac{1}{2} y_0 = 0.$$

They are linearly dependent when $c_0 = 1$. The standard choice is to impose the following normalizing condition for the constant terms:

$$x_0 - y_0 = 1 \quad \text{and} \quad A_{0l} - B_{0l} = 0 \quad \text{for } l = 1, 2, \ldots$$

so that

$$x_0 = \frac{1}{4} \quad \text{and} \quad y_0 = -\frac{3}{4}.$$

Next consider the coefficient of m in (18):

$$c_1 - 1 + \frac{9}{2} x_1 + \frac{3}{2} y_1 = 0$$

$$\frac{3}{2} x_1 + \frac{1}{2} y_1 = 0.$$

Also, these terms are independent of time, and with $c_1 = 1$ and the normalizing condition we find

$$x_1 = 0 \quad \text{and} \quad y_1 = 0.$$

In order to express the terms of m^l with $l > 1$ it is convenient to introduce the notation

$$c + 1 + m + D = \sum_{l=0}^{\infty} \tilde{c}_l m^l,$$

$$c - 1 - m - D = \sum_{l=0}^{\infty} \hat{c}_l m^l,$$

so that
$$c_i = \tilde{c}_i = \hat{c}_i \qquad \text{for} \quad i > 1$$

but $\tilde{c}_0 = D + 2$, $\hat{c}_0 = -D$, $\tilde{c}_1 = 2$, and $\hat{c}_1 = 0$. The square of these series are

$$(c + 1 + m + D)^2 = \sum \tilde{d}_i m^i \qquad \text{with} \quad \tilde{d}_i = \sum_{j=0}^{i} \tilde{c}_j \tilde{c}_{i-j}$$

and

$$(c - 1 - m - D)^2 = \sum \hat{d}_i m^i \qquad \text{with} \quad \hat{d}_i = \sum_{j=0}^{i} \hat{c}_j \hat{c}_{i-j}$$

so that the terms at order l are

$$\sum_{i=0}^{l} (\tilde{d}_i x_{l-i} + \tilde{M}_i x_{l-i} + \tilde{N}_i \overline{y}_{l-i}) = 0,$$

$$\sum_{i=0}^{l} (\hat{d}_i y_{l-i} + \tilde{N}_i \overline{x}_{l-i} + \tilde{M}_i y_{l-i}) = 0.$$

Extracting from it the unknown terms x_l, y_l, and c_l we obtain

$$c_l + \left((2 + D)^2 + \frac{1}{2} \right) x_l + \frac{3}{2} \overline{y}_l + \alpha_l = 0,$$

$$\frac{3}{2} \overline{x}_l + \left(D^2 + \frac{1}{2} \right) y_l + \beta_l = 0.$$

Terms that depend on those of lower order are α_l and β_l. They have the form

$$\alpha_l = \sum_{k=-(l+1)/2}^{l/2} \alpha_{kl} \zeta^{2k}, \qquad \beta_l = \sum_{k=-l/2}^{(l+1)/2} \beta_{kl} \zeta^{2k}$$

and are given by

$$\alpha_l = \frac{1}{4} \tilde{M}_l - \frac{3}{4} \tilde{N}_l + \frac{1}{4} \sum_{j=1}^{l-1} \tilde{c}_j \tilde{c}_{l-j} + \sum_{i=1}^{l-1} (\tilde{d}_i x_{l-i} + \tilde{M}_i x_{l-i} + \tilde{N}_i \overline{y}_{l-i}),$$

$$\beta_l = -\frac{3}{4} \tilde{M}_l + \frac{1}{4} \tilde{N}_l - \frac{3}{4} \sum_{j=2}^{l-2} c_j c_{l-j} + \sum_{i=l}^{l-1} (\hat{d}_i y_{l-i} + \tilde{N}_i \overline{x}_{l-i} + \tilde{M}_i y_{l-i}).$$

With

$$x_l = \sum_{k=-(l+1)/2}^{l/2} A_{kl} \zeta^{2k}, \qquad y_l = \sum_{k=-l/2}^{(l+1)/2} B_{kl} \zeta^{2k},$$

we find

$$c_l + \sum_{k=-l/2}^{l/2} \left(\left((2+2k)^2 + \frac{1}{2} \right) A_{kl} + \frac{3}{2} B_{-kl} + \alpha_{kl} \right) \zeta^{2k} = 0,$$

$$\sum_{k=-l/2}^{l/2} \left(\frac{3}{2} A_{kl} + \left(4k^2 + \frac{1}{2} \right) B_{-kl} + \beta_{-kl} \right) \zeta^{-2k} = 0.$$

The general solution for the individual system of equations is

$$A_{kl} = \frac{\frac{3}{2}\beta_{-kl} - (4k^2 + \frac{1}{2})\alpha_{kl}}{4k(k+1)(2k+1)^2},$$

$$B_{-kl} = \frac{\frac{3}{2}\alpha_{kl} - ((2+2k)^2 + \frac{1}{2})\beta_{-kl}}{4k(k+1)(2k+1)^2},$$

which is valid when $k \neq 0, -1$. For the case $k = 0$ we find that the two equations are linearly dependent when

$$c_l = 3\beta_{0l} - \alpha_{0l},$$

so that together with the normalizing condition we have

$$A_{0l} = -\beta_{0l}/2, \qquad B_{0l} = -\beta_{0l}/2.$$

When $k = -1$ the terms are more difficult to determine, as then the corresponding equations are linearly dependent. It means that terms at order $l - 1$ have to be used to satisfy the condition at order l. The resulting equations are

$$\frac{1}{2}A_{-1,l} + \frac{3}{2}B_{1l} + A_{-1,l-1} + 3B_{1,l-1} = -\tilde{\alpha}_{-1,l},$$

$$\frac{3}{2}A_{-1,l} + \frac{9}{2}B_{1l} + 3A_{-1,l-1} + B_{1,l-1} = -\tilde{\beta}_{1l}$$

with $\tilde{\alpha}_{-1,l}$ and $\tilde{\beta}_{1l}$ now denoting the remaining known terms. These equations show that these terms at order l are determined except for a solution to the homogeneous equation $A_{-1,l} + 3B_{1l} = 0$. We will set $B_{1l} = 0$ so that we have the following set of equations:

$$\frac{1}{2}A_{-1,l} + A_{-1,l-1} + 3B_{1,l-1} = -\tilde{\alpha}_{-1,l},$$

$$\frac{3}{2}A_{-1,l} + 3A_{-1,l-1} + B_{1,l-1} = -\tilde{\beta}_{1l}, \tag{18}$$

$$A_{-1,l-1} + 3B_{1,l-1} = 0,$$

with the solutions

$$A_{-1,l} = -2\tilde{\alpha}_{-1,l}, \quad A_{-1,l-1} = \frac{9\tilde{\alpha}_{-1,l} - 3\tilde{\beta}_{1l}}{8}, \quad B_{1,l-1} = \frac{\tilde{\beta}_{1l} - 3\tilde{\alpha}_{-1,l}}{8}.$$

The MACSYMA program that implements the above formulas is given in Appendix D, and it produced the following results for the solution in Cartesian coordinates:

$$x = \frac{1}{4} - \frac{45}{32}\zeta^{-2}m + \left(\frac{3}{16}\zeta^2 + \frac{3}{16} - \frac{555}{128}\zeta^{-2}\right)m^2$$

$$+ \left(\frac{27}{64}\zeta^2 + \frac{651}{1024} - \frac{22441}{2048}\zeta^{-2} - \frac{15}{256}\zeta^{-4}\right)m^3$$

$$+ \left(\frac{177}{1024}\zeta^4 + \frac{697}{1536}\zeta^2 + \frac{2817}{4096} - \frac{63583}{3072}\zeta^{-2} - \frac{193}{512}\zeta^{-4}\right)m^4 + \cdots,$$

$$y = -\frac{3}{4} + \frac{15}{32}\zeta^2m + \left(\frac{277}{128}\zeta^2 + \frac{3}{16} - \frac{1}{32}\zeta^{-2}\right)m^2$$

$$+ \left(\frac{45}{128}\zeta^4 + \frac{36329}{6144}\zeta^2 + \frac{651}{1024} + \frac{29}{192}\zeta^{-2}\right)m^3$$

$$+ \left(\frac{137}{64}\zeta^4 - \frac{87599}{9216}\zeta^2 + \frac{2817}{4096} + \frac{35}{4608}\zeta^{-2} + \frac{5}{512}\zeta^{-4}\right)m^4 + \cdots,$$

and for the mean motion of the perigee

$$c = 1 + m - \frac{3}{4}m^2 - \frac{201}{32}m^3 - \frac{2367}{128}m^4 - \frac{1\,11749}{2048}m^5$$

$$- \frac{40\,95991}{24576}m^6 - \frac{3325\,32037}{5\,89824}m^7 - \frac{1\,51062\,11789}{70\,77888}m^8$$

$$- \frac{597\,53329\,16861}{6794\,77248}m^9 - \frac{1\,54777\,54421\,75567}{4\,07686\,34880}m^{10}$$

$$- \frac{818\,42933\,65560\,24967}{489\,22361\,85600}m^{11}$$

$$- \frac{2\,18559\,43284\,86055\,04951}{29353\,41711\,36000}m^{12} + \cdots.$$

Of interest is the growth of the coefficients. One has to go fairly far in this expansion to get the desired accuracy of 14 decimal digits. Performing these calculations in floating point arithmetic shows that for $m = 0.08084\ 89375\ 3667$, one has to go to order 30 in m.

Appendix A: Canonical Transformation to Jacobi Coordinates

Jacobi coordinates for the N-body problem are related to binary trees with N leaves in a very natural way. As shown in [8] the transformation matrix can then be generated in a straightforward manner. The following iterative method describes the construction of this binary tree from the bottom up. The tree will contain all the information needed to write down the transformation to the desired set of Jacobi coordinates.

Let m_j and m_k be the masses of the two bodies that at least conceptually will be replaced by a new virtual body of mass $m_j + m_k$. In terms of coordinate transformations, the position coordinates q_j and q_k are to be replaced by the vector giving their relative position $q_j - q_k$ and by the vector to their center of mass $v = (m_j q_j + m_k q_k)/(m_j + m_k)$. The node in the binary tree corresponding to this virtual body will have m_j as its left child and m_k as its right child. The order of the children indicates that the relative coordinates are from q_j to q_k and not in the opposite direction. Repeat the above step for $N - 1$ bodies, that is, the $N - 2$ original bodies plus the new virtual body.

Since the leaves of the binary tree corresponding to the N bodies will be numbered from 1 to N, we will number the $N - 1$ internal nodes from $N + 1$ to $2N - 1$ in the order in which they were generated. The root of the binary tree thus has the number $2N - 1$. Next construct an N by N matrix H whose N columns correspond to the N bodies and whose rows 1 through $N - 1$ correspond to the internal nodes $N + 1$ through $2N - 1$.

The entries h_{ij} for $1 \leq i \leq N$, $1 \leq j \leq N - 1$ of this matrix are defined by

$$h_{ij} = \begin{cases} -1, & \text{if } m_j \text{ is in the left subtree of the node } N + i, \\ +1, & \text{if } m_j \text{ is in the right subtree of the node } N + i, \\ 0, & \text{otherwise.} \end{cases}$$

The matrix is completed by setting $h_{Nj} = +1$ for $1 \leq j \leq N$. It indicates that one of the coordinates is the vector to the center of all masses.

Let $A = (a_{ij})$ be the matrix for the transformation from the old position coordinates q_i to the new Jacobi coordinates Q_i. Each entry a_{ij} is a 3 by 3 diagonal matrix, as the transformation is from $\mathbf{R}^{3N}$ to $\mathbf{R}^{3N}$. Similarly, $B = (b_{ij})$ is the matrix for the transformation of the old momenta to the new ones. In order to simplify the discussion we will only write down the entries of these matrices as scalars and not as diagonal block matrices.

For each node i of the binary tree let M_1 be the sum of the masses in the left subtree and M_2 the sum of all masses in the right subtree, that is,

$$M_1 = \sum_{h_{ij}=-1} m_j, \qquad M_2 = \sum_{h_{ij}=1} m_j.$$

Then the entries for the transformation matrices are

$$a_{ij} = \begin{cases} -m_j/M_1, & \text{if } h_{ij} = -1, \\ m_j/M_2, & \text{if } h_{ij} = +1, \\ 0, & \text{otherwise,} \end{cases}$$

and

$$b_{ij} = \begin{cases} -M_2/(M_1 + M_2), & \text{if } h_{ij} = -1, \\ M_1/(M_1 + M_2), & \text{if } h_{ij} = +1, \\ 0, & \text{otherwise.} \end{cases}$$

The above formulas hold for all rows $i = 1, \ldots, N$ with M_1 and M_2 recomputed for each row. The formulas hold for the last row since the root of the binary tree, node $2N - 1$, should be right child of an additional node $2N$, indicating the fact that the center of mass of the N bodies is one of the coordinates. Actually, the above formulas for the last row simplify and are more easily written down from

$$a_{Nj} = m_j/M_2, \qquad b_{Nj} = 1, \qquad 1 \leq j \leq N.$$

The fact that the above construction leads to a canonical transformation is proven in [8] so we restrict ourselves here to present a MACSYMA program that implements the above ideas. The program is to be called via Jacob(N) with N the number of bodies. The program will then request in an interactive manner the construction of the binary tree. In the end the transformation matrices are available in the MACSYMA variables A and B

```
Jacob(n) := block([b1,l,l1,l2,i,j,s],
    h : zeromatrix(n-1,n),
    a : zeromatrix(n,n),
    b : zeromatrix(n,n),
    s : [],
    for i thru n do s : endcons(i,s),
    print("1 thru ",n," represent the ",n," masses,",n+1, " thru ",
         2*n-1," will be the internal nodes of the binary tree"),
    print("The transformation matrix for the coordinates will be A,"),
    print("for the momenta is B"),
    print("Select a pair of numbers from the displayed list"),
    print("Enter the two number as a list, i.e. [1,2]"),
    for i from 1 thru n-1 do
        (b1 : true,
        while(b1) do
            (l:read("Select two values from ",s),
            l1 : first(l),
            l2 : last(l),
            if member(l1,s) and member(l2,s) and not l1=l2 then
                (b1 : false,
                s : delete(l2,delete(l1,s)),
                s : endcons(i+n,s),
                if l1<= n then h[i,l1]:-1
                else (l1: l1 - n,
                   for j thru n do h[i,j]:- abs(h[l1,j])),
                      if l2<= n then h[i,l2]: 1
                      else (l2 : l2 - n,
                          for j thru n do
                              h[i,j]: h[i,j] + abs(h[l2,j]))))),

    for i thru n-1 do
       ( m1:m2:0,
       for j thru n do
          (if h[i,j]<0 then m1:m1 + m[j]
          else if h[i,j]>0 then m2:m2 + m[j]),
          m3 : m1 + m2,
          for j thru n do
             (if h[i,j]<0 then
                (a[i,j]:-m[j]/m1,
                b[i,j]:-m2/m3)
             else if h[i,j]>0 then
                (a[i,j]: m[j]/m2,
                b[i,j]: m1/m3))),

    for j thru n do (a[n,j] : m[j]/m3, b[n,j]:1) ) ;
```

Appendix B: MACSYMA Program for the Intermediate Orbit

```
Hill(n) := block(local(a),
    array([u,v,r2,r32],n),
    globalsolve : true ,
    ratmx : true,              /* use canonical form */
    u[0]:v[0]:r2[0]:r32[0] : 1 ,
    for k : 1 thru n do
      (
      n2 : entier(k/2),

      r2[k] : sum(u[k-1]*v[1],1,1,k-1),    /* u*v   */
                                           /* (u*v)^(-3/2) */
      r32[k] : -3/2*r2[k]
            + sum((-3/2*(k-1)-1)/k*r2[k-1]*r32[1],1,1,k-1),
      eq : - 2*diff(u[k-1]*zeta,zeta)
            + sum(r32[k-1]*u[1],1,0,k-1),

      if k > 1 then
         eq : eq - 3/2*(u[k-2] + zeta^-2 * v[k-2]) ,

      /* compare coefficients of zeta^l and zeta^-1 l=0,-2,+2,... */

      a[k,0]:coeff(eq,zeta,0)/3,
      for l : 1 thru n2   do
        (
        a1 :ratcoef(eq,zeta,2*l),
        a2 :ratcoef(eq,zeta,-2*l),
        a[k,l]:(((2*l-1)^2+1/2)*a1-3/2*a2)/(4*l^2*(4*l^2-1)),
        a[k,-l]:(((2*l+1)^2+1/2)*a2-3/2*a1)/(4*l^2*(4*l^2-1))
        ),

      u[k] : sum(a[k,l]*zeta^(2*l),l,-n2,n2),
      v[k] : sum(a[k,l]*zeta^(-2*l),l,-n2,n2),
      print("u[",k,"]=",expand(u[k])) ,
      r2[k] : rat(r2[k]+u[k]+v[k]),    /*update r2[k] and r32[k] */
      r32[k] : rat(r32[k]-3/2*(u[k]+v[k]))
      ) );
```

Once the above program has been compiled it is called by `Hill(n)` with n the order to which the intermediate orbit is desired. In designing a first program one might start out with u_k having undetermined coefficients, then evaluate the corresponding differential equation (8) and let MACSYMA solve the resulting linear equations by calls to its built-in routines. One finds out soon that it is very time consuming to work with undetermined coefficients. It is much faster to calculate the solution as given by the formulas at the end of section 4.

The evaluation of $(u^{-1/2}v^{-3/2})_k$ remains the time-consuming aspect of the entire computation, even when we initially set $u_k = v_k = 0$. The method that we use to find the terms of order k for the series

$$y = \sum y_j m^j \qquad \text{so that} \qquad y = x^\alpha \quad \text{when} \quad x = \sum x_j m^j$$

is based on the trick of differentiating the defining relation with respect to m. By comparing coefficients of order k in

$$x\frac{\mathrm{d}y}{\mathrm{d}m} = \alpha y \frac{\mathrm{d}x}{\mathrm{d}m},$$

one obtains

$$y_k = \frac{\alpha x_k y_0}{x_0} + \sum_{l=1}^{k-1} \frac{(\alpha(k-l)-l)x_{k-l}y_l}{kx_0}.$$

Appendix C: MACSYMA Program for Inclination

The program below is a direct coding of the formulas given in section 5. The program uses the array MM, which equals the function $2M(\zeta)$ of the text. Once the program is compiled it is called via Cowell(m) with m not greater than the argument used in the previous program.

```
cowell(n):=    /* compute first order terms for inclination */
  ( array([g,w,MM],n),
   for i from 0 thru n do mm[i] : r32[i],
        /* factor 1/2 still missing */
   mm[2] : mm[2] + 1,     /* add m^2    */
   g[0]:w[0]:1,
   for l thru n do
     (
     temp : sum(g[j]*g[l-j],j,1,l-1)
          + sum( 2*g[j]*diff(zeta*w[l-j],zeta)
             +(sum(g[i]*g[j-i],i,1,j-1)-MM[j])*w[l-j],j,1,l-1)
          - MM[l] ,
```

```
temp : rat(temp) ,
w[1] : 0 ,
for i from -1 thru 1  do  if (i=0 ) then
    g[1] : - ratcoef( temp,zeta,0 )/2
  else if (i=-1) then
    w[1-1] : w[1-1]
            + ratcoef(temp,zeta,-2)/(zeta^2*(2+2*g[1]))
    else
      w[1] : w[1]
            - ratcoef(temp,zeta,2*i)/(4*i*(i+1)) * zeta^(2*i),
  print ("g[",1,"]=",g[1]) ) ) ;
```

Appendix D: MACSYMA Program for First-Order Terms in e

The program below is a direct coding of the formulas given in section 6. The program uses the array M2, which corresponds to the function $M(\zeta)$, and the array NN corresponds to the function $N(\zeta)$. These functions are computed from the series found in Hill(n). To call the MACSYMA program after it is compiled, enter Adams(m) with the value of the parameter not greater than what was used for the intermediate orbit.

```
Adams(n) :=  /* compute first order terms in e */
  ( array( [c,M2,NN,R25,RT,x,y],n ) ,
  for i from 0 thru n do M2[i] : R32[i]/2,
  M2[2] : M2[2] + 1/2,
  R52[0] : RT[0] : 1 ,
  NN[0] : 3/2 ,
  for k : 1 thru n do
    (
    R52[k] : -5/2*r2[k]
            + sum((-5/2*(k-1)-1)/k*r2[k-1]*R52[1],1,1,k-1),
    RT[k] : sum(u[1]*R52[k-1],1,0,k) ,
    NN[k] : 3/2*sum(u[1]*RT[k-1],1,0,k)
    ),
  NN[2] : NN[2] + 3/2 *zeta^-2,
  for l : 0 thru n do M2[l] : MM[l]/2,
  c[0] : c[1] : 1 ,
  x[0] : 1/4 ,
  y[0] :-3/4 ,
  x[1] : y[1] : 0 ,
```

```
for l from 2 thru n do
  (
  l2 : entier((l+1)/2),
  c[0] : c[1] : 2 ,
  alpha : ( M2[l] - 3* NN[l] + sum( c[j]*c[l-j],j,1,l-1 ) )/4
        + sum(sum(c[j]*c[i-j],j,0,i)*x[l-i]
              + 2*c[i]*diff(x[l-i],zeta)*zeta
        + M2[i]*x[l-i]+NN[i]*ev(y[l-i],zeta=1/zeta),i,1,l-1) ,
  c[0] : c[1] : 0 ,
  beta : 1/4*(-3*M2[l]+NN[l]  - 3* sum(c[j]*c[l-j],j,1,l-1))
       + sum(sum(c[j]*c[i-j],j,1,i-1)*y[l-i]
             - 2*c[i]*diff(y[l-i],zeta)*zeta
             + NN[i]*ev(x[l-i],zeta=1/zeta)+M2[i]*y[l-i],i,1,l-1) ,

  x[l] : y[l] : 0 ,
  for k : -l2 thru l2 do
    (
    alphakl : ratcoef( alpha,zeta,2*k ),
    betamkl : ratcoef( beta,zeta,-2*k ) ,
    if (k=0) then
      (
      x[l] : x[l] - betamkl/2,
      y[l] : y[l] - betamkl/2,
      c[l] : 3 * betamkl - alphakl,
      print("c[",l,"]=",c[l])
      )
    else if (k=-1) then
      (
      x[l] : x[l] - 2*alphakl * zeta^(-2),
      x[l-1] : x[l-1] + (3*alphakl-betamkl)*3/8*zeta^(-2),
      y[l-1] : y[l-1] + (betamkl-3*alphakl)/8*zeta^2
      )
    else /* general case */
      (
      x[l]:x[l]+(3/2* betamkl-(4*k*k+1/2)*alphakl)*zeta^(2*k)
            /(4*k*(k+1)*(2*k+1)^2),
      y[l]:y[l]+(3/2*alphakl-((2+2*k)^2+1/2)*betamkl)*zeta^(-2*k)
            /(4*k*(k+1)*(2*k+1)^2)
      )
    )
  )
) ) ;
```

References

[1] E. Brown. Theory of the motion of the moon. *Memoirs of the Royal Astronomical Society of London*, 53:39–116, 1897.

[2] P. H. Cowell. On the inclinational terms in the moon's coordinates. *Amer. J. Math*, 18:99–127, 1896.

[3] M. Gutzwiller and D. S. Schmidt. *The motion of the moon as computed by the method of Hill, Brown and Eckert.* Volume 23, Nautical Almanac Office, U.S. Naval Observatory, U.S. Government Printing Office, Washington, DC: 1986.

[4] M. C. Gutzwiller. The numerical evaluation of Eckert's lunar ephemeris. *Astron. J.*, 84:889–899, 1979.

[5] G. W. Hill. Literal expression for the moon's perigee. *Ann. of Math*, 9:31–41, 1894.

[6] G. W. Hill. On the part of the motion of the lunar perigee which is a function of the mean motions of the sun and the moon. *Acta. Math.*, 8:1–36, 1886.

[7] G. W. Hill. Researches in lunar theory. *Amer. J. Math*, 1:5–16, 129–147, 245–260, 1878.

[8] C. C. Lim. Binary trees, symplectic matrices, and canonical transformations for classical n body problems. *IMA Preprint Series*, 480:1–22, 1989.

[9] D. S. Schmidt. Computing the motion of the moon accurately. In H. S. Dumas, K. R. Meyer, and D. S. Schmidt, editors, *Proceedings on Hamiltonian Dynamical Systems: History, Theory and Applications*, pages 341–361, Springer-Verlag, New York, 1994.

Photograph: Ivan Feitosa

Lectures on Geometrical Methods

in Mechanics

Mark Levi

DEPARTMENT OF MATHEMATICS
PENN STATE UNIVERSITY
UNIVERSITY PARK, PA 16803, USA
e-mail: levi@math.psu.edu

1   Variational Principles: Their Explanation and Applications

Variational principles lie in the foundation of classical mechanics and modern physics. Most texts on mechanics start with variational principles taken as axioms. Our goal in this section is to "explain" why mechanics and optics obey variational principles. We will do that by considering the following four topics.

1. Explanation of Fermat's principle.

2. High-frequency asymptotics: from waves to rays.

3. Maupertuis–Euler–Lagrange principle.

4. Optimal control and the geometry of the Euler–Lagrange equations.

1.1   Explanation of Fermat's Principle

In the first part of these notes we give a physical motivation of the governing variational principles, first in optics and then in mechanics. Our exposition itself follows its own variational principle of minimizing technicalities, sometimes at the expense of rigor. Fermat was the first to discover the applicability of Voltaire's principle ("we live in the best of all possible worlds") to optics, by observing that the light travelling from point A to point B chooses the path of least time (provided A and B are close enough).

How does the light have the "intelligence" to choose the fastest path? Does the light "try out" other paths? The short answer is yes[1].

The light actually does follow *every* path from A to B, and what arrives at B is indeed the sum of all the contributions from all paths. All these contributions, except for the ones that came along the quickest path, tend to cancel! This is a kind of Darwinian survival of the fastest. The following simple model shows what is going on.

---

[1] See the marvelous book QED by Feynman, [1, p. 39]. In our discussion here we violate the truth somewhat by treating the light as a wave, rather than as probabilistic particles from quantum electrodynamics. Nevertheless, the discussion captures the essential mechanism of Fermat's least time principle.

Instead of the uncountable variety of all possible paths we consider only the simplest broken paths ACB, as shown in Figure 1. We will write down the contribution at the receiver B due to the sum of arrivals along all the paths ACB as an integral over ds, and then we will estimate this integral for shortwave light by the method of stationary phase. We will see that the intensity of light at B is due mostly to the path of shortest time *because the sum of contributions from all nonminimal paths ACB cancel each other*[2].

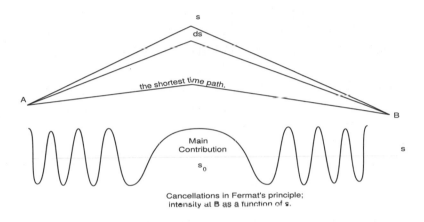

Figure 1. A heuristic explanation of Fermat's principle

We want to find the intensity at B delivered via a particular path ACB. Assume that the light's electric field at the source A is given by $E_0 \cos \omega t$; then the electric field at B, arriving via ACB, will oscillate with a delay $T(s) = $ travel time along ACB. The intensity of electric field at B carried by the wave passing through the segment ds (see Figure 1) is

$$a(s) \cos(\omega(t - T(s))), \tag{1}$$

where $a(s)$ characterizes the amplitude of the arriving signal; it depends on the length of ACB, but its value is irrelevant for our explanation. The

----

[2]A better approximation would have been to take not just these simplest broken paths, but all possible paths $AC_1C_2...C_nB$, for all n and have estimated an appropriate infinite sum. Actually, this has to be done not just in physical space but in space–time, but we do not go into this here.

intensity of electric field at B is thus

$$E_B = \int_{-\infty}^{\infty} a(s) \cos(\omega(t - T(s))) \, ds. \tag{2}$$

The method of stationary phase.   What is the approximate value of this integral, assuming that the light waves are short, $\omega \gg 1$? A look at the graph of the integrand suggests the answer: the contribution must come from the value of s giving T a minimum! Physically, this means that the waves arriving along a nonstationary path are cancelled by their neighbors arriving 180° out of phase. By contrast, the shortest-time paths are enhanced by their neighbors since their phases are nearly equal.

Problem 1.1 *(The stationary phase method). Find an approximate value of the integral (2). The solution will show, as expected, that the leading term of (2) is given in terms of stationary value of the phase T. Hint: Single out a small neighborhood of the extremum of $T(s)$. The integral over the rest of the interval is small, and the integral over the neighborhood can be almost explicitly evaluated for small ω.*

1.2   High-Frequency Asymptotics: From Waves to Rays

Having given a heuristic explanation of Fermat's least time principle, we show how the concept of the ray arises from wave equations.

Consider the wave equation

$$E_{tt} = c^2(x)\Delta E, \tag{3}$$

describing the propagation of light in an inhomogeneous medium. We study the special case of standing waves, which are the solutions of the form

$$E = a(x)\mathrm{Re}\, e^{\frac{i}{\epsilon}(t - S(x))}, \tag{4}$$

in the high-frequency limit of small ϵ. The asymptotic approach based on (4) is sometimes referred to as the WKB method (after Wentzel, Kramers, and Brillouin)[3]. We note that S has a very concrete physical meaning:

---

[3]Although it was used earlier by Jeffreys [2].

it is the *phase function*: along the level surfaces of S the phase of the standing wave is constant. Since the standing wave can be viewed as the superpositon of two countermoving waves, we can think of these surfaces as *wave fronts*.

To learn more about S, we substitute the ansatz (4) into the wave equation (3); equating the terms of the highest order ϵ^{-2}, we obtain

$$S_x^2 = n^2(x), \tag{5}$$

where $n(x) = \frac{1}{c(x)}$. This is a nonlinear first-order partial differential equation, called the *eikonal* equation. This equation specifies the magnitude of ∇S at every x. The motivation behind the beautiful classical theory of first-order nonlinear PDEs comes from optics. Before presenting some of this theory we show that the quantum mechanical phase satisfies a similar eikonal equation.

1.2.1   The Quantum Mechanical Eikonal Equation

We consider the eigenstates Ψ for the quantum mechanical particle of mass m in potential V with energy E:

$$-\frac{h}{2m}\Delta\Psi + V(x)\Psi = E\Psi. \tag{6}$$

Seeking the solutions in the form

$$\Psi = e^{\frac{i}{h}S(x)}, \tag{7}$$

we substitute this in (6). Equating the highest order terms gives the partial differential equation (PDE) for S:

$$S_x^2 = E - V(x). \tag{8}$$

This equation is equivalent to the eikonal equation (5) (with $n^2(x) = E - V(x)$) we had obtained from the wave equation.

1.3   Solving First-Order PDEs Using an Optical Analogy

Both equations (5) and (8) are of the form

$$|S_x| = n(x), \tag{9}$$

where $n(x)$ is a given function that we interpret as the index of refraction.

How does one solve for S? The key is to recall that S is the time it takes the light (emitted from some initial surface Σ) to reach x.

Problem 1.2 *What is the physical meaning of $|S_x|$?*

Answer: $|S_x|$ is the speed of light; PDE (9) asks us to recover the time function S from the knowledge of the speed of light at every point: $\langle \dot{x}, \dot{x} \rangle = \frac{1}{n^2}$.

If the medium is anisotropic, that is, if the speed of light depends on the direction $\langle B(x)\dot{x}, \dot{x} \rangle = 1$, where B is a positive matrix, then S satisfies a more general PDE,

$$\langle AS_x, S_x \rangle = 1, \tag{10}$$

where $A = A(x) = B^{-1}$.

Here is a physically motivated solution of the PDE (10). Let us start with an inhomogeneous anisotropic optical medium with the velocity of light $\dot{x}$ satisfying[4]

$$\langle B(x)\dot{x}, \dot{x} \rangle = 1, \tag{11}$$

where $B = B(x)$ is a symmetric positive definite matrix. Let $S(x)$ be the least time required to reach the point x from a given initial surface Σ, subject to the velocity constraint (11). We show the following lemma.

Lemma 1.1 *The least travel time $S(x)$ in an optical medium with the indicatrix (11) satisfies the generalized eikonal equation (10) with $A = B^{-1}$.*

Proof. Let $x(t)$ be the ray, that is, the path from Σ to x reaching x in least time. To reiterate, our ray is singled out from the infinity of functions satisfying the velocity constraint (11) by the optimality condition: it has to be the first to arrive at x. Now comes the key idea. The ray that arrives first has this crucial property: its direction $\dot{x}(t)$ maximizes the normal velocity $\text{proj}_{\nabla S} \dot{x}$ to the front! Maximizing $\langle \nabla S, \dot{x} \rangle$ over all $\dot{x}$ subject to (11) gives $\nabla_{\dot{x}} \langle S_x, \dot{x} \rangle = \lambda \nabla_{\dot{x}} \langle B(x)\dot{x}, \dot{x} \rangle$, with a scalar factor λ. Equivalently, we have

$$S_x = \lambda B \dot{x}. \tag{12}$$

---

[4]The set of such $\dot{\mathbf{x}}$ is called an *indicatrix*.

In fact, $\lambda = 1$. Indeed, differentiating the identity $S(x(t)) = t$ (S is the arrival time!) along the ray, we obtain

$$1 = \langle S_x, x \rangle = \lambda \langle B\dot{x}, \dot{x} \rangle - \lambda. \tag{13}$$

Substituting $\dot{x} = B^{-1}S_x$ into (11), we obtain (10) for S with $A = B^{-1}$. To summarize, we showed that *the time S of arrival at x of light with the indicatrix (11) satisfies (10)*. This physical interpretation gives an equivalence between the PDE (10) on the one hand and the ray dynamics on the other. This completes the proof of the lemma.

We will now show that the ray dynamics is Hamiltonian.

Rewriting (11) in the form $\langle \sqrt{B}\dot{x}, \sqrt{B}\dot{x} \rangle = 1$ and denoting $\sqrt{B}\dot{x} = u$, we turn the optical problem into a control problem

$$\dot{x} = \sqrt{A}u, \ |u| = 1, \ A = B^{-1}, \tag{14}$$

where the control u has to be chosen according to the Fermat principle. Here $\sqrt{A}$ denotes the positive matrix square root of A. The adjoint linearized equation is given by

$$\dot{p} = -(\sqrt{A}u)_x^T p. \tag{15}$$

The system (14)–(15) is Hamiltonian with $H(x, p, u) = \langle \sqrt{A(x)}u, p \rangle$. For the solutions to satisfy the Fermat's principle, we must maximize the velocity normal to the front; that is, we must choose $u = u(x, p)$ so that

$$\mathcal{H}(x, p) = \max_{|v|=1} H(x, p, v) = \langle \sqrt{A(x)}v, p \rangle = |\sqrt{A(x)}p|. \tag{16}$$

We observe that when the maximizing u is substituted in (14)–(15), the resulting system is still Hamiltonian with the Hamiltonian $\mathcal{H}(x, p) = |\sqrt{A(x)}p|$, since the u-derivatives of $H(x, p, u)$ vanish when u is a maximizer.

We conclude that the rays satisfy the Hamiltonian system with the Hamiltonian $\mathcal{H} = |\sqrt{A(x)}p|$; moreover, p has the interpretation of the normal to the front, and the unknown time S can be found from $S_x = p$ by integration of $\frac{dS}{dt} = S_x \dot{x} = p\dot{x} = \mathcal{H}$.

Equation (10) is the Hamilton–Jacobi equation for the Hamiltonian system (14)–(15).

Equation (10) is quadratic in ∇S and thus is of a rather particular form; the most general first-order nonlinear PDE can be written in the form

$$F(x, S, S_x) = 0. \tag{17}$$

This PDE can be reduced to solving a system of ODEs, using the optical analogy, in a similar way, with the resulting equations for rays (see Figure 2):

$$\dot{x} = F_p(x, p), \quad \dot{p} = -F_x(x, p). \tag{18}$$

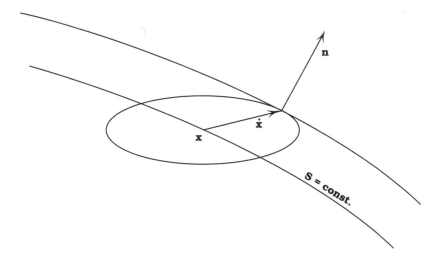

Figure 2. The evolution of (x, S_x) along the ray is governed by a Hamiltonian system.

1.4 From Optics to Mechanics by Bernoulli

In this section we present, as a footnote, Bernoulli's beautiful derivation of the Maupertuis variational principle from Fermat's principle. Since Bernoulli, this optical-mechanical analogy has proved to be extremely fruitful in the work of Hamilton, Lagrange, and others. Some relatively recent developments, such as Bellman's equation and Pontryagin's maximum principle of optimal control, are based on this analogy as well.

To derive the Maupertuis least action principle $\delta \int v \, ds = 0$, we consider a point mass moving in a potential force field. Conservation of energy determines the speed in terms of position. We discretize the medium (Figure 3) by assuming that the potential is constant within strips so that the velocity within strips is constant; with no force acting in the tangential direction to the interace between layers, that velocity is preserved:

$$v_i \sin \alpha_i = v_{i+1} \sin \alpha_{i+1} = \text{const.} \tag{19}$$

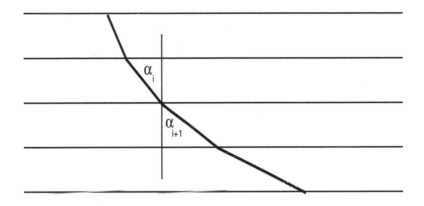

Figure 3. From Fermat's principle in optics to the Maupertuis principle in mechanics.

This is a discretization of the differential equation for the shape of the smooth path. What is the variational formulation of this equation? To answer this, consider an optical medium superimposed on Figure 3 and choose the speed of light $= c_i$ in this optical system so that the rays will follow the same path as the particles (if sent from the same point in the same direction); that is, we want Snell's law,

$$\frac{\sin \alpha_i}{c_i} = \frac{\sin \alpha_{i+1}}{c_{i+1}} = \text{const.}, \tag{20}$$

to be equivalent to Newton's law (19). Taking $c_i = \frac{1}{v_i}$ guarantees such equivalence. Now the ray minimizes $\int dt = \int \frac{ds}{c} = \int v \, ds$. Hence, so does the mechanical particle! Thus Bernoulli's argument gave a variational formulation of a mechanical problem.

1.5   The Maupertuis–Euler–Lagrange Principle

One of the drawbacks of the Maupertuis principle is the fact that it is only applicable to systems with conserved energy, and consequently does not work for time-dependent systems.

The more general principle of Lagrange can be formulated as follows. Consider a mechanical system with the n-dimensional configuration space, that is, one in which the system's position is determined completely by n coordinates $\mathbf{q} = (q_1, \ldots, q_n)$. One can think of finitely many particles, such as planets, or masses connected by rods such as pendula, or rigid bodies (the latter can be viewed as finitely many particles with the rigidity constraint). We consider only the constraints on the *coordinates*:

$$\mathbf{f(q)} = \mathbf{0}. \tag{21}$$

Definition 1.1 *A mechanical system of the type described above is called* holonomic.

Given any holonomic system, let $L = T(\mathbf{q}, \dot{\mathbf{q}}) - V(\mathbf{q})$ be the difference between its kinetic and potential energies. The variational principle of Lagrange states that if $\mathbf{q}(t)$ describes the motion with

$$\mathbf{q}(t_0) = \mathbf{q}_0 \text{ and } \mathbf{q}(t_1) = \mathbf{q}_1, \tag{22}$$

then $\mathbf{q}(t)$ is a critical point of the action

$$\int_{t_0}^{t_1} L(\mathbf{q}, \dot{\mathbf{q}}) dt, \tag{23}$$

with $\mathbf{q}$ subject to (22).

Problem 1.3 *Prove that (23) is invariant under coordinate transformations* $\mathbf{q} = f(\mathbf{Q})$*, that is, that it is equivalent to* $\int_{t_0}^{t_1} \mathcal{L}(\mathbf{Q}, \dot{\mathbf{Q}}) dt$*, where* $\mathcal{L}(\mathbf{Q}, \dot{\mathbf{Q}}) = L\left(f(\mathbf{Q}), f'(\mathbf{Q})\dot{\mathbf{Q}}\right)$*. Hint:* $\mathcal{L}(\mathbf{Q}, \dot{\mathbf{Q}}) = L(\mathbf{q}, \dot{\mathbf{q}})$*.*

The mere statement of Lagrange's principle (23)–(22) as an axiom is unsatisfactory and calls for an explanation. Such an explanation can in fact be given, on the basis of quantum electrodynamics, in the spirit of our earlier discussion.

1.6 Hamiltonian Mechanics via Optical Motivation

In this section we show how the main concepts of Hamiltonian mechanics, such as the Legendre transform, the Hamiltonian function, the Hamilton–Jacobi equations, the Hamilton variational principle, and more come from the optical analogy. The key is to treat the $(\mathbf{q}, t)$-space as an optical medium and to follow the evolution of rays just as we did in the section on geometrical optics. The only difference from that section is that we *deprive time of its privileged role and treat it as another coordinate* in the extended configuration space (q, t) (see Figure 4).

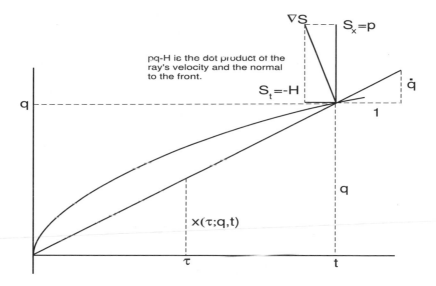

Figure 4. Geometric optics and Lagrange's principle.

We define the action integral

$$S(t, \mathbf{q}) = \min_{\mathbf{x}(\cdot)} \int_0^t L(\mathbf{x}, \dot{\mathbf{x}})dt, \quad \mathbf{x}(0) = 0, \quad \mathbf{x}(t) = \mathbf{q}. \tag{24}$$

The action $S(t, \mathbf{q})$ is the analog of the time function in optics, giving the "cost" of reaching $(t, \mathbf{q})$. Level sets of $S(t, \mathbf{q})$ are the mechanical analogs of the optical wave fronts.

Problem 1.4 *Find the action $S(t, \mathbf{q})$ for the free particle Lagrangian* $L = \frac{m\dot{\mathbf{q}}^2}{2}$. *Take $m = 1$.*

Answer: $S(t, q) = \frac{q^2}{2t}$. The level sets of S are shown in Figure 4.

Problem 1.5 *Find the action $S(t, \mathbf{q})$ for the Lagrangian of the harmonic oscillator $L = \frac{\dot{\mathbf{q}}^2}{2} + \frac{\mathbf{q}^2}{2}$ and sketch the level curves of S.*

We now ask the key question that leads automatically to all the concepts mentioned above: *What is the normal to the boundary of the reachable set?* The answer is given by the following simple theorem.

Theorem 1.1

$$\nabla S(t, \mathbf{q}) \equiv \langle S_t, S_\mathbf{q} \rangle = (L - \mathbf{p}\dot{\mathbf{q}}, \ \mathbf{p}), \tag{25}$$

where $\mathbf{p} = L_{\dot{\mathbf{x}}}(t, \mathbf{q}, \dot{\mathbf{q}})$ *and* $\dot{\mathbf{q}} = \frac{\partial}{\partial \tau}\big|_{\tau = t}\mathbf{x}(\tau; t, \mathbf{q})$, *where* $\mathbf{x}(\tau) = \mathbf{x}(\tau; t, \mathbf{q})$ *is the minimizer of the action (23) connecting the origin with $(t, \mathbf{q})$:*

$$\mathbf{x}(0; t, \mathbf{q}) = \mathbf{0} \quad and \quad \mathbf{x}(t; t, \mathbf{q}) = \mathbf{q}. \tag{26}$$

Proof. Differentiating (24) with respect to $\mathbf{q}$, we obtain

$$
\begin{aligned}
\frac{\partial S}{\partial \mathbf{q}} &= \frac{\partial}{\partial \mathbf{q}} \int_0^t L\left(\mathbf{x}(\tau; t, \mathbf{q}), \frac{\partial}{\partial \tau}\mathbf{x}(\tau; t, \mathbf{q})\right) d\tau \\
&= \int_0^t \left(L_{\mathbf{x}}\mathbf{x}_\mathbf{q} - \frac{d}{d\tau}L_{\dot{\mathbf{x}}}\mathbf{x}_\mathbf{q}\right) dt + L_{\dot{\mathbf{x}}}\mathbf{x}_\mathbf{q}\big|_0^t = L_{\dot{\mathbf{q}}},
\end{aligned}
\tag{27}
$$

where we integrated by parts; we then used the Euler–Lagrange equation and, finally, the fact that

$$\mathbf{x}_\tau(\tau; t, \mathbf{x})\big|_{\tau=t} = -\mathbf{x}_t(\tau; t, \mathbf{x})\big|_{\tau=t}. \tag{28}$$

The latter equality follows by differentiation of the second equation in (26) by t.

Now to find $\frac{\partial S}{\partial t}$, we use a quick shortcut (instead of a direct computation): observe that

$$\frac{d}{dt}S(t, \mathbf{x}(t)) = L(\mathbf{x}, \dot{\mathbf{x}}); \tag{29}$$

since $\frac{d}{dt}S\left(t,\mathbf{x}(t)\right) = S_t + S_\mathbf{x}\dot{\mathbf{x}}$, we obtain the desired

$$S_t = L - S_\mathbf{x}\dot{\mathbf{x}} = L - L_{\dot{\mathbf{x}}}\dot{\mathbf{x}}. \tag{30}$$

∎

Remark 1.1 *Theorem 1.1 leads automatically to the* Legendre transform *as follows: wishing to trace the evolution of the normal to the front, we use, instead of the ray velocity* $\mathbf{v} = \langle 1, \dot{\mathbf{x}} \rangle$, *the normal to the front* $\nabla S = \langle S_t, S_\mathbf{x} \rangle \equiv$ *(by the definition)* $\langle -H, \mathbf{p} \rangle$, *just as we did in the discussion in section 1.3. Choosing* $\mathbf{p}$ *as the independent variable instead of* $\dot{\mathbf{x}}$ *leads to the Legendre transform:*

$$\mathbf{p} = L_{\dot{\mathbf{x}}} \quad and \quad H = \dot{\mathbf{x}}L_{\dot{\mathbf{x}}} - L, \tag{31}$$

where $\mathbf{p}$ *is now a new phase variable and* $H = H(\mathbf{x}, \mathbf{p}, t)$ *is the t-coordinate of* ∇S. *It is now most natural to ask, What is the equation satisfied by the new phase variables* $\mathbf{x}, \mathbf{p}$? *We observe that the motions are critical points of the functional* $\int dS = \int \mathbf{p}d\mathbf{x} - Hdt$ *and thus satisfy the variational equations* $\dot{\mathbf{x}} = H_\mathbf{p}$, $\dot{\mathbf{p}} = -H_\mathbf{x}$.

Remark 1.2 *By rotating Figure 4 by* 90°, *that is, by switching the roles of* H, t *with those of* p, x *(we assume throughout this remark that* x *is scalar), we observe at once that the system is Hamiltonian with* $(H, t) =$ *(position, momentum) and* $(x, p) =$ *(time, Hamiltonian).*

Here is another corollary of Theorem 1.1.

Remark 1.3 *S satisfies the equation*

$$S_t + H(\mathbf{x}, S_\mathbf{x}) = 0, \tag{32}$$

called the Hamilton–Jacobi equation.

Proof. By Theorem 1.1, $S_t(t, \mathbf{x}) = L - \dot{\mathbf{x}}L_{\dot{\mathbf{x}}} = -H\left(\mathbf{x},\ p(\dot{\mathbf{x}})\right) = -H(\mathbf{x}, S_\mathbf{x})$. ∎

2   Geometrical Basics of Optimal Control

Optimal control theory deals with finding the "best" way to drive a system from one state to another. Some examples: how to launch a rocket into an orbit with a minimal expenditure of fuel or in the shortest possible time; how to reorient a satellite with the least amount of fuel or in shortest time, etc. We will look at problems modelled by systems of ODEs

$$\dot{\mathbf{x}} = \mathbf{f}(\mathbf{x}, u), \tag{33}$$

where $u \in U$ is the control parameter whose time dependence $u = u(t)$ is to be determined subject to two conditions: (1) there must exist a solution connecting the initial state $\mathbf{x}_0$ to the final state $\mathbf{x}_1$, and (2) this solution must minimize the given cost functional

$$F(u(\cdot), \ \mathbf{x}_0, \mathbf{x}_1) = \int_{t_0}^{t_1} f_0(\mathbf{x}(\tau), u(\tau)) d\tau. \tag{34}$$

The case of $f_0 \equiv 1$ amounts to the problem of minimizing the time.

Our goal is to obtain a sufficient condition for the control function $u(\cdot)$ to be optimal in the sense specified above. We begin with a heuristic discussion.

2.1   Heuristics of the Maximum Principle

Consider first the time-optimization problem $f_0 \equiv 1$. We already saw in section 1.1 that nature "knows" how to solve the time-minimization problem by sending light along the path of shortest time; to minimize the time, each ray follows, according to Huygens, the following simple recipe.

The Maximum Principle *"To arrive in the shortest time, always choose the direction to maximize the velocity's normal component to the front."* Imitation of this recipe for the nonoptical case (33) leads to the maximum principle. To that end, we interpret the problem (33) optically as follows.

Imagine an anisotropic optical medium where the set of possible velocities of rays at $\mathbf{x}$ is given by $\mathbf{f}(\mathbf{x}, u)$ and where u is thought of as parametrizing the ray direction[5]. We thus have a physical interpretation of (33).

---

[5]Recall that Huygens's principle can be formulated precisely in such a way that the

With this interpretation, the time-minimizing control u is singled out via the above maximum principle. To express it explicitly, we introduce the reachable set R_T consisting of points in phase space that can be reached from $\mathbf{x}_0$ in time $\leq T$ with all possible control functions $u(t)$. In other words, the set R_T is "painted" by all possible points $\mathbf{x}(T)$, where $\mathbf{x}$ satisfy (33) with all possible (say, summable) $u(t)$ and with $\mathbf{x}(0) = \mathbf{x}_0$. Now the maximum principle states:

For a solution to $\dot{\mathbf{x}} = \mathbf{f}(\mathbf{x}, u(t))$ to minimize the time from $\mathbf{x}_1$ to $\mathbf{x}_1$, it must lie on the boundary of R_t for all t, and to that end must maximize its velocity normal to that boundary.

Translating this into a computationally useful formula is a simple matter. Assume that the boundaries of R_t are smooth surfaces foliating the space, and let $\mathbf{N}(\mathbf{x})$ be a normal at $\mathbf{x}$ to the surface through $\mathbf{x}$. The maximum principle characterizes the optimal choice u via

$$\mathbf{f}(\mathbf{x}, u) \cdot \mathbf{N}(\mathbf{x}) = \sup_{v \in U} \mathbf{f}(\mathbf{x}, v) \cdot \mathbf{N}(x). \tag{35}$$

Rewriting (35) by expressing $\mathbf{N}(x) = \nabla T(\mathbf{x})$, where $T(\mathbf{x})$ is the minimal travel time from $\mathbf{x}_0$ to $\mathbf{x}$, we get the equation for the best u, known as *Bellman's equation*:

$$\mathbf{f}(\mathbf{x}, u) \cdot \nabla T(\mathbf{x}) = \sup_{v \in U} \mathbf{f}(\mathbf{x}, v) \cdot \nabla T(). \tag{36}$$

This equation suffers from several drawbacks, the main one being the difficulty of finding the function T. In fact, finding it requires knowledge of the optimal trajectory. Furthermore, $T(\mathbf{x})$ fails to be differentiable in many physically interesting examples (see below). We circumvent these difficulties by following the evolution of the ray and the normal to the front along the ray, just as we did in sections 1.3 and 1.6.

A key observation. Any solution $\eta(t)$ to the adjoint linearized system

$$\dot{\eta} = -\mathbf{f}_{\mathbf{x}}^T \left(\mathbf{x}(t, u(t)) \right) \eta \tag{37}$$

along $\mathbf{x}(t, u(t))$ such that $\eta(t_0)$ is normal to the front at $t = t_0$, remains normal to the front at all times t (as long as the front is smooth).

---

disturbance along the ray travels in every direction and thus propagates according to $\dot{\mathbf{x}} = \mathbf{f}(\mathbf{x}, u)$; the actual propagation follows the envelope of infinitesimal indicatrices.

Proof. Let $\xi(t)$ be the solution of the linearized equation

$$\dot{\xi} = \mathbf{f_x}(\mathbf{x}(t, u(t)))\xi, \tag{38}$$

with $\xi(t^0)$ tangent to the front for some t^0. The solution $\xi(t)$ thus remains tangent for all t, since any tangent to the front is the limit of the scaled difference between two nearby solutions; such a limit satisfies the linearized equations. Now, the solution $\eta(t)$ of the adjoint system (37) remains perpendicular to ξ for all t if we choose $\eta \perp \xi_0$ for some t.

We observe now that (33) and (37) form a Hamiltonian system with

$$H(\mathbf{x}, \mathbf{N}, u(t)) = \sup_{v \in U} H(\mathbf{x}, \eta, v), \tag{39}$$

where $H(\xi, \mathbf{N}, u) := \mathbf{f}(\mathbf{x}, u) \cdot \mathbf{N}$. This is the formulation of Pontryagin's maximum principle.

In our heuristic discussion we had made an implicit assumption that the front is smooth. In fact, in most applications it is not; consider, for instance, the free particle subject to a bounded control force u:

$$\ddot{x} = u, \quad |u| \le 1; \tag{40}$$

the front is shown in Figure 5.

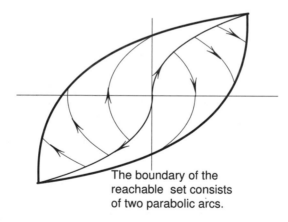

The boundary of the
reachable set consists
of two parabolic arcs.

Figure 5. Reachable set for the particle driven by a bounded force.

The maximum principle is still valid for nonsmooth reachable sets. The precise formulation is given by the following theorem.

Theorem 2.1 *Equation (39) above is a necessary condition for the control u(t) to be time-optimal.*

Example. Find the time-optimal control to drive the harmonic oscillator

$$\ddot{x} + x = u, \quad |u| \leq 1, \tag{41}$$

to the origin $x = \dot{x} = 0$ in the shortest time.

Solution (see Figure 6). The equivalent system is

$$\dot{x}_1 = x_2,$$
$$\dot{x}_2 = -x_1 + u, \tag{42}$$

together with the linearized adjoint system

$$\dot{\eta}_1 = \eta_2,$$
$$\dot{\eta}_2 = -\eta_1. \tag{43}$$

We maximize the Hamiltonian

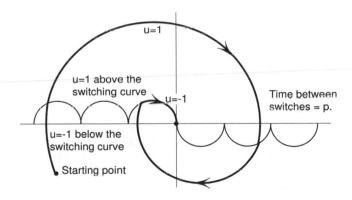

Figure 6. Stopping the pendulum in shortest time.

$$H(x, \eta, u) = x_2\eta_2 + (-x_1 + u)(-\eta_1) = -u\,\eta_1 + \cdots, \tag{44}$$

where $\cdots$ denotes terms independent of u. Since $|u| \leq 1$, the maximum is achieved by choosing $u = -\text{sgn}\,\eta$, where sgn is the signum function. Since

η_1 changes sign at times exactly π apart, so does u. Now this information is sufficient to analyze the problem completely. First, we note that the only trajectories leading into the origin are the two semicircles shown in figure. Therefore, for any trajectory to reach the origin it must reach one of these semicircles. In other words, if a trajectory reaches a semicircle, the u must be switched at that moment. Consequently, the previous switch must be π earlier, and thus control must be switched at the curves obtained by flowing the two semicircles backwards in time by π, as shown in Figure 6.

2.2   A Geometric Derivation of the Euler–Lagrange Equations

The familiar derivation, using integration by parts, of the Euler–Lagrange equations

$$\frac{d}{dt}L_{\dot{\mathbf{x}}} - L_{\mathbf{x}} = 0 \tag{45}$$

from the variational principle

$$S([\mathbf{x}]) = \int_{t_0}^{t_1} L(\mathbf{x}, \dot{\mathbf{x}})d\tau = 0, \quad \text{where} \quad \mathbf{x}(t_0) = \mathbf{x}_0, \; \mathbf{x}(t_1) = \mathbf{x}_1 \tag{46}$$

has the advantage of being very short, but it hides the geometry of the problem. In this section we give a geometrical derivation of (45) from (46) using the optical idea.

Let us turn (46) into a geometrical problem: equation (46) is equivalent to the problem of finding u such that

$$\begin{cases} \dot{s} = L(\mathbf{x}, u), \\ \dot{\mathbf{x}} = u, \end{cases} \quad \text{or} \quad \dot{\mathbf{X}} = \mathbf{F}(\mathbf{X}, u), \tag{47}$$

with the boundary conditions

$$s(t_0) = 0, \; \mathbf{x}(t_0) = \mathbf{x}_0, \; \mathbf{x}(t_1) = \mathbf{x}_1, \tag{48}$$

and most importantly, such that

$$s(t_1) \text{ is minimal.} \tag{49}$$

As an example, the boundary of the reachable set for (47) with $L = \frac{\dot{\mathbf{x}}^2}{2}$ is shown in Figure 7.

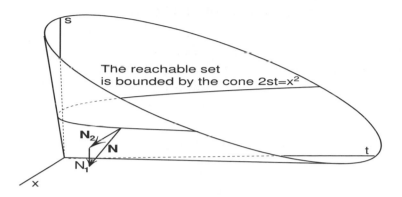

Figure 7. Boundary of the reachable set in the case of $L = \frac{\dot{x}^2}{2}$ is a cone $2st = x^2$.

The optimal solution to (47) must maximize the normal velocity to the front:

$$\frac{\partial}{\partial u} \mathbf{F} \cdot \mathbf{N} = \frac{\partial}{\partial u} \left(L(\mathbf{x}, u) N_1 + u \cdot \mathbf{N}_2 \right) = L_u N_1 + \mathbf{N}_2 = 0, \qquad (50)$$

where $\mathbf{N} = (N_1, \mathbf{N}_2)$ is a normal to the front at $\mathbf{X}$ (Figure 7). We recall that a solution to the adjoint system

$$\dot{\mathbf{N}} = -(D\mathbf{F})^T \mathbf{N}, \qquad (51)$$

produces such a normal. More explicitly, (51) is rewritten as

$$\begin{cases} \dot{N}_1 = 0, \\ \dot{\mathbf{N}}_2 = -L_{\mathbf{x}} N_1. \end{cases} \qquad (52)$$

Since the Hamiltonian in (50) is constant along solutions, we have

$$\frac{d}{dt} \left(L(\mathbf{x}, u) N_1 + u \cdot \mathbf{N}_2 \right) = 0. \qquad (53)$$

Using (52) gives

$$\frac{d}{dt} \left(L(x, u) \right) N_1 - L_{\mathbf{x}} N_1 = 0, \qquad (54)$$

which, upon cancelling $N_1 = \text{const.}$ and substituting $\dot{\mathbf{x}} = u$ becomes the Euler–Lagrange equation!

3   Geometric Phases in Mechanics

In this section we discuss some physical manifestations of the geometrical effect of parallel transport. Our goal is to give the student a flavor of a variety of applications, most of which can be observed in everyday life. Here is a partial list of examples where the geometric phases manifest themselves.

1. Reorienting a shopping cart or parking a car.

2. Propagation of polarized light through an optical fiber or of the polarization of a transversal wave.

3. A cat reorienting its body to land on its feet; a spaceship reorienting itself with rotors.

4. Amoebas' propulsion in viscous fluids.

5. Berry's quantum mechanical phase.

6. Extra rotation in Poincaré maps in periodic orbits.

7. Composition of noncommuting rotations.

8. Twisting a hose causes it to spiral.

9. The writhe of a curve.

10. Berry's particle in a hoop.

11. Hannay's angles.

This list can be continued. Rather than putting all these examples in a unified framework, we give separate treatment of some selected problems.

3.1   A Shopping Cart Problem

One late evening in a supermarket, waiting for my turn at the register, I was guiding a front wheel of my shopping cart around the square outline of the floor tile, about 1 by 1 foot square. It transpired that while the front wheel came back to its starting point, the rear wheel did not. By how much did the cart rotate after the front end executed a closed path? Figure 8 illustrates the situation. In that figure the cart has been replaced by an idealized bike, that is, by a directed segment RF, the velocity of whose rear end R is restricted to the direction RF (no skidding).

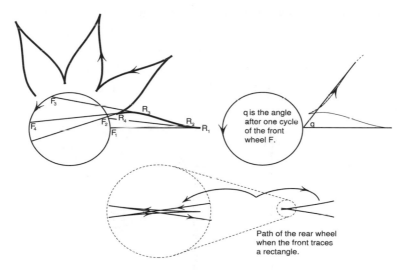

Figure 8. The bike rotates by $\theta \approx \frac{A}{L^2}$.

Theorem 3.1 *If the front end F of the bike RF traces out a curve enclosing a domain D of area A, then the bike rotates through the angle $\theta = A/l^2 + O(A^2)$, where l is the length of RF. Here we assume that D is not "too stretched", that is, that the ratio of the area to the length is bounded from infinity and from zero.*

Proof. Let (x, y) be the position of F, and let θ be the angle between the positive x-axis and RF. The triple (x, y, θ) specifies the position of the bike. The no-skidding constraint is given by

$$l\dot{\theta} + \dot{x} \sin \theta + \dot{y} \cos \theta = 0. \tag{55}$$

Geometrically, (55) amounts to the restriction of velocity vectors in $\mathbf{R}^3 = \{(x, y, \theta)\}$ to the planes

$$\omega = l\,d\theta + dx \sin \theta - dy \cos \theta = 0 \tag{56}$$

in the tangent space at (x, y, θ). Now the distribution of these planes is nonintegrable; that is, there exists no surface all of whose tangent planes belong to the distribution (56). Indeed, $\omega \wedge d\omega \neq 0$ by a simple calculation, which implies nonintegrability by Frobenius's theorem.

To compute the change $\Delta\theta$ in θ as the front end F traverses its cycle, we consider the lift $ABA'A$ of this cycle in the phase space (Figure 9). More explicitly, the curve ABA' is such that its projection onto the xy-plane is the path traced out by F while the tangents to ABA' lie in the planes of the distribution and $A'A$ is the vertical segment (Figure 9).

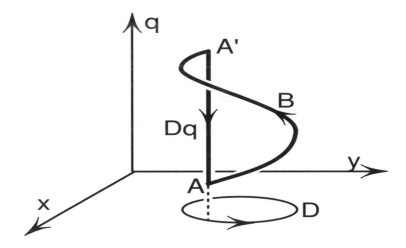

Figure 9. Computing the holonomy of the bike.

Integrating over a surface S spanning the contour $ABA'A$, we obtain

$$\oint_{ABA'A} \omega = \int_S d\omega, \tag{57}$$

or in more detail:

$$\oint_{ABA'A} l\,d\theta + \sin\theta\,dx - \cos\theta\,dy = \int_{ABA'} + \int_{A'A} l\,d\theta + \sin\theta\,dx - \cos\theta\,dy$$

$$\equiv 0 - l\Delta\theta = \int\int_S \cos\theta\,dx \wedge d\theta + \sin\theta\,d\theta \wedge dy, \quad \text{by Stokes.} \tag{58}$$

Now if the region D is small, then the tangent planes to S are close to the planes $\omega = 0$ when we are not too close to the line AA', so that, apart from this small neighborhood $l\,d\theta = -\sin\theta\,dx + \cos\theta\,dy + \text{small}$; using this in (58), we obtain

$$\Delta\theta = l^{-2} \int\int \left(\cos^2\theta + \sin^2\theta\right) dx \wedge dy + \text{small} = \frac{A}{l^2} + \text{small}. \qquad (59)$$

■

3.2   The Holonomy of a Bike Wheel

You are holding a bicycle wheel by its axis. The wheel is at rest and there is no friction in the bearings, so that no rotation around the axis can be imparted. Still, can the wheel be turned around its axis? The wheel is perfectly balanced, so the gravity is irrelevant. The answer to this question turns out to be equivalent to the proof of the Gauss–Bonnet formula relating the Gaussian and the geodesic curvatures!

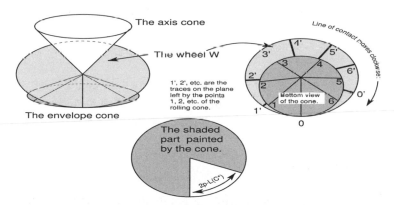

Figure 10. If the wheel's axis sweeps a solid angle A, the wheel ends up rotated through the angle A.

Assume for simplicity that the geometrical center of the wheel is fixed in space. Let the wheel's axis sweep a closed cone C, not necessarily circular (Figure 10). In the process, the plane of the wheel remains tangent to another cone C^*, which is the envelope of the planes normal to the generators of C. Such a cone will be called dual. Let c and c^* be the curves given by C and C^* on the unit sphere centered at the common vertex of the two cones.

Theorem 3.2 *If the wheel's axis sweeps out a simple closed curve c on the unit sphere, then the wheel rotates through the angle*

$$\theta = 2\pi - L(c^*), \tag{60}$$

where $L(c^*)$ is the length of dual curve c^*.

Proof. During the motion, the wheel's plane W rolls without sliding along the dual cone C^*. Indeed: (rolling) at all times W is tangent to C^* by the definition of C^*, and (no sliding) the spoke of W that is in contact with C^* has zero velocity, so that there is no sliding. ∎

Theorem 3.3 *If the axis of the wheel sweeps out a closed path bounding the solid angle A, then the wheel ends up rotated through the angle*

$$\theta = A. \tag{61}$$

Proof. The proof follows from Theorem 3.4 on dual cones given in the next subsection. This theorem, applied to the situation at hand, states that $A(c) + L(c^*) = 2\pi$, which implies (61) via (60).

3.3   Dual Cones

In this section we exploit the fruitful idea of duality suggested by the wheel with a moving axis. We restate the definition of duality. As the wheel's axis sweeps out a cone C, the wheel's plane[6] remains tangent to the cone C^*, which we will call the dual of C. More precisely, we treat the wheel's axis as a projective line, since there is no natural assignment of direction.

Definition 3.1 *Given a projective cone C, we define its dual C^* as the envelope of normal planes to the generators of C.*

---

[6]The geometric center of the wheel is fixed, as before.

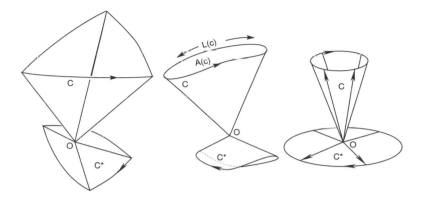

Figure 11. Examples of dual cones.

Remark 3.1 *In the case of a convex cone, we choose one of its "halves,"*
which we will still call C, with the corresponding half of the dual cone chosen
so that it is separated from C by a plane (Figure 11).

From now on we assume that the cone C is convex (and thus, so is C^*).

Problem 3.1 *Prove that duality is reflexive:* $(C^*)^* = C$.

Let $A(c)$ be the area bounded by the curve $c = C \cap S^2$ on the unit
sphere, and let $L(c)$ be the length of c. We now state the main theorem.

Theorem 3.4 *(Lemma on dual cones).*

$$A(c) + L(c^*) = 2\pi. \qquad (62)$$

We next state the infinitesimal version of this theorem, and then prove
both versions. Let $k(s)$ be the geodesic curvature of c at $c(s)$, with s being
an arclength parameter on c, and let s^* be the arclength parameter on c^*.
The two values s and s^* correspond to each other if the generators $C(s) \perp$
$C^*(s^*)$. In the case of strictly convex smooth cones this correspondence is
a diffeomorphism.

Theorem 3.5 *The geodesic curvature k_c of a spherical curve $c(s)$ is given by the length of the dual curve:*

$$k_c(s) = \frac{ds^*}{ds},\tag{63}$$

where s^* and s correspond to each other as described in the preceding paragraph.

Proof of Theorem 3.4 on dual cones from the infinitesimal version (Theorem 3.5). Assume that (63) holds. Let us deform the curve c into a point, through a one-parameter family c_μ of convex curves. The dual curve undergoes a corresponding deformation through the family c_μ^* into a great circle. Wishing to prove that $A(c_\mu) + L(c_\mu^*) = $ const., we differentiate the left-hand side by μ. Note that

$$\frac{d}{d\mu}A(c_\mu) = \int v\,ds, \quad \frac{d}{d\mu}L(c_\mu) = \int kv\,ds,\tag{64}$$

where v is the normal velocity of the curve with respect to the parameter μ and where k is the geodesic curvature of c. Leaving the proof of this intuitively rather plausible statement to the reader, we obtain

$$\partial_\mu\left(A(c_\mu) + L(c_\mu^*)\right) = \int_c v\,ds + \int_{c^*} k^*v^*\,ds^*,\tag{65}$$

where * denotes the quantities related to c^*. We note that $v(s) = -v^*(s^*)$, as is clear from the duality of c and c^*. By (63) $k^*ds^* = ds$, which proves that the expression on the left in (62) is constant. By shrinking c to a point we identify this constant as 2π, thus completing the proof of Theorem 3.4 modulo (63).

Proof of the infinitesimal version (Theorem 3.5). Equation (63) is a restatement of the definition of the geodesic curvature: Indeed, $c^*(s^*)$ is the binormal vector[7] to the curve $c(s)$, and thus

$$k_c = |\partial_s c^*(s^*)| = |\partial_{s^*} c^*(s^*)|ds^*/ds = ds^*/ds.\tag{66}$$

■

---

[7] We recall that $s^* = s^*(s)$ is the function of s.

Corollary 3.1 *The geodesic curvatures of dual curves are reciprocal:*

$$k_c k_{c^*} = 1. \tag{67}$$

Proof of the corollary. Duality is reflexive, so that $k_{c^*} = \frac{ds}{ds^*}$, which together with (63) implies (67).

3.4 Proof of the Gauss–Bonnet Formula Using Dual Cones

The Gauss–Bonnet formula is the generalization of the fact that the tangent vector to a simple planar closed curve makes a turn of $\int k \, ds = 2\pi$ during one cycle around the curve, k being the curvature. For a curve γ on a *curved* surface, the turn of the tangent vector is measured similarly by $\int_\gamma k ds$, where k is the geodesic curvature[8], but this turn is no longer 2π. The Gauss–Bonnet formula states that the mismatch of this angle with 2π is precisely the integral of the Gaussian curvature over the domain enclosed by the curve (Figure 12).

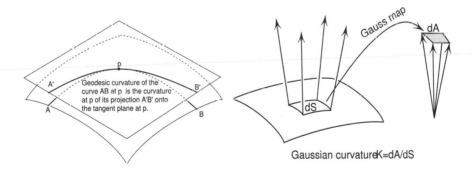

Figure 12. Definition of the geodesic and of Gaussian curvatures.

Theorem 3.6 *(The Gauss–Bonnet formula).* *Let D be a two-dimensional disc with a smooth boundary curve γ, embedded in $\mathbf{R}^3$. We also assume D*

---

[8]indeed, k measures, by its definition, the angular velocity of the tangent vector within the tangent plane.

to be convex[9]. Then

$$\int_\gamma k\,dt + \int_D K\,dS = 2\pi,\tag{68}$$

where k is the geodesic curvature of γ, t is the arclength parameter along γ, and K is the Gaussian curvature of D.

Proof. The proof consists in the observation that Gauss–Bonnet formula (68) is the preimage, under the Gauss map, of the dual cones theorem $L(c^*) + A(c) = 2\pi$. In other words, (68) is the same, term-by-term, as the dual cones theorem $L(c^*) + A(c) = 2\pi$, where c is the image of γ under the Gauss map. For the proof, we map D via the Gauss map G onto the unit sphere; that is, we drag each unit outward normal vector to D, to a common point[10]. The curve γ maps onto its Gauss image $c = G(\gamma)$ on the unit sphere (Figure 13). Let c^* be the dual curve to c.

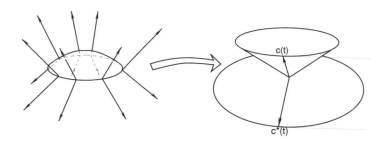

Figure 13. Gauss map of D.

By the definition of Gaussian curvature, $A(c) = \int_D K\,dS$, so that the second terms in (68) and (62) are equal, and it remains to prove the equality of the first terms:

$$\int_\gamma k\,dt = L(c^*).\tag{69}$$

Consider to that end the angle $\alpha(t) = \angle(T(t), c_t^*)$, where c_t^* is computed[11] at the point $\gamma(t)$.

---

[9]Not out of necessity, but to minimize technicalities.

[10]Suggesting a new name "the mouse map," for the Gauss map.

[11]To interpret c_t^* physically, one can think of transporting a wheel along γ holding the wheel's axis normal to D. The wheel's angular velocity in the axial direction remains zero. Then c_t^* is *that spoke of the wheel which serves as an instantaneous axis of rotation.*

Differentiation gives $\dot{\alpha}(t) = k(t) - |\dot{c}_t^*|$, since the angular velocity of $T(t)$ within the tangent plane is $k(t)$ and since both $T(t)$ and c_t^* lie in the tangent plane at $\gamma(t)$. Integrating over t and using periodicity of α we obtain $0 = \int_\gamma k\,dt - \int |\dot{c}_t^*|\,dt \equiv \int_\gamma k\,dt - L(c^*)$, which proves (69) and thus completes the proof of the Gauss–Bonnet theorem.

Remark 3.2 *Note that $L(c^*) = \int ds^* = \int k(s)\,ds$ is the total geodesic curvature of c (Theorem 3.5). The total geodesic curvature of a closed curve is an invariant of the Gauss map, by (69). The same is true of the total Gaussian curvature by the definition of the latter.*

Thus our proof of the Gauss–Bonnet theorem can be restated as follows:

By the invariance of the total curvatures, both geodesic and the Gaussian, under the Gauss map G,

$$\int_D K\,dS + \int_\gamma k_\gamma(t)\,dt = \int_{G(D)} 1\,dS + \int_c k_c(s)\,ds$$

$$= A(c) + \int ds^* = A(c) + L(c^*) = 2\pi,$$

$$(70)$$

where $c = G(\gamma)$.

3.5 Duality of External and of Internal Billiards on the Sphere

There is a beautiful relationship between the internal and the external billiards (to be defined shortly) on the sphere: the two are dual via the duality we just used in the proof of the Gauss–Bonnet theorem.

We recall the definition of billiards on the sphere. Consider a sphere S with a curve $K \subset S$ bounding a spherical domain D. The (usual) billiard trajectories in D are defined as broken geodesics inside D with vertices on K and such that the incidence and reflection angles with $K = \partial D$ are equal. The *external* billiard trajectory on $D^c = S - D$ is defined as a broken geodesic with vertices in D^c with every segment tangent to K and bisected by the tangency point.

Theorem 3.7 *Let K be a geodesically convex curve on the unit sphere, and let K^* be its dual curve. Then any billiard trajectory in K is dual to an external billiard trajectory on K^* and vice versa.*

Proof. The proof is almost immediate from Figure 14. Let l_1 and l_2 be two consecutive segments of the billiard trajectory in K and let T be the (great circle) tangent to K at P. The dual P^* of P is a great circle, while the duals l_i^* are points. Moreover, $l_i^* \in P^*$ since $P \in l_i$. Now $\alpha = \beta$ (see Figure 14) since the angle between the planes of l_1 and T equals the angle between their normals Ol_1^* and OT^*. Thus l_1^* and l_2^* are equidistant from T^* (by α), proving that the external billiard is dual to the internal one. The proof of the reverse is left as an exercise. ∎

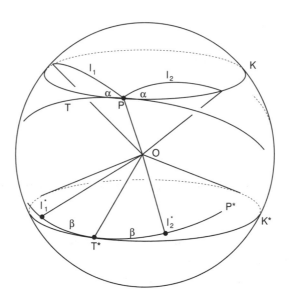

Figure 14. Duality of internal and external spherical billiards.

Remark 3.3 *Note that simple closed period n orbits of internal billiards on the sphere maximize the inscribed n-gon's perimeter. By Theorem 3.4, any simple closed n-orbit of the dual billiard on the sphere minimizes the area of circumscribed n-gons.*

3.6 Geometric Phases in the Motion of Free Rigid Bodies

3.6.1 The Energy and Momentum.

First we recall that the angular velocity vector $\boldsymbol{\Omega}$ of a rigid body is, by definition, the vector pointing along the instantaneous axis of rotation and whose magnitude is the instantaneous angular speed of rotation. In the absence of all external forces the center of mass of the body moves with zero acceleration, and we place our inertial frame at that center. Thus the only nontrivial dynamics is rotational. Our goal is to

describe the apparent motion of the angular momentum **m**
viewed from the reference frame of the body.

Since the angular momentum is constant in space, this almost determines the motion of the body relative to the inertial frame.

Choosing a frame attached to the body, we let $\mathbf{M} = (M_1, M_2, M_3)$ be the angular momentum vector expressed in this frame; in other words, the components M_i are the projections of **m** onto the axes of the body frame (Figure 15).

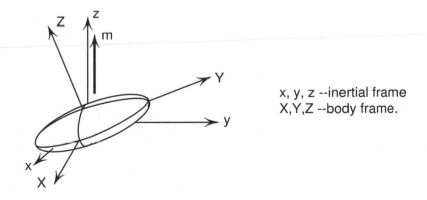

x, y, z --inertial frame
X,Y,Z --body frame.

Figure 15. Space frame and the body frame.

Let $\boldsymbol{\Omega}$ be the angular velocity of the rigid body expressed in the body frame[12]. The relationship between $\mathbf{M}$ and $\boldsymbol{\Omega}$ is linear:

$$\mathbf{M} = I\boldsymbol{\Omega}, \tag{71}$$

via a symmetric matrix I, called the *tensor of inertia*. Indeed, an infinitesimal element of the body of mass dm positioned at $\mathbf{r}$ has the angular momentum

$$d\mathbf{M} = dm(\mathbf{r})\mathbf{v} \times \mathbf{r} = dm(\mathbf{r})(\omega \times \mathbf{r}) \times \mathbf{r} \equiv dI(\mathbf{r})\omega; \tag{72}$$

integration in $\mathbf{r}$ gives (71).

Since the axes of the body frame can be chosen along the eigendirections of I, we can assume that $I = \mathrm{diag}(I_1, I_2, I_3)$.

The kinetic energy E is given by the dot product $\frac{1}{2}\langle \mathbf{M}, \boldsymbol{\Omega} \rangle$, and it is a conserved quantity of the motion:

$$\langle I\boldsymbol{\Omega}, \boldsymbol{\Omega} \rangle = \langle \mathbf{M}, I^{-1}\mathbf{M} \rangle = \frac{M_1^2}{I_1} + \frac{M_2^2}{I_2} + \frac{M_3^2}{I_3} = 2E = \text{const.} \tag{73}$$

3.6.2   Phase Portrait on the Momentum Sphere

Since $|\mathbf{M}| = |\mathbf{m}| = \text{const.}$, the vector $\mathbf{M}$ moves on a sphere $|\mathbf{M}| = \mu = \text{const.}$ Moreover, by the conservation of energy, $\{\mathbf{M}(t)\}$ is confined to an ellipsoid (73). Consequently, any trajectory $\mathbf{M}(t)$ lies on an energy curve on the momentum sphere. We thus have a simple phase portrait (Figure 16).

3.6.3   A Tennis Racket Puzzle

Consider the following experiment: Toss up a tennis racket and catch it by the handle after it makes one "handle over the head" tumble. When tossing, release the racket when the net's plane is horizontal. You may notice that not only does the long axis of the racket make a full turn in flight, but in addition, the racket turns around its *long* axis during its flight. Now, the surprising observation is this: in most tosses that angle of turn around the long axis is a multiple of 180°! This manifests itself in the fact that the racket is usually caught with the plane of its head horizontal, despite the

---

[12]$\boldsymbol{\Omega}$ should not be confused with the angular velocity *with respect to* the body frame, which is, of course, zero.

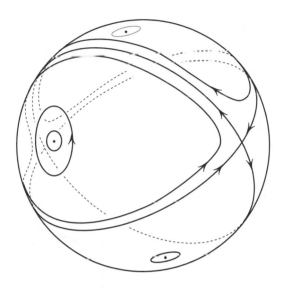

Figure 16. Phase flow on the momentum sphere.

fact that during the flight the racket is seen rotating around the long axis. (When I had noticed this effect, only 1 out of 20 tosses didn't conform to the rule.)

Here is the explanation of this phenomenon. The racket is launched so that $\mathbf{M}(0)$ is close to the intermediate axis of inertia, and thus $\mathbf{M}(0)$ *is close to the saddle point on the momentum sphere.* From Figure 16 it is clear that $\mathbf{M}(t)$ spends most of its time near the two saddles, with quick transitions between the saddles. Since we are watching from the inertial frame, we see not $\mathbf{M}$ but the racket itself making the 180° flip.

3.6.4 Poinsot's Description

Definition 3.2 *The* energy ellipsoid *of the rigid body is the ellipsoid of all angular velocity vectors* $\mathbf{\Omega}$ *for which the kinetic energy* $\frac{1}{2}\langle I\Omega, \Omega\rangle = E$.

We note that the energy ellipsoid is rigidly attached to the body. The energy ellipsoid imitates the shape of the body. Indeed, the largest moment of inertia I_3 in a rigid body is around the body's shortest dimension; the semiaxis $\sqrt{2E/I_i}$ is also shortest for the largest I_i (we assume $I_1 < I_2 < I_3$). The following theorem presents the beautiful result of Poinsot.

Theorem 3.8 *(Poinsot). The free rigid body in the inertial frame of its center of mass moves so that its energy ellipsoid rolls without sliding on the plane P fixed in the inertial frame of the center of mass, given in the inertial frame by $\{\mathbf{x} : \langle \mathbf{m}, \mathbf{x} \rangle\} = 2E$ (Figure 17).*

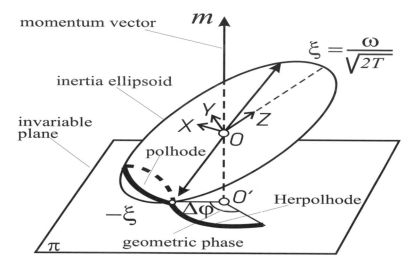

Figure 17. Poinsot description of the motion of the free rigid body.

This theorem gives a virtually complete description of the motion.

Proof of Poinsot's theorem.

1. The body angular velocity $\boldsymbol{\Omega}$ lies in the invariable plane. Proof: The equation of that plane in the *body* frame is $\{\mathbf{X} : \langle \mathbf{M}, \mathbf{X} \rangle = 2E\}$. The energy relation $\langle \mathbf{M}, \boldsymbol{\Omega} \rangle = 2E$ proves that $\boldsymbol{\Omega}$ lies in the invariable plane, that is, that the ellipsoid has a nonempty intersection (at $\boldsymbol{\Omega}$) with the invariable plane. In fact, it is a tangency point.

2. The plane is tangent to the ellipsoid at $\boldsymbol{\Omega}$. Proof: The normal $\mathbf{M}$ to

the plane (we are working in the body frame) has the same direction as the normal $\nabla\langle I\Omega, \Omega\rangle$ to the ellipsoid: $\nabla\langle I\Omega, \Omega\rangle = 2I\Omega = 2\mathbf{M}$.

3. There is no sliding between the ellipsoid and the plane. Proof: The instantaneous velocity of any point of the body on the line of the angular velocity vector is zero.

■

3.7 Euler's Equations of Motion: An Instantaneous Derivation

The equations of motion of the rigid body in the body frame are

$$\dot{\mathbf{M}} = \mathbf{M} \times (I^{-1}\mathbf{M}); \tag{74}$$

here is a one-paragraph proof. Since the body's angular velocity relative to the inertial frame is Ω, the inertial frame has the angular velocity $-\Omega$ relative to the body frame; therefore, any point $\mathbf{x}$ stationary in the inertial frame appears to have the velocity

$$\dot{\mathbf{X}} = -\Omega \times \mathbf{X} \tag{75}$$

in the body frame, where $\mathbf{X} = (X_1, X_2, X_3)$ is the point's body position. The angular momentum is just such a vector, fixed in the inertial frame! It thus satisfies $\dot{\mathbf{M}} = -\Omega \times \mathbf{M}$, which is equivalent to (74).

Problem 3.2 *Show that the flow of (74) conserves the energy and momentum and that it is area-preserving.*

Problem 3.3 *Write the equations of motion for M in the Hamiltonian form $\dot{\mathbf{M}} = J_{\mathbf{M}}\nabla H(\mathbf{M})$, where J_M is a skew-symmetric operator on the tangent space to the momentum sphere.*

Solution: Let $H = \frac{1}{2}\langle \mathbf{M}, I^{-1}\mathbf{M}\rangle = $ the energy, and let $J_{\mathbf{M}}\mathbf{v} = \mathbf{M} \times \frac{\mathbf{v}}{|\mathbf{v}|}$, so that $J_{\mathbf{M}}$ rotates by $\frac{\pi}{2}$ in the tangent plane to the sphere at $\mathbf{M}$. This description is completely analogous to the familiar Hamiltonian flow $\dot{\mathbf{z}} = JH_{\mathbf{z}}(\mathbf{z})$ in the plane!

3.7.1   Poinsot Description, Parallel Transport and the Gauss–Bonnet Theorem

According to the Poinsot description, the free rigid body finds itself rotated through a certain angle θ around $\mathbf{m}$ every time the polhode rolls out precisely once on the invariable plane. What is θ?

The answer is particularly easy to see in the body frame. Posed in the body frame, our problem is *to find the angle θ through which an xy-frame lying in the invariable plane turns with respect to the rigid body during one period T.*

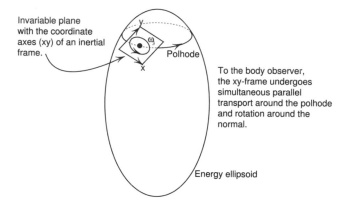

Figure 18. Parallel transport on the energy ellipsoid along the polhode.

Here is a quick solution. As the body executes the Poinsot motion, our xy-frame is transported along the polhode (Figure 18), while rotating with the angular velocity ω_3. Had ω_3 been zero, we would have had parallel transport of the frame along the polhode, and the holonomy angle would have been $\mp \int K \, dS$ (with the sign depending on one's orientation), where K is the Gaussian curvature of the energy ellipsoid. The effect of ω_3 is simply additive, so that (note that $\omega_3 = $ const.)

$$\theta = \mp \int K + \omega_3 T. \tag{76}$$

It is not hard to see that the sign is $+$ when $E < E_2$, where $E_2 = \frac{\mu^2}{2I_2}$ is the energy of the unstable rotation around the intermediate axis.

Lemma 3.1 *The phase orbits on the momentum sphere (Figure 16) are (scaled) images of the polhode on the energy ellipsoid under the Gauss map.*

Proof. $\mathbf{M}$ is orthogonal to the energy ellipsoid at Ω:

$$\nabla \frac{1}{2} \langle I\Omega, \Omega \rangle = \mathbf{M}, \tag{77}$$

and $\mathbf{M}$ stays on a sphere, so that $\mathbf{M}$ is the image (up to scaling by $|\mathbf{M}|^{-1}$) of Ω under the Gauss map. ∎

We now reformulate (76) in terms of momenta, rather than the Gaussian curvature, by observing that $\int K \, dS = A$, where A is the solid angle subtended by the closed orbit of $\mathbf{M}$ on the momentum sphere. With this remark, (76) is equivalent to

$$\theta = \mp A + \omega_3 T. \tag{78}$$

An alternative derivation of (76). From the Poinsot description it is clear that θ equals the angle (modulo 2π) by which the tangent to the herpolhode turns as the point sweeps out one fundamental segment of the herpolhode. Denoting by γ one such segment, we have $\theta = \int_\gamma k_\gamma ds \mod 2\pi$, where k_γ is the curvature of the polhode. Now it is not hard to show that

$$k_\gamma(t) = k_\Gamma(t) + \frac{\omega_3}{v}, \tag{79}$$

where k_Γ is the geodesic curvature of the herpolhode, v is the speed of the point of contact and t is time. Multiplying by v and integrating by t, we obtain

$$\theta = \int k_\Gamma ds + \omega_3 T. \tag{80}$$

This is equivalent to the previous answer via the Gauss–Bonnet theorem. The $\pm$ sign in (76) is not seen in this formula because it is hidden in the sign of k_Γ. Of course, (80) is also immediate from our first derivation, by observing that the inertial xy-plane turns, from the body observer's point of view, through the angle equal to $\int_\Gamma ds + \omega_3 T$.

3.8   Fluid in a Tube

Consider a fluid-filled pipe forming a closed curve K in the plane. Assume that the pipe executes a 2π turn around a fixed center O. We are to treat the fluid as a one-dimensional object; assuming the pipe to be of constant thickness, we have all particles going around the pipe with the same speed relative to the pipe. What position will the fluid particles occupy after one 2π turn, assuming that everything starts at rest?

The key to the answer is the observation that the circulation of the fluid is a conserved quantity:

$$C \equiv \int_K \mathbf{v} \cdot \mathbf{T} ds, \tag{81}$$

here $\mathbf{v} = \mathbf{v}(s, t)$ is the velocity of the particles in the *inertial frame*, while $\mathbf{T} = \mathbf{T}(s, t)$ is the unit tangent to the curve. Here s measures arclength from a reference point on the curve.

The conservation of circulation follows from the invariance of the Lagrangian of the system under the shift $s \to s + a$ of the arclength parameter[13]. Let u be the (tangential) speed of a water particle *relative to the tube*; u is the same for all particles, but may change in time if the tube's rotation rate changes. We have to find the distance $\int u dt$ travelled by each particle over the time of one 2π revolution of K.

We have

$$\mathbf{v} = u\mathbf{T} + \omega \mathbf{r}^\perp, \tag{82}$$

where $\mathbf{r}^\perp$ is the $\frac{\pi}{2}$-rotation of the position vector $\mathbf{r}$, and $\omega = \omega(t)$ is the angular velocity of the tube, with $\int \omega(t) dt = 2\pi$. Substituting (82) into (81) and using $C = 0$, we obtain

$$0 = \int \mathbf{v} \cdot \mathbf{T} ds = \int \left(u\mathbf{T} + \omega \mathbf{r}^\perp \right) \cdot \mathbf{T} ds. \tag{83}$$

---

[13]Indeed, pick a preferred particle, denoting its arclength coordinate by $s = s(t)$. The position of this particle determines the positions of all the rest of the particles since the fluid is incompressible. The position of an arbitrary particle is $R_\alpha \mathbf{r}(s(t) + \sigma)$, where R_α is the α-rotation around O, $\mathbf{r}(\cdot)$ is the parametric representation of K in its initial position, and σ is the arclength distance from the particle in question to the reference particle. The Lagrangian of the system is then given by

$$L(s, \dot{s}) = \int_0^L \mathbf{v}(s + \sigma, t)^2 d\sigma,$$

where $\mathbf{v}$ is the velocity: $\mathbf{v}(s + \sigma, t) = \frac{d}{dt} R_{\alpha(t)} \mathbf{r}(s(t) + \sigma) = \dot{\alpha} \mathbf{r} + \dot{s} \mathbf{r}'$. This Lagrangian is invariant under $s \to s + a$ by periodicity of the integrand. The conserved quantity is $L_{\dot{s}} = 2 \int_0^L \langle \dot{\alpha} \mathbf{r} + \dot{s} \mathbf{r}', \mathbf{r}' \rangle d\sigma = 2 \int_0^L \langle \mathbf{v}, \mathbf{T} \rangle d\sigma = 2C$, as claimed.

Since $\int \mathbf{r}^\perp \cdot \mathbf{T}ds = 2A$, with A being the area enclosed by K, and since $u = u(t)$ is independent of s, we obtain

$$u - -2\omega\frac{A}{L},\qquad(84)$$

where L is the length of the curve K.

Integrating over T and using $\int_0^T \omega\,dt = 2\pi$ we obtain the displacement of each particle along K:

$$\Delta s = -4\pi\frac{A}{L}.\qquad(85)$$

In terms of the angular coordinate $\theta = 2\pi s/L$ on K, the displacement is given by $\Delta\theta = 2\pi\Delta s/L$ or

$$\Delta\theta = -8\pi^2\frac{A}{L^2}.\qquad(86)$$

Problem 3.4 *Consider the interior of a simple closed curve K filled with a two-dimensional irrotational incompressible fluid, initially at rest. Describe the result of rotating K around a fixed center O through the angle 2π.*

3.9   A Spherimeter

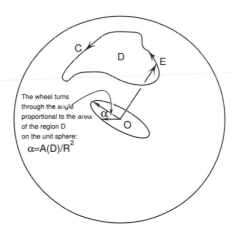

Figure 19. A spherimeter: the area of a country is given by the angle of the turn of the gyroscope.

The mechanical device in Figure 19 measures the area of a spherical domain D. The device is just a small wheel, that is, a gyroscope mounted at the end O of the axis OE with the plane of the wheel perpendicular to OE. The wheel encounters no rotational friction. The end O of the axis is held fast at the origin, while the other end E of the wheel's axis can be moved freely along the sphere. To measure the area $A(D)$ of a country D, that is, of a spherical domain, we place E at some point on the boundary C of D; the wheel must be at rest. We then guide E along C, bringing it back to the original point, and record the angle α by which the wheel has turned as the result of this circumnavigation. The area of D is given by

$$A(D) = \alpha R^2,$$

R being the radius of the sphere. We leave the proof as an exercise.

3.10   Optical Fibers

Consider an object in the plane and its image transmitted through an optical fiber to a parallel plane, as in Figure 20.

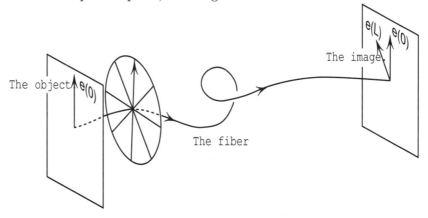

Figure 20. The image is transmitted by parallel, or inertial, transport.

The angle α between the object and its image is given by the solid angle A formed by the cone of unit tangent vectors to the fiber (considered as a one-dimensional curve), all carried to a common origin. If a polarized beam is sent through the fiber, the polarization is carried along the fiber by the same parallel transport rule and thus undergoes the same rotation.

The explanation of this effect lies in the fact that *the image is transmitted by parallel transport*: a moving plane perpendicular to the curve carries the image so that the latter does not rotate about the tangential direction. In other words, the image is transported precisely as a spoke of a wheel whose axis is tangent to the curve (Figure 20). As we have seen before, the wheel turns through the angle described by its axis during the motion, which is precisely the solid angle formed by unit tangents. ∎

Example. We illustrate these remarks for a helical waveguide

$$\mathbf{r}(t) = (R\cos t,\ R\sin t,\ at).\tag{87}$$

The unit tangent vector

$$\mathbf{T} = \frac{1}{|\dot{\mathbf{r}}|}\dot{\mathbf{r}} = \frac{(-R\sin t,\ R\cos t,\ a)}{\sqrt{R^2+a^2}}\tag{88}$$

traces out a circle on the unit sphere. The area A of the enclosed spherical cap equals 2π times its height, which gives $A = 2\pi(1 - a/\sqrt{R^2+a^2})$. Thus

$$\theta = A = 2\pi\left(1 - \frac{1}{\sqrt{1+(\frac{R}{a})^2}}\right).\tag{89}$$

For small $\frac{R}{a} \approx 0$ the helix is close to a straight line and consequently θ is small, an intuitively obvious result. Fixing R and decreasing a towards zero, we make the spiral tight and thus $\theta = 2\pi\left(1 - \frac{1}{\sqrt{1+(\frac{R}{a})^2}}\right) \approx 2\pi$.

This shows a familiar effect: if you pull a spring straight, not allowing the ends of the wire to slip in your hold, then the straightened wire is twisted with a twist of 2π per each straightened coil.

Acknowledgments

It is a pleasure to thank Hildeberto Cabral for his hospitality and in particular for providing a stimulating environment that enabled me to write down these notes.

References

[1] Richard Feynman, *QED*. Princeton University Press, Princeton, NJ, 1985, p. 39.

[2] H. Jeffreys, On certain approximate solutions of linear differential equations of the second order. Proc. London Math. Soc., 23(1924), pp. 428–436.

[3] V. G. Boltyanskii, *Mathematical Methods of Optimal Control*. Holt, Rinehart and Winston, New York, 1971.

Momentum Maps and Geometric Phases

Jair Koiller (coordinator)
LABORATÓRIO NACIONAL DE COMPUTAÇÃO CIENTÍFICA
AV. GETÚLIO VARGAS 333
PETRÓPOLIS R.J 25651-070, BRAZIL
e-mail: jair@lncc.br

with collaborators

Earnest W. Coli (Planet Earth)
Richard Montgomery (U.C., Santa Cruz)
Joaquin Delgado Fernandez (UAM, Iztapalapa, Mexico)
Kurt Ehlers (TMCC, Nevada)
Teresinha J. Stüchi (UFRJ, Brazil)
Maria de Fátima L. B. P. Almeida (UERJ, Brazil)

Foreword

Momentum maps are one of the cornerstones of geometric mechanics. In these lectures, we present three related situations in which they are instrumental, summarized in Part 1 (an overview). In Part 2, we study adiabatic phases for integrable mechanical systems in a (slowly) moving frame. *The momentum map of the group is averaged over the invariant tori.* The simplest examples are Foucault's pendulum or any system with S^1 symmetry. A nontrivial example is the rotating elliptic billiard. In Part 3, we give an example of nonadiabatic phases, extending work by Montgomery and Levi on Euler's rigid body motion. We find the geometric phase around the angular momentum vector of a *gyrostat*. Here the momentum acts as a "pillar" around which the geometric phases are depicted. In Part 4, we make the restriction of zero momentum. Typically, the configuration space is a principal bundle with a connection; the base is called the "shape space." *A robot (or an organism) can control its shape and by so doing can navigate in the configuration space.* After one cycle of shape deformations, the new position differs from the original by an element of the Lie group. We assume the reader has had an undergraduate-level analytic mechanics class and knows the basic facts about symplectic forms, canonical transformations, and momentum maps of groups acting symplectically. Part 4 requires knowledge of principal bundles and connections (actually, it can be used to motivate that theory).

Our work, a companion to sections 2 and 3 of Mark Levi's lectures (Chapter 7 of this volume), is based on a set of four lectures given in January 1999 at the Mathematics Department, Universidade Federal de Pernambuco, Brazil. I would like to thank Hildeberto Cabral for his friendship and support for almost 35 years!

Misrepresentations are the coordinator's responsibility. To excuse ourselves, we quote Mark Twain: "Adults don't really lie, they just exaggerate" (*Tom Sawyer*) and Oscar Wilde "It is a terrible thing for a man to find out suddenly that all his life he has been speaking nothing but the truth." *(The importance of being Earnest.)*

Earnest W. Coli is an intelligent Escherichia coli which is the maskot for our collective work. See the site: http://web.bham.ac.uk/bcm4ght6/res.html, the E. Coli homepage.

The collaboration of the participants is as follows:
- Part 2: Classical Adiabatic Angles
  With Richard Montgomery, Mathematics Department, University of California, Santa Cruz 95064, USA.

- Part 3: Holonomy for Gyrostats
  With Maria de Fatima L. B. de Paiva Almeida, PUC-RJ, and Teresinha J. Stuchi, Instituto de Física da UFRJ, Brazil.

- Part 4: Microswimming
  With Kurt M. Ehlers, Truckee Meadows Community College, Reno, Nevada 89512, USA, and Joaquin Delgado Fernandez, UNAM-Iztapalapa, Mexico.

Part 1: Overview

1 Adiabatic Phases: Overview

Berry [4] and Hannay [10] showed, around 1985, that geometrical and topological effects are ubiquitous in classical and quantum mechanical systems subject to *adiabatic* variations of parameters. Since then, the theme "geometric phases in physics" has exploded[1].

Our examples in Part 2 will fit in the following framework[2]. Consider a configuration space S of dimension n and a natural mechanical system on TS with Lagrangian $L = T - V$. S (the "laboratory") is immersed on a bigger Riemannian manifold W (the "universe") of dimension N. W is acted upon by a Lie group G so that every map $i_g : S \to g \cdot S$ is an isometry. Using the metric, we have the identifications $TS \equiv T^*S$, $TW \equiv T^*W$, and the embedding $TS \to TW$ yields the corresponding *symplectic embedding* $T^*S \to T^*W$ (not to be confused with the *projection* $T^*W \to T^*S$ corresponding to the dual of $TS \hookrightarrow TW$).

Given a curve $t \to g(t) \in G$,

$$L(Q, \dot{Q}, t) = T\left(\frac{d}{dt}(g(t)Q) \right) - V(Q), \quad Q \in S, \tag{1}$$

is a moving Lagrangian system inside W. As is well known, Euler–Lagrange equations for the generalized coordinates Q contain "fictitious" forces, such as Coriolis and centripetal ([2, chapter 4]). It is more convenient for our purposes to work directly in the underlying inertial frame W. We set

$$q(Q, t) = g(t)Q \tag{2}$$

so the Lagrangian (1) becomes[3]

$$L(q, \dot{q}, t) = T(\dot{q}) - V(g(t)^{-1} \cdot q). \tag{3}$$

The Lie group G can be thought of as a parameter space.

---

[1] For a comprehensive introduction, see [24].

[2] See [12] for a souped up version of the theory.

[3] One of the nicest features of the formalism of calculus of variations is the invariance of Euler–Lagrange equations [2, section 12D]. Lagrangian (3), in which the time-dependence is transferred to the potential energy, is *mathematically* equivalent to Lagrangian (1) through the coordinate change (2). Notice that (3) *can also represent a physically distinct problem.*

The potential V contains three types of terms:

$$V = V_{\text{s.c.}} + V_{\text{int}} + V_{\text{ext}}.$$

$V_{\text{s.c.}}$ are *infinitely constraining* potentials (see, e.g., [23] or [13]) to S. This means that $g(t)^{-1}q \in S$ throughout the motion. Terms in V that are completely equivariant under G, $V_{\text{int}}(g^{-1} \cdot q) = V_{\text{int}}(q)$ form the *internal* or *interaction* potential. The remaining terms define the *external* potential.

Given a slow variation $g = g(\epsilon t)$, $g(0) = I$ describing a closed curve $C \subset G$, we want to compare the moving system (1) with a system evolving in the inertial frame ($g \equiv I$).

As a basic example, we consider a spherical pendulum orbiting a planet[4]. To simplify matters, we assume that our spaceship is the spherical pendulum itself. Taking the origin at the planet, the center of the sphere describes a curve $r(t) \in W = \mathbb{R}^3$. In the inertial frame W we have

$$H(p,q,t) = \frac{1}{2}|p|^2 - \frac{k}{|q|} + \mu(|q - r(t)| - 1)^2, \quad \mu \to \infty, \tag{4}$$

where k is the planet's attraction constant and units are such that the length of the rod and the mass of the particle are 1. The usual Foucault pendulum corresponds to $r(t)$, describing a parallel (constant latitude circle) of the planet. See Figure 1.

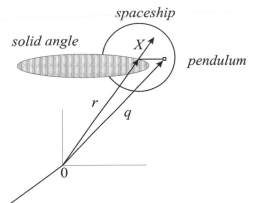

Figure 1. The "Star Wars" Foucault pendulum. Holonomy is given by a solid angle.

[4]This is a "Star Wars" Foucault pendulum.

We do not need to linearize gravity, nor to require that the pendulum is in the small oscillation regime. We just use the S^1 symmetry around the axis

$$x = \frac{r(t)}{|r(t)|}.$$

We consider any moving frame $R(t) \in SO(3)$ whose third axis is $x(t) = r(t)/|r(t)| \in S^2$. Here the change of coordinates (2) is

$$q = r + R(x)Q, \quad r = |r| \cdot x, \; R(x) e_3 = x = r/|r|. \tag{5}$$

Given any integrable system with S^1 symmetry around e_3, we use the angle θ around this axis as one of the generalized coordinates[5]. We do not care about the remaining generalized coordinates, because there will be no effect on them to first order on ϵ. Let the system evolve in an inertial frame $(r(t) \equiv r_o)$; the function

$$\theta = \theta_{dyn}(t) \tag{6}$$

is called the *dynamic phase*.

Now, let the spaceship go, as it actually does; notice that $r = r(\epsilon t)$ is slow compared to the frequencies of the system. We will show that when the spaceship makes a complete cycle, coordinate θ will have acquired an extra *geometric phase* (of order $1 = \epsilon^0$) given by

$$\Delta\theta_{\text{geometric}} = \text{solid angle subtended by } r(t). \tag{7}$$

Note that the parameter space $G = SE(3)$ reduces here to $X = \mathbb{R}^3 - 0 \equiv S^2 \times \mathbb{R}^*$. The curve $r(t)$ of the pendulum pivot can be retracted to its S^2 projection x, and the geometric phase is precisely the spherical angle. See Figure 1 again.

The dynamic phase is of order ϵ^{-1}, so the geometric phase usually can only be extracted [6] by interference with a twin system evolving in the inertial frame.

---

[5]Caveat: This does not necessarily imply that θ must be one of the uniformizing (i.e., part of an action-angle coordinate system) coordinates. In fact, for the spherical pendulum θ is not an angle coordinate.

[6]In Foucault's pendulum, which swings vertically, the dynamic phase vanishes identically, so the geometric phase effect can be seen directly!

2   Nonadiabatic Phases of Gyrostats

Recall Euler's rigid body motion with a fixed point. With the help of Lie groups, one can describe the motion as follows. The configuration space is the group $SO(3)$ of orthogonal matrices. Rigid body motions $R = R(t)$ are *geodesics* of a *left-invariant Riemannian metric* on $SO(3)$, since the Lagrangian consists of kinetic energy alone[7].

Due to Noether's theorem, it follows that the angular momentum (written on the inertial frame coordinates) is a conserved vector,

$$\mathbf{m}(t) \equiv \mathbf{m}. \tag{8}$$

Euler's reduced equations[8] describe the motion $\mathbf{M}(t)$ of the angular mo-

Momentum sphere $\|\mathbf{M}\| = J$

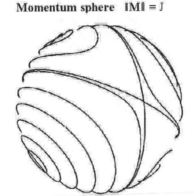

Figure 2. Euler's reduced system for the phase curves $\mathbf{M}(t)$.

mentum $\mathbf{m}$ *as viewed from the rotating body.* Following Arnold's notation [2], every capital vector means an object written in the body frame:

$$R(t)\mathbf{M}(t) = \mathbf{m}.$$

---

[7]*Left-invariance* is the mathematical wording of the fact that physical behavior does not depend on the choice of coordinate axis x, y, z for the inertial frame.

[8]*Reduction* goes back to Jacobi's "elimination of a node" and was cast in modern language by Marsden and Weinstein [17]. If the reduced system is integrable, the problem of finding the time-dependence of the "ignorable" coordinates was often dismissed on the argument that it can be done easily by quadratures. But the reconstruction is not always easy. A bit of geometry can be of great help, especially when the symmetry group is nonabelian. For a general theory for *reconstruction* of the complete solutions of Hamiltonian systems with group symmetry, see [18].

Differentiating, after some simple manipulations, one gets

$$\frac{d\mathbf{M}}{dt} = \mathbf{M} \times \nabla_M H(\mathbf{M}) \qquad (9)$$

with $H(\mathbf{M}) = \frac{1}{2}(\mathbf{M}, \mathbf{A}^{-1}\mathbf{M})$. The inertia matrix $\mathbf{A}$ is symmetric and positive definite. The reduced phase portrait on the sphere $\|\mathbf{M}\| = J$ is sketched in Figure 2.

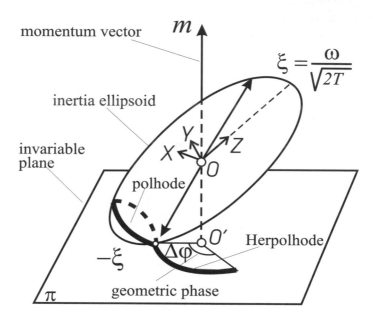

Figure 3. Poinsot's description of the rigid body motion. The geometric phase is $\Delta\phi$.

Reconstruction can be geometrically visualized in Poinsot's description: the inertia ellipsoid (carrying the body frame $R(t) : XYZ$) rolls without slipping over the invariable plane [2]. Figure 3 depicts a *polhode* in the ellipsoid and the corresponding *herpolhode*.

Recall that $R \in SO(3)$ is parametrized by the three Euler angles (see Figure 4)

$$\varphi = \angle(x, \text{nodes}), \quad \psi = \angle(\text{nodes}, X), \quad \sigma = \angle(Z, z). \qquad (10)$$

What is the geometric phase $\Delta\phi$ around $\vec{\mathbf{m}}$?

Using a bit of symplectic abstract nonsense, Montgomery [19] found

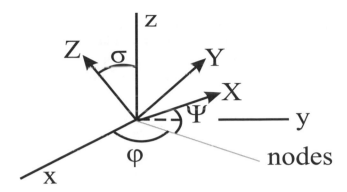

Figure 4. Euler angles. $\vec{\mathrm{m}}$ along the z-axis. Holonomy is measured using ϕ.

the following formula[9]:

$$\Delta\phi = \frac{2hT}{J} - \Upsilon. \tag{11}$$

Here h is the energy of the trajectory, J is the modulus of the angular momentum vector, T is the period of the polhode, and Υ is the (signed) solid angle, swept by the polhode. This solid angle is also the (normalized, signed) area enclosed by the curve $\mathbf{M}(t)$ in Euler's phase portrait!

Thus, we do not need elliptic functions to get the most important information about the full rigid body motion. The herpolhode angular shift is given directly in terms of the reduced Euler system (the polhode spherical angle).

More generally, take any system with symmetry and consider a *periodic* solution of the reduced system. Reconstructing the full solution, some variables usually acquire *holonomies*. In many cases, differential geometric insight allows one to split a phase into two parts, one of which is called "dynamic phase," as is the first term on the right-hand side of (11), while the other is called "geometric phase," as is the solid angle in (11).

Here we will extend (11) to *gyrostats*, mechanical systems that are composed of more than one body yet have the rigid body property that its inertia components are time independent constants [25, 16]. Such systems consist of a main rigid body, called the *carrier*, together with one or more

---

[9]Also obtained by M. Levi using differential geometry (see Levi's chapter 7 in this book).

rigid symmetric rotors, which we may call *flywheels*, supported by rigid bearings on the carrier. See Figure 5. Gyrostats have important technological applications, particularly in the attitude control of artificial satellites [15, 11]. Rotations of the n flywheels relative to the carrier do not change

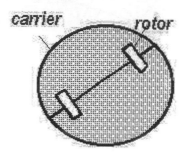

Figure 5. A gyrostat with one rotor or flywheel.

the mass geometry of the whole system. For a gyrostat with n flywheels, the configuration space is $SO(3) \times S^1 \times \cdots \times S^1$. We denote the additional degrees of freedom $\theta_1, \ldots, \theta_n$. In addition to *left (or spatial) SO(3)* invariance, there is also *right (or material) $S^1 \times \cdots \times S^1$* symmetry. In this case Noether's theorem implies the existence of n additional conserved scalar momenta $I_1, \ldots, I_n$, besides the conserved vector **m**.

The reduced equations of motion also give rise to trajectories on the sphere $\|\mathbf{M}\| = J$. They are given by

$$\frac{d\mathbf{M}}{dt} = \mathbf{M} \times \nabla_M H(\mathbf{M}, I), \quad \frac{d\theta}{dt} = \nabla_I H(\mathbf{M}, I). \tag{12}$$

The reduced Hamiltonian $H(\mathbf{M}, I)$ is given by a quadratic function of $3 + n$ variables $M_1, M_2, M_3, I_1, \ldots, I_n$. In the first set of equations, the conserved momenta $I_1, \ldots, I_n$ are thought of as parameters. We have shortened the notations, $\theta = (\theta_1, \ldots, \theta_n)$ and $I = (I_1, \ldots, I_n)$ for the flywheel coordinates and momenta. Once a closed trajectory $C : \mathbf{M} = \mathbf{M}(t)$ is found, say with period T, *the phases associated to the flywheels can be found by quadratures*:

$$\Delta\theta_j = \int_0^T \nabla_{I_j} H(\mathbf{M}(\mathbf{t}), I) dt. \tag{13}$$

What about the geometric phase $\Delta\phi$ of the main body? Using Montgomery's approach, we prove

$$\Delta\phi = \frac{2hT}{J} - \Upsilon - \frac{1}{J}\sum_{j=1}^{n} I_j \Delta\theta_j. \tag{14}$$

We obtained the same result following Levi's method; see [1].

3   Holonomy with Zero Momentum

In this section we present the abstract framework in a nutshell[10]. The reader not acquainted with connections on principal bundles should read Part 4 first.

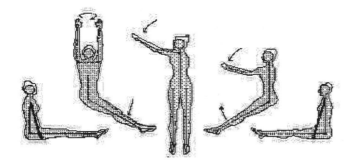

Figure 6   A cat performing holonomy (adapted from [9]).

The configuration space Q for self-locomotion problems[11] is a principal bundle $\pi : Q \to S$ over a base manifold S, with group G. That is, G acts on Q and $S = Q/G$ is the space of "shapes." Q has a G-invariant Riemannian metric, inherited by S. Denote $V_q = T_q\pi^{-1}(s)$ the vertical distribution. The statement

$$H = V^{\perp} \text{ (with respect to the metric)} \tag{15}$$

defines a *mechanical connection.*

We assume that this connection is *fat* (has lots of curvature). By the Ambrose and Singer theorem [7, p.389], there are admissible paths between any two points $q_o, q_1 \in Q$. Take them in the same fiber, $q_1 = g \cdot q_o$.

---

[10]See [20] for details.

[11]Examples include the connections for falling cats or gymnasts, deforming molecules, satellites with moving panels, swimming in a tar pool, and many more (Figure 6).

For low Reynolds number swimming, $\mathcal{Q}$, the set of "located shapes," is the set of embeddings of a given manifold (say, the sphere) on $\mathbb{R}^3$. $\mathcal{S}$, the set of "unlocated shapes," is the quotient of $\mathcal{Q}$ by the group $G = SE(3)$ of euclidean motions. An infinitesimal deformation of a given shape in $\mathcal{Q}$ gives a vector field along the shape, and the set of infinitesimal deformations defines the tangent bundle $T\mathcal{Q}$. In Part 4 we show how the connection can be computed.

If a group G acts on a manifold Q and $\langle, \rangle$ is a G-invariant Riemannian metric, recall that the momentum map $\mu : TQ \to \mathcal{G}^*$ is given by

$$\mu(v_q) \cdot \xi = \langle \xi_q, v_q \rangle, \quad \xi_q = \frac{d \exp(t\xi)}{dt}\Big|_{t=0} \cdot q, \quad \xi \in \mathcal{G}, \, v_q \in T_q Q, \qquad (16)$$

where we have identified $TQ \equiv T^*Q$ via the metric.

There is an intrinsic formulation for the connection form. Define the *locked inertia tensor, $I : \mathcal{G} \to \mathcal{G}^*$,* by

$$I_q(\xi) \cdot \eta = \langle \xi_q, \eta_q \rangle. \qquad (17)$$

The connection form is

$$A(v_q) = I_q^{-1}(\mu(v_q)). \qquad (18)$$

The horizontal distribution is the kernel of the momentum map[12] μ.

For a cycle $s(t)$ of infinitesimal deformations of a shape s spanned by vectors $\epsilon u_1, \epsilon u_2, -\epsilon u_1, -\epsilon u_2 x \in T_s$, the holonomy is given by

$$g \sim \exp(\epsilon^2 \Omega(u_1, u_2)),$$

where Ω is the curvature of the connection.

Problem. With prescribed holonomy $g \in G$, find the shortest loop $s(t)$ in S such that its horizontal lift connects q_o to $q_1 = g q_o$. The elements $\dot{s} \in TS$ are viewed as *controls*.

This variational problem with constraints is a special *sub-Riemannian geometry problem*. Montgomery [20] studied the beautiful structure of the corresponding Euler–Lagrange equations, given by

$$\frac{ds_j}{dt} = p_j, \quad \frac{dp_j}{dt} = -\frac{1}{2}\frac{\partial g^{ik}}{\partial x_j}p_i p_k + F_{jk}^I p^k \xi_I, \quad \frac{d\xi_I}{dt} = -c_{IJ}^K A_j^J p^j \xi_k, \qquad (19)$$

---

[12]In the case of microswimming, zero momentum means that the organism can exert neither net force nor torque on the fluid.

where A_j^J and F_{jk}^I are, respectively, the coefficients of the connection and the curvature forms. Here j runs over the dimension of S and I over the dimension of the group. The intrinsic formulation of these equations, known in the Yang–Mills theories as Wong's equations, are

$$\langle \nabla_{\dot\gamma}\dot\gamma, \bullet \rangle = \lambda \cdot F(h\dot\gamma, h\bullet), \quad \frac{D\lambda}{dt} = 0, \qquad (20)$$

where $\lambda(t)$, the Lagrange multiplier, belongs to $Ad^*Q = Q \times_{Ad^*} \mathcal{G}^*$, an associated fiber bundle over S with fiber $\mathcal{G}^*$. Here h denotes the horizontal lift via the connection. There is an Ad-ambiguity in the value of F, choosing different elements $q \in Q$ over the vertical fiber over $s \in S$. This ambiguity is cancelled with the corresponding Ad^*-ambiguity of λ. The optimal trajectories in the total space Q are obtained from the projections $s(t) \in S$ by horizontal lift.

Not developed further in these notes due to space limitations, we briefly discuss *nonholonomic systems*, a theme regarded until a few years ago as only curious or bizarre, but that has recently become of great interest in robotics[13]. One example is certains stones, known to the Celts and supposedly possessing magical properties, that prefer to rotate in one direction (experiment performed in class). This and other examples show that the momentum map is often *not conserved* in nonholonomic systems!

For nonholonomic mechanical systems such as those encountered in engineering applications, using d'Alembert's principle is believed to be the correct way to eliminate the constraints (Sommerfeld [22])[14].

In the case of nonholonomic systems with symmetry, eliminating the constraints leads to an interesting reduced system (for details, we refer to our work [14]). In the simplest situation, the same ingredients are used: a principal bundle Q over a base manifold S, with group G, a G-invariant Riemannian metric (or more generally, a G-equivariant Lagrangian L on Q), and finally a connection, whose horizontal spaces define the constraints. Here, however, the horizontal spaces do not need to be orthogonal to the

---

[13]Notwithstanding a somewhat dubious reputation, nonholonomic systems attracted the interest of important scientists in the past. Hertz advocated replacing forces by constraints and Cartan made an interesting address at the 1928 IMU Congress [6]. See [8] for the state of the art of nonholonomic systems.

[14]As Hertz already knew, a different set of equations results if one uses the rules of calculus of variations (as in (20)). But their solutions do not fit the results of experiments. Paraphrasing Leibniz, the engineer's world is not the best (nor the worst), so it is not described by variational principles.

vertical spaces.

Using the horizontal distribution, the equivariant Lagrangian L on TQ projects to a Lagrangian L^* on TS, but with an extra "strange" force!

We denote (following Arnold [3]) by $[L^*](v_s) \in T_s^*S$ the Euler–Lagrange derivative[15]. The reduced equations of motion on T^*S are written as (see [14])

$$[L^*](v_s)(\bullet) = \mu(\mathrm{FL}h_q(v_s)) \cdot \Omega(h_q(v_s), h_q(\bullet)). \tag{21}$$

Here $q \in Q$ is any point on the fiber over s, h_q is the horizontal lift operator, FL is the Legendre transform associated to L, $\mathrm{FL} : TQ \to T^*Q$, and Ω is the curvature of the connection. Figure 6 exemplifies reorientation manouvers performed by cats, gymnasts, and astronauts (without violating the constraint of total zero angular momentum). Not surprisingly, the momentum map μ is the key ingredient in this strange force.

Figure 7. Snakeboard (from [5], with permission from the authors). Shape variables: (ψ, ϕ_1, ϕ_2). Lie group variables (x, y, θ).

Once the reduced equations are solved, the full motion is recovered on Q by horizontally lifting the trajectory $s(t)$.

---

[15]In coordinates, the familiar $\frac{d}{dt}(\frac{\partial L^*}{\partial \dot{s}_i}) - \frac{\partial L^*}{\partial s_i}$, which transforms as a covector.

To finish, we would like to mention a hybrid situation, in which the nonholonomic system depends on parameters that can be used as control variables. This new feature, which allows potential applications in robotics, especially in biomechanics, was first studied by Jerry Marsden and his associates; see [5]. A delightful example is the snakeboard, (see Figure 7) which is a souped-up version of a skate. The rider can twist his body (coordinate ψ) and turn the front and back pairs of wheels (coordinates ϕ_1, ϕ_2).

The structure of such *D'Alembert control systems* is as follows: denote g the group variables, μ the momentum, and s the parameters that describe the shape of the system. At a glance, the equations of motion have the form

$$g^{-1}\dot{g} = -A(s)\dot{s} + B(r)\mu \qquad \text{(geometric phase)}, \qquad (22)$$

$$\dot{\mu} = \dot{s}^{\dagger}\left(\alpha(s)\dot{s} + \beta(s)\mu\right) + \mu^{\dagger}\gamma(s)\mu \quad \text{(momentum)}, \qquad (23)$$

$$M(r)\ddot{s} = -C(s, \dot{s}) + N(s, \dot{s}, \mu) + \tau \quad \text{(shape dynamics)}. \qquad (24)$$

If in (24) the forcing $\tau \equiv 0$, the shape varies solely by the nonholonomic dynamics. The rider can impose a full control over s. Equation (24) is deleted and a prescribed $s(t)$ is imposed in the first two equations. The reader should study Figure 8 and convince himself (herself) that indeed μ (momentum about p) is *not* conserved.

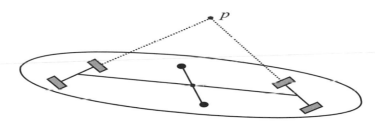

Figure 8. Angular momentum about p is not conserved. Not even p is conserved.

Summarizing: We have described three (four, if we include nonholonomic systems) situations where momentum maps and geometric phases interplay through adiabatic change of parameters, reconstruction, or geometric control. We anticipate that interesting examples will be found in which these features appear simultaneously.

References

[1] Almeida, M. F. L. B. P., Koiller, J., Stuchi, T. J., *Toy tops, gyrostats and Gauss-Bonnet*, Matem. Contemp. **9**, 1–14 (1993).

[2] Arnold, V., *Méthodes mathematiques de la mécanique classique*, MIR, Moscow (1978).

[3] Arnold, V. ed., *Dynamical Systems III*, Springer Encyclopaedia of Math. Sciences vol. 3, Springer-Verlag, New York (1988).

[4] Berry, M. V., *Classical adiabatic angles and quantal adiabatic phase*, J. Phys. A **18**, 15–27 (1985)

[5] Bloch, A. M., Krishnaprasad, P. S., Marsden, J. E., Murray, R. M., *Nonholonomic systems with symmetry*, Arch. Rational Mech. Anal. **136**, 21–99 (1996).

[6] Cartan, E., *Sur la réprésentation géometrique des systèmes matériels non holonomes*, Proc. Int. Cong. Math. vol 4, Bologna, 253–261 (1928).

[7] Choquet-Bruhat, Y., DeWitt-Morette, C., and Dillard-Bleick, M., *Analysis, manifolds and physics*, North Holland, Amsterdam (1982).

[8] Cushman, R., Snyaticki, J., eds., Proc. Workshop on Nonholonomic Constraints in Dynamics, Calgary 1997, *Reports on Mathematical Physics*, **42**, 1/2 (1998).

[9] Frohlich, C., *The physics of somersaulting and twisting*, Sci. American **263** 155–164 (1980).

[10] Hannay, J. H., *Angle variable holonomy in adiabatic excursion of an integrable hamiltonian*, J. Phys. A **18**, 221–230 (1985).

[11] Hubert, C. H., *An attitude acquisition technique for dual-spin spacecraft*, Ph.D. thesis, Cornell University, Ithaca, NY (1980).

[12] Koiller, J., *Classical adiabatic angles for slowly varying mechanical systems*, Contemp. Math. **97**, 159–185 (1989).

[13] Koiller, J., *A note on classical motions under strong constraints*, J. Phys. A 23, L521–527 (1990).

[14] Koiller, J., *Reduction of some classical non-holonomic systems with symmetry*, Arch. Rational Mech. Anal. 118, 113–148 (1992).

[15] Krishnaprasad, P. S., *Lie Poisson structures on dual spin spacecraft and asymptotic stability*, Nonlinear Anal. **9:10**, 1011–1035 (1985).

[16] Leimanis, E., *The General Problem of the Motion of Coupled Rigid Bodies about a Fixed Point*, Springer-Verlag, New York (1965).

[17] Marsden, J., Weinstein, A., *Reduction of symplectic manifolds with symmetries*, Rep. Math. Phys. **5**, 121–130 (1974).

[18] Marsden, J. E., Montgomery, R., Ratiu, T., *Reduction, symmetry, and phases in mechanics*, Mem. Amer. Math. Soc. **88, 436** (1990).

[19] Montgomery, R., *How much does the rigid body rotate? A Berry's phase from the 18th century*, Am. J. Phys. **59**, 394–398 (1991).

[20] Montgomery, R., *The isoholonomic problem and some applications*, Commun. Math. Phys. **128**, 565–592 (1990).

[21] Montgomery R., *Optimal control of deformable bodies and its relation to gauge theory*, in T. Ratiu, ed., *The geometry of Hamiltonian systems*, MSRI 1998 Workshop, Springer-Verlag, New York (1991).

[22] Sommerfeld, A., *Mechanics*, Academic Press, New York (1952).

[23] Takens, K., *Motion under the influence of a strongly constraining force*, in Lecture Notes in Math., vol. **819**, Springer-Verlag, New York (1979).

[24] Shapere, A., Wilczek, F., *Geometric Phases in Physics*, World Scientific, Singapore, Teaneck, NJ (1989).

[25] Wittenburg, J., *Dynamics of Systems of Rigid Bodies*, B.G. Teubner, Stuttgart (1977).

Part 2: Classical Adiabatic Angles

Geometry glitters, Birkhoff used to say, but to find gold one has to dig harder into analysis. So we begin with a disclaimer: we will omit here important issues related to resonance phenomena[16].

4   Averaging Heuristics

This section presents a geometrical derivation of Hannay's classical adiabatic angles. It is a pleasure to acknowledge conversations with R. Montgomery and A. Weinstein; the reader will find their papers [8, 11, 12] very inspiring.

Let $\theta = (\theta_1, \ldots, \theta_n)$ denote angle variables in $T^n = S^1 \times \cdots \times S^1$ (n times), and $I = (I_1, \ldots, I_n)$ the corresponding action variables. Consider a family of canonical transformations, parametrized by $x \in X$:

$$F : (I, \theta, x) \to (\, p = p(I, \theta, x)\,,\ q = q(I, \theta, x)\,). \tag{25}$$

More abstractly, we may think of a family, parametrized by $x \in X$, of Lagrangian foliations on a symplectic manifold (M^{2n}, ω):

$$\{\mathcal{L}_x\}_{x \in X}\,. \tag{26}$$

We consider a family (again parametrized by X) of completely integrable systems $H(p, q; x) = F(I; x)$. Now take $x = x(t)$ and extend phase space and Hamiltonian in the usual way:

$$\mathcal{H}(p, q, E, t) = H(p, q, t) + E, \quad \omega = dp \wedge dq + dE \wedge dt.$$

Here E is a dummy moment associated to a time "coordinate," such that $\dot{H} = -\dot{E} = \partial H / \partial t$ (since $\mathcal{H}$ is constant). In the extended space the map $(E, t, I, \theta) \to (E, t, p, q)$ is not canonical: the symplectic form pulls back as

$$\omega = dE \wedge dt + dI \wedge d\theta + (I, t)dI \wedge dt + (\theta, t)d\theta \wedge dt. \tag{27}$$

---

[16]See Lochak and Meunier [8], Golin, Knauf, and Marmi [4], and (for time estimates of the adiabatic approximation) Neishtadt [10].

Here, for $u, v = I, \theta$, or t, (u, v) denotes the *Lagrange bracket*

$$(u, v) = \frac{\partial p}{\partial u} \frac{\partial q}{\partial v} - \frac{\partial p}{\partial v} \frac{\partial q}{\partial u}. \tag{28}$$

Lagrange brackets are not as famous as Poisson brackets, so, to warm up, we propose a simple exercise.

Exercise 4.1 *Let (M^{2n}, ω) be a symplectic manifold, and $H : M \rightarrow \mathbb{R}$ be a Hamiltonian. Given any coordinate system $u = (u_1, \ldots, u_{2n})$ Hamilton's equation is written as*

$$L(\dot{u}_1, \ldots, \dot{u}_{2n})^\dagger = (\partial H/\partial u_1, \ldots, \partial H/\partial u_{2n})^\dagger,$$

where L is the matrix of Lagrange brackets

$$L_{ij} = (u_i, u_j) = \omega\left(\frac{\partial}{\partial u_i}, \frac{\partial}{\partial u_j}\right).$$

The relationship with Poisson brackets is transparent: $L^{-1} = P$, where

$$P_{ij} = \{u_i, u_j\}.$$

Exercise 4.2 *In our case, the matrix of ω with respect to the basis*

$$\partial/\partial E, \, \partial/\partial t, \, \partial/\partial I, \partial/\partial \theta$$

is $L = J + K$, where

$$J = \begin{bmatrix} 0 & 1 & 0 & 0 \\ -1 & 0 & 0 & 0 \\ 0 & 0 & 0 & 1 \\ 0 & 0 & -1 & 0 \end{bmatrix}, \quad K = \begin{bmatrix} 0 & 0 & 0 & 0 \\ 0 & 0 & (t, I) & (t, \theta) \\ 0 & (I, t) & 0 & 0 \\ 0 & (\theta, t) & 0 & 0 \end{bmatrix}. \tag{29}$$

Proposition 4.1 *In the action-angle coordinates, Hamilton's equations for the time-dependent system $H(I, \theta, x(t))$ are given by*

$$L \frac{d}{dt} (E, t, I, \theta)^\dagger = \text{grad } \mathcal{F} = (1, F_t, F_I, 0)^\dagger. \tag{30}$$

Exercise 4.3 *Let the parameter x vary slowly in time, $x = x(\epsilon t)$, and K be $O(\epsilon)$. Show that $L^{-1} = -J - JKJ + O(\epsilon^2)$.*

Proposition 4.2 *If $x = x(\epsilon t)$, then the Hamiltonian vector field*

$$\frac{d}{dt}(E, t, I, \theta) = L^{-1} grad F$$

is given by

$$\dot{\theta} = \partial F/\partial I + (t, I) + O(\epsilon^2), \quad \dot{I} = -(t, \theta) + O(\epsilon^2), \tag{31}$$

$$\dot{E} = -\partial F/\partial t + (t, \theta)\partial H/\partial I + O(\epsilon^2), \tag{32}$$

where the Lagrange brackets are $O(\epsilon)$.

Notice that the angle coordinates are *fast* (due to the term $\partial H/\partial I$), and all the other terms are $O(\epsilon)$ or higher and therefore *slow*.

Given any function $f(\theta)$ of the angle variables, we denote by $\langle f \rangle$ the average

$$\langle f \rangle = \frac{1}{(2\pi)^n} \int_{T^n} f(\theta) \, d^n\theta.$$

The "averaging principle," in its outermost heuristical form (see Arnold, [2, chapter 10, section 52]), states that it is "reasonable" to replace (31) by the averaged system[17]

$$\dot{\theta} = \partial H/\partial I + \langle t, I \rangle, \quad \dot{I} = -\langle t, \theta \rangle. \tag{33}$$

Caveat: It is not easy to transform the word "reasonable" into sound mathematics. See Neishtadt [10].

Definition 4.4 *Family (25) is free if there is a globally defined generating function $S(I, \theta; x)$ such that*

$$pdq - Id\theta = d_{(I, \theta)}S. \tag{34}$$

From now on, we assume all families of Lagrangian foliations to be free. The reason for this hypothesis is as follows.

Lemma 4.5 *If (26) is free, then $\langle t, \theta \rangle = 0$. So in the averaged system (33),*

$$\dot{I} = 0.$$

---

[17]The idea goes back to work of Gauss on celestial mechanics; he proposed replacing a perturbing object by an annulus of same total mass.

Proof. For fixed x the 1-form $pdq - Id\theta$ is locally exact; that is, locally there exists a function $S(I, \theta; x)$ with $pdq - Id\theta = d_{(I,\theta)}S$. In particular,

$$p\partial q/\partial \theta = I + \partial S/\partial \theta .$$

Differentiating with respect to t, we have

$$pq_{\theta t} + p_t q_\theta = S_{\theta t}.$$

Now,

$$(t, \theta) = p_t q_\theta - p_\theta q_t = [-pq_t]_\theta + pq_{\theta t} + p_t q_\theta = [-pq_t + S_t]_\theta. \qquad (35)$$

Since the average of a $\partial/\partial \theta$ is zero, we are done. ∎

We say that I is an *adiabatic invariant*. Heuristically, one expects that the solution $I(t)$ of the full system stays close to $O(\epsilon)$ from the initial value, at least for times of order $O(1/\epsilon)$. However, due to resonances, this is not necessarily true for systems with more than one degree of freedom [10].

Having said that, we will proceed bluntly with the averaged equations,

$$\dot\theta = \partial H/\partial I + \langle t, I \rangle , \quad \dot I = 0. \qquad (36)$$

5   The Classical Adiabatic Phases

Curiously, it was only around 1985 that attention was given, by Hannay and Berry, to the extra term in the equation for $\dot\theta$. The solution of the averaged system (36) is given as follows.

Theorem 5.1

$$\theta(T) - \theta(0) = \int_0^T H_I(I, t)\, dt + \int_C \rho\, dx, \quad \rho = \langle x, I \rangle\, dx, \qquad (37)$$

where

$$\langle x, I \rangle = \frac{1}{(2\pi)^n} \int_{T^n} (p_x q_I - p_I q_x)\, d^n\theta. \qquad (38)$$

The first term in (37) gathers the *dynamic phases*, while the second term gathers the *geometric phases*. Geometric phases are indeed geometric: their value does not depend on reparametrizations of the curve $C : x =$

$x(\epsilon t) \in X$. They depend solely on the family of Lagrangian foliations, not on the specific time-dependent Hamiltonian.

Definition 5.2 *The 1-form ρ in parameter space, depending on I, with values on $\mathbb{R}^n$, is called the Hannay–Berry 1-form.*

Remark 5.3 *For those familiar with principal bundles: since ρ has vector values, one suspects that it defines a connection on a torus bundle over X (fixing the values of I). See Montgomery [9]. More generally, can one define a connection for adiabatic variations of general systems, not necessarily integrable? See Weinstein [13].*

Exercise 5.4 *Prove Berry's original expression*

$$\langle x, I \rangle = grad_I \langle -pq_x + S_x \rangle.$$

Although C can be an open curve, in most cases we are interested in finding the geometric phase for a *closed* curve C in parameter space. If C bounds a disk inside X, we can use Stokes's theorem.

Theorem 5.5

$$d\rho = -\text{grad}_I \beta. \tag{39}$$

Here β is a real-valued 2-form, given by

$$\beta(I) = \frac{1}{(2\pi)^n} \int_{\theta \in T^n} \sum_{i<j} (x_i, x_j) \, dx_i \wedge dx_j, \tag{40}$$

and where (x_i, x_j) are the Lagrange-like brackets

$$(x_i, x_j) = \sum_k \left(\frac{\partial p_k}{\partial x_i} \frac{\partial q_k}{\partial x_j} - \frac{\partial p_k}{\partial x_j} \frac{\partial q_k}{\partial x_i} \right). \tag{41}$$

Exercise 5.6 *Prove Theorem 5.5.*

Hint. Use Exercise 5.4. You can also start directly with (38). When computing $d\rho$, interchange some indices. For instance,

$$\sum p_{I\,x_i} q_{x_j} \, dx_i \wedge dx_j = -\sum p_{I\,x_j} q_{x_i} \, dx_i \wedge dx_j.$$

Terms like $\sum p_{x_i x_j} q_I \, dx_i \wedge dx_j$ vanish. Notice the minus sign in (39) and the sum $\sum_{i<j}$ in (40). ∎

Definition 5.7 *β is called the Hannay–Berry 2-form, a 2-form in parameter space X depending on the value I of action variables.*

Exercise 5.8 *For fixed values of (I, θ), define the function*

$$\lambda_{(I,\theta)} : X \to M^{2n}$$

using the family (25). Then

$$\sum_{i<j}(x_i, x_j) \, dx_i \wedge dx_j = \lambda^*_{(I,\theta)}\omega, \tag{42}$$

where ω is the symplectic form on M.

Given a closed curve C in parameter space, bounding a disk $D \subset X$, denote by $D(I, \theta) \subset M$ the image of D under $\lambda_{(I,\theta)}$:

$$D(I, \theta) - \lambda_{(I,\theta)}(D).$$

Proposition 5.1 *The geometric phases can be written in an almost intrinsic way as*

$$-\mathrm{grad}_I \left\langle \int_{D(I,\theta)} \omega \right\rangle. \tag{43}$$

Remark 5.9 (i) *To make this expression completely intrinsic, one needs to study its dependence on changes of action-angle variables $(I, \theta, x) \to (J, \phi, x)$.* (ii) *Weinstein used (42) as a starting point to define new invariants for loops of symplectomorphisms on compact symplectic manifolds [12].*

We finish this section with a question: Relax the freeness hypothesis (Definition 4.4) for (26). Is there any new term, of topological origin, present in the classical adiabatic angles?

6   Pendulum with Slowly Varying Gravity

This example is closely related to the Foucault pendulum. Take units so that mass, gravity constant, and length are all equal to 1. We have

$$H(p, q, x) = \frac{1}{2}|p|^2 + (q, x) \tag{44}$$

with $q \in S^2, (p, q) = 0$, and $x \in X = S^2$ gives the direction of the force of gravity.

The spherical pendulum is obviously integrable because of the S^1 symmetry. Fix $x = e_3$, the north pole, so the gravity vector is $\vec{g} = -ge_3$. One of the momenta, say I_1, is the angular momentum about e_3. The other, I_2, is the area enclosed by the energy curve in the reduced one-degree-of-freedom system (a complete elliptic integral).

The solutions require elliptic functions (of time). In the case of the simple pendulum, three different sets of action-angle variables are needed, one for the libration regime and the other two (essentially the same) for the circulation regimes. It was recently found that it is impossible to find global action-angle coordinates for the spherical pendulum![18].

Exercise 6.1 *Make a bibliographical search and get the solutions of the simple (planar) pendulum and of the spherical pendulum in terms of elliptic functions. In other words, find explicitly the canonical transformations*

$$P = P(I, \theta), \quad Q = Q(I, \theta), \tag{45}$$

where Q are the Cartesian coordinates and $P = \dot{Q}$ the corresponding momenta.

For the parametrized system (44), we may write

$$q = R(x)Q(I, \theta), \quad p = R(x)P(I, \theta), \tag{46}$$

where (P, Q) are given by (45) and $R : S^2 \to SO(3)$ satisfies $R(x) \cdot e_3 = x$; that is, $R(x)$ has x as the third column.

Exercise 6.2 *R does not exist globally!*

---

[18]There is a *monodromy* phenomenon in any domain $I = (I_1, I_2) \in D$ around the unstable equilibrium. We will not discuss this beautiful issue here; see [3].

Hint. The sphere is not "parallelizable"; even worse, it is not "combable."

Exercise 6.3 *Try the mission impossible of constructing R globally.* (i)
*Consider stereographic projection $T_p S^2 \sim \mathbb{R}^2 \to S^2 - \{-p\}$, with rays
emanating from $-p$. Via the differential, map the Euclidian (parallel) frame
on $T_p S^2 \equiv \mathbb{R}^2$ to a frame on $S^2 - \{-p\}$. Then apply Gram–Schmidt
ortogonalization. Sketch a picture; how does it look near $-p$? (ii) Obtain
a more familiar picture mapping the polar coordinates frame on $\mathbb{R}^2 - 0$ to
the meridian-parallel frame on $S^2 - \{p, -p\}$. What happens at p and $-p$?*

At first sight, performing the averages over the tori in (40) using the
action-angle system (45) seems to be a byzantine exercise involving elliptic
functions. However, from the special "isotropic" form of the canonical
transformation (46), we have the following.

Exercise 6.4

$$\beta = f(I_1, I_2) \quad area\ element\ of\ \ S^2. \tag{47}$$

Exercise 6.5 *To find $f(I_1, I_2)$, it suffices to compute the Lagrange bracket
(x_1, x_2) in(40) at $x_o = e_3$.*

Exercise 6.6 *Use stereographic coordinates x_1, x_2 at the tangent plane at
e_3, as in Exercise 6.3 (i). Observe that $R(e_3) = I$ and*

$$\partial R / \partial x_1 = e_2 \times \cdot, \quad \partial R / \partial x_2 = -e_1 \times \cdot.$$

Draw a picture. Show that the Jacobian at the origin is 1.

Exercise 6.7 *Fill in the blanks below. With spherical coordinates*

$$Q = \begin{bmatrix} \sin\alpha \cos\beta \\ \sin\alpha \sin\beta \\ \cos\alpha \end{bmatrix}, \quad \dot{Q} = \begin{bmatrix} \dot\alpha \cos\alpha \cos\beta - \dot\beta \sin\alpha \sin\beta \\ \dot\alpha \cos\alpha \sin\beta + \dot\beta \sin\alpha \cos\beta \\ -\dot\alpha \sin\alpha \end{bmatrix}, \tag{48}$$

compute

$$\partial p / \partial x_1 = e_2 \times \dot{Q} = \cdots,$$
$$\partial p / \partial x_2 = -e_1 \times \dot{Q} = \cdots,$$

$$\partial q / \partial x_1 = e_2 \times Q = \cdots,$$
$$\partial q / \partial x_2 = -e_1 \times Q = \cdots,$$

$$\sum_k \frac{\partial p_k}{\partial x_1} \frac{\partial q_k}{\partial x_2} - \frac{\partial p_k}{\partial x_2} \frac{\partial q_k}{\partial x_1} = \cdots = -\sin^2 \alpha \, \dot{\beta}. \tag{49}$$

Miracle! The quantity to be averaged over the tori is precisely minus the angular momentum of the spherical pendulum with respect to the axis $x_o = e_3$.

(Here's a shortcut. For the special parametrized family of the form (46),

$$\sum_k \frac{\partial p_k}{\partial x_1} \frac{\partial q_k}{\partial x_2} - \frac{\partial p_k}{\partial x_2} \frac{\partial q_k}{\partial x_1} = -(e_2 \times p)(e_1 \times q) + (e_1 \times p)(e_2 \times q) = p_1 q_2 - p_2 q_1.)$$

■

The argument works for any integrable system with symmetry S^1 (Hannay's top [5] is the simplest example), and parameter space is $X = S^2$. We get $f = -I_1$ in (47) so

$$\beta = -I \text{ area element of } S^2, \tag{50}$$

where I is the momentum associated with the S^1 symmetry. Taking into account the minus sign in (39), we conclude as follows.

Proposition 6.1 *For systems $H(\bullet, x)$ with S^1 symmetry around $x \in X = S^2$, the holonomy is the spherical angle in S^2 enclosed by the tip of the symmetry axis x.*

7   Slowly Moving Integrable Mechanical Systems

Recall the framework in section 1. We have a natural mechanical system with Lagrangian $L = T - V$ on TS, where S is immersed on a bigger Riemannian manifold W. W is acted upon by a Lie group G so that every map $i_g : S \to g \cdot S$ is an isometry.

We set

$$q(Q,t) = g(t)Q, \ p = g(t)^{-1*}P, \tag{51}$$

so that in the Hamiltonian formalism,

$$H(q,p,t) = T(p) + V(g(t)^{-1} \cdot q). \tag{52}$$

The parameter space X is the Lie group G.

As an example, we presented the "Star Wars" Foucault pendulum. Recall (4). As a warmup, we propose a simple exercise.

Exercise 7.1 *The linear approximation for $-k/|q|$ is*

$$g(t)(q - r(t)) \cdot \frac{r(t)}{|r(t)|}, \ g(t) = k/|r(t)|^2.$$

As we mentioned before, we need no linearization, nor do we require that the pendulum is in the small oscillation regime. We just use the S^1 symmetry around the axis $x = \frac{r(t)}{|r(t)|}$. The parameter space reduces to $X = \mathbb{R}^3 - 0 = S^2 \times \mathbb{R}^*$. The canonical transformation (46) is now replaced by

$$q = r + R(x)Q, \ p = R(x)P, \tag{53}$$

where $r = |r| \cdot x$, $R(x) e_3 = x - r/|r|$.

Exercise 7.2 *Show that*

$$\beta = -I \text{ element of solid angle.}$$

Hint. Show that the affine part in (53) produces no holonomy.

8 Hannay–Berry's 1-Form for Moving Systems

For concreteness, we suppose that $W = \mathbb{R}^{3N}$, with the familiar metric[19]

$$T = \sum_i^N m_j \dot{Q}_j^2$$

---

[19]This is not too restrictive due to Nash's embedding theorem.

on $W = \mathbb{R}^{3N}$. Assume that the mechanical system $H = T + V$ in T^*S is completely integrable: there are action-angle coordinates $(I, \theta) \in \mathbb{R}^n \times T^n$ such that $H(P, Q) = K(I)$ under the canonical immersion of $T^*S \to T^*W$:

$$Q_j = Q_j(I, \theta), \; P_j = P_j(I, \theta) = m_i \dot{Q}_j = m_j \frac{\partial Q_j}{\partial \theta} \cdot K_I, \quad j = 1, \ldots, N. \quad (54)$$

The parameter space X is the Lie group $SE(3) = SO(3) \times \mathbb{R}^3$ acting diagonally on $W = \mathbb{R}^3 \times \cdots \times \mathbb{R}^3$. The parametrized family of canonical transformations is given by

$$p_j = RP_j(I, \theta), \quad q_j = RQ_j(I, \theta) + r, \quad j = 1, \ldots, N \quad (55)$$

where $g = (R, r) \in SE(3)$.

Theorem 8.1 *The 1-form ρ on $X = SE(3)$ is left invariant and vanishes in the translation subgroup. For any $v \in SO(3)$,*

$$\rho(v) = grad_I \langle -M(I, \theta) \rangle \cdot v. \quad (56)$$

Here $M \in \mathbb{R}^3$ is the total angular momentum vector, $\langle \, , \, \rangle$ is the averaging operator, and we identify $sO(3) \equiv \mathbb{R}^3$ in the usual way:

$$\begin{bmatrix} 0 & -v_3 + v_2 \\ +v_3 & 0 & -v_1 \\ -v_2 + v_1 & 0 \end{bmatrix} \leftrightarrow \begin{bmatrix} v_1 \\ v_2 \\ v_3 \end{bmatrix}. \quad (57)$$

Proof. The reader should decompress our shortened notation. Theorem 5.1 gives

$$\rho = \langle p_r q_I - q_r p_I \rangle \, dr + \langle p_R q_I - q_R p_I \rangle \, dR.$$

The first Lagrange bracket averages to zero. This is because (55) gives $p_r \equiv 0$, q_r = identity operator, and (54) yields

$$p_I = \frac{\partial}{\partial \theta} (mK_I \cdot Q(I, \theta))_I.$$

For the second bracket,

$$p_R q_I - q_r p_I = (dRP, RQ_I) - (dRQ, RP_I) = (R^{-1}dRP, Q_I) - (R^{-1}dRQ, P_I),$$

where we changed the inner product notation to $(\, , \,)$. The presence of the

$R^{-1}dR$ implies left invariance. Using the correspondence (57), we get

$$\rho = dv \cdot \left\langle \sum_j P_j \times \partial/\partial I Q_j - Q_j \times \partial/\partial I P_j \right\rangle,$$

where $\times$ is the vector product in $\mathbb{R}^3$. In short notation,

$$\rho = dv \cdot \langle P \times Q_I - Q \times P_I \rangle.$$

It can also be written

$$\rho = dv \cdot \frac{\partial}{\partial I} \langle P \times Q \rangle = dv \cdot \frac{\partial}{\partial I} \langle -M \rangle,$$

as we wanted to prove. ∎

Example: Hannay's Slowly Rotating Planar Hoop

A unit mass is travelling along a hoop $C : s \to (x(s), y(s))$ of length L. Action-angle coordinates are given by

$$\theta = \frac{2\pi s}{L}, \ I = \frac{L\dot{s}}{2\pi}.$$

The hoop is slowly rotated through a full turn (parameter space $G = S^1$). Compute the geometric phase (the 2π factor in front is the integral over S^1):

$$\Delta\theta = 2\pi \frac{d}{dI} \frac{1}{2\pi} \int_0^{2\pi} (y\dot{x} - x\dot{y})d\theta.$$

One gets

$$\Delta\theta = 2\pi \frac{d}{dI} \frac{1}{L} \int_0^L (y\dot{x} - x\dot{y})ds$$

$$= 2\pi \frac{1}{L} \frac{d}{dI} \int_0^L (ydx - xdy)\dot{s}$$

$$= (\frac{2\pi}{L})^2 \frac{d}{dI} \left[I \oint_C ydx - xdy \right]$$

so that

$$\Delta\theta = -8\pi^2 \frac{A}{L^2},$$

where A is the area enclosed by the hoop. In terms of the arc length,

$$\Delta s = -4\pi \frac{A}{L}. \tag{58}$$

If C is a circle of radius r,

$$[\Delta s]_{\text{circle}} = -2\pi r,$$

which is simply an artifact, since the origin of the arc length advances by $2\pi r$ as the hoop makes a circle. Thus we must add L to (58) to cure this "Jules Verne's syndrome":

$$\Delta s = L - 4\pi \frac{A}{L}, \tag{59}$$

which is ≥ 0 in view of the isoperimetric inequality.

9 The Main Theorem

The astute reader will suspect that something deeper is going on, involving the momentum map.

Theorem 9.1 *Let N a symplectic submanifold of a symplectic manifold M. Assume that a Lie group G acts symplectically on M with momentum map $J : M \to \mathcal{G}^*$. Fix a Lagrangian foliation on N. Then $\pi : G \cdot N \to G$ is a symplectic fiber bundle, and its torus subbundles $\mathcal{T}$ (locally defined by fixing a set of action variables) are T^n fiber bundles over G. Moreover, the Hannay–Berry 1-form*

$$\rho(\xi) = grad_I \langle -J_\xi(I, \theta) \rangle \tag{60}$$

defines a parallel transport operator on the torus subbundles. The parallel transport does not depend on the choice of action-angle variables on N.

Exercise 9.2 *Prove Theorem 9.1 for $N = T^*S$, $M = T^*W$, where $S \subset W$. The Lie group G acts on W by isometries; the action is lifted to T^*W via $g \cdot p_q = (g \cdot q, (g^{-1})^* p_q)$.*

Hint. Take local coordinates (P, Q) on T^*S and action-angle coordinates $P = P(I, \theta), Q = Q(I, \theta)$ such that $P dQ - I d\theta = dS(I, \theta)$. The lifted action $p_q = g \cdot P_Q$ satisfies $pdq = PdQ$. Therefore, for the family of symplectic immersions $p = p(I, \theta, g)$, $q = q(I, \theta, g)$, we have

$$pd_{(I,\theta)}q - Id\theta = dS(I, \theta),$$

with generating function S independent of parameter $g \in G$. In particular, $\frac{\partial S}{\partial t} \equiv 0$. From Exercise 5.4 we get

$$\rho = \langle g, I \rangle dg = grad_I \langle -pq_t \rangle dt,$$

where

$$pq_{t\,|t=0} = P\left(\frac{d}{dt}e^{t\xi}Q_{|t=0}\right) = J_\xi(P, Q), \; \dot{g}(0) = \xi. \tag{61}$$

■

Problem 9.1 *Are there global obstructions to define the torus subbundles $\mathcal{T}$?*

9.1   Invariance under a Maximal Torus T^r

The reader unfamiliar with Lie algebra theory can stick to $r = 1$, $T = S^1$, $G = SO(3)$. N is the phase space of a one- or two-degrees-of-freedom Hamiltonian with S^1-symmetry. An extra degree of freedom is allowed, still maintaining integrability; there will be no holonomy for the latter angle. He or she can go directly to formula (63), and in the next section we compute the holonomy $\Delta\theta_1$ for several examples.

Suppose G is compact semisimple, acting on M with momentum map J. Suppose that N is invariant under the action of a maximal torus T^r of G, with dim $N = 2r$ or $2(r+1)$. Choose a basis $\{\xi_i, \}_{i=1,\dots,\dim G}$ of the Lie algebra with the first $\xi_1, \dots, \xi_r \in t^d$, the Lie algebra of T^r, the last ones in the Killing-perpendicular of $\mathcal{T} \subset \mathcal{G}$.

Corollary 9.1 *The Berry–Hannay 1-form is*

$$\rho(\xi) = -projection \; of \; \xi \; over \; t^r. \tag{62}$$

Proof. Choose action-angle variables for N such that the first r actions are the momenta associated with $\xi_1, \ldots, \xi_d \in t^r$:

$$J_{\xi_j} = I_j \, , \ j = 1, \ldots, r.$$

If dim $N = 2r$, we have the Lagrangian foliation. If dim $N = 2(r+1)$, complete the set of canonical coordinates for N with any pair of conjugate coordinates (I_{r+1}, θ_{r+1}).

Claim. The average over T^r of the momenta associated with Lie algebra elements in the Killing-perpendicular of t^d is zero.

Then (62) follows immediately from (60).   ∎

Proof of the claim. Decompose ξ in (61) in its t^d component ξ and the Killing orthogonal component ζ. Use coordinates $(p(t), q(t), \theta(t))$ on N as above. By the equivariance of the momentum map, the claim boils down to the following fact from Lie group theory (see [1, section 4.10]):

Let ζ be in the Killing orthogonal to t^d. Then $\int_T Ad_g \, \zeta \, dg = 0$.   ∎

Remark 9.3 *Impose no restrictions on* $\dim N = 2(r+s)$. *Introduce coordinates* (p, q, I, θ) *on* N *so that* $\theta \in T^r$ *corresponds to the maximal torus and* I *are the associated momenta as above. We have s extra degrees of freedom,* q, *with conjugate momenta* p. *A* T^r *invariant Hamiltonian on* N *writes as* $H_o(p, q, I)$. *In general,* H_o *is nonintegrable, but it has the invariant measure* $d^s p \wedge d^s q \wedge d^r I \wedge d^r \theta$. *Assuming ergodicity, an (even more) outrageous averaging principle can be invoked:*

A perturbation $\epsilon H_1(p, q, I, \theta)$ *can be replaced by its average along* H_o.

Since $d^r \theta$ *is a factor in the invariant measure, any perturbation whose average over* T^r *vanishes does not contribute (to first order) to the perturbed dynamics. In our case, we have the time-dependent Hamiltonian* $H_o(g(\epsilon t)^{-1} \cdot n)$ *constrained to* $g(\epsilon t) \cdot N \subset M$. *Using the symplectic coordinates* (p, q, I, θ), *it writes as* $H_o - \epsilon J_{g^{-1}\dot{g}}$, *where* J *is the momentum of the G-action on M. It is reasonable to conclude that (62) remains valid, and there is no holonomy for the (p, q) component.*

Summarizing: Let $C : g = g(\epsilon t)$ a curve in G (not necessarily closed). The non-vanishing geometric phases $\Delta\theta_i$, $i = 1, \ldots, d$, associated with the maximal torus T^d are given by

$$\Delta\theta_i = -\text{proj}_{\xi_i} \int_C g^{-1} \, dg. \tag{63}$$

The integral $\int_C g^{-1} \, dg$ is the *hodograph* of $C \subset G$ to the Lie algebra $\mathcal{G}$.

9.2   Isotropic Oscillator

Let a planar isotropic harmonic oscillator or any one- or two-degrees-of-freedom system with S^1 symmetry be adiabatically transported along a curve C in $\mathbb{R}^3$. Recall that the Frenet frame $R = (t, n, b)$ satisfies the Frenet–Serret equations

$$t' = \kappa n, \; n' = -\kappa t + \tau b, \; b' = -\tau n, \quad \left(' = \frac{d}{ds}\right),$$

which in our notation writes (using the correspondence (57)) as

$$R^{-1} dR \leftrightarrow (\tau, 0, \kappa) ds. \tag{64}$$

- If the oscillator is transported along the oscullating plane (t, n), then we project $R^{-1} dR$ on the third axis. Equation (63) gives

$$\Delta\theta = -\int_C \kappa \, ds. \tag{65}$$

- If the oscillator is transported along the normal plane (n, b), we project over the first axis, so

$$\Delta\theta = -\int_C \tau \, ds. \tag{66}$$

This phenomenon has been observed in optical fibers (see [11]).

- For a surface curve, we use Darboux's frame $R = (t, n, N)$, where t is tangent to the curve and N is normal to the surface. The structure equations are $R^{-1} dR = (\tau_g, -\kappa_n, \kappa_g)$, where κ_n is the normal curvature, κ_g the geodesic curvature, and τ_g the geodesic torsion. If the system is transported along the tangent plane, then *the holonomy is minus the total geodesic curvature along C.* Using Gauss–Bonnet, we get

$$\Delta\theta = -\int \kappa_g\,ds = -2\pi + \int\int K\,dS, \qquad (67)$$

where K is the Gaussian curvature[20].

Exercise 9.4 *The term* -2π *is missing in Proposition 6.1. Is there a mistake?*

Answer. Both results are correct. We made two mistakes that cancelled out. In Proposition 6.1 we used the parallel-meridian frame on S^2. β is singular at the poles, which produces an extra 2π (which we ignored there). Well, this oversight is compensated by Hopf's umlaufsatz (the tangent turns by 2π around a simple closed curve). Since the angle is measured from the tangent vector, a 2π should be discounted, which we did not do.

9.3   Foucault's Pendulum Revisited

We take the moving frame $(r(\varphi), R(\varphi))$ along the circle of latitude $\frac{\pi}{2} - \psi$.

Exercise 9.5 *Here* $R^{-1}dR = (-\sin\psi, 0, \cos\psi)d\varphi$.

Thus

$$\Delta\theta = -2\pi cos\psi. \qquad (68)$$

Exercise 9.6 *Transport adiabatically Hannay's top along the center circle of the Moebius strip.*

10   Final Remarks

At least two issues related to Corollary 9.1 may have disturbed the reader. First, it does not apply to our examples! $SE(3)$ is not compact, and the symplectic submanifold N is only invariant under $S^1 \times 0$, that is, if there is no translation part. Not to worry: by Theorem 8.1 the translation part $r(t) \in \mathbb{R}^3$ of $(r, R) \in SE(3)$ produces no holonomy. The reader should not

---

[20]According to Klein [6], this result was first discovered by Radon as a "mechanical" demonstration of parallel transport.

have difficulty to modify the statement of Corollary 9.1 in order to handle these cases.

The second issue is not only embarrassing, it is actually frightening: our proof of Corollary 9.1 has a flaw when $dimN = 2(r+1)$! Given a T^r-invariant Hamiltonian $H_o(p, q, I)$, in general (as we observed in footnote 5) $\dot{\theta} = H_I(I, p, q) \neq$ const. We see that θ is not uniform in time (even if H is integrable). Retracing all logical steps to Theorem 9.1, we see that (in principle) we should replace θ by uniformizing variables, which we denote α. In other words, we should average over the latter, not over θ. What comes to our rescue is Remark 9.3. It is *declared* that time averages can be replaced by space averages with respect to the measure $dp \wedge dq \wedge d^r I \wedge d^r \theta$. A justification of this more general averaging principle is in order[21].

To close this chapter, we list some examples, not discussed here for lack of space. They are discussed in our paper [7]. We leave them for further reading:

- slowly rotating elliptic billiard,
- rigid body with slowly changing inertia matrix,
- systems subjected to a strong constraining, nonhomogeneous, force,
- coupled slow–fast mechanical systems.

References

[1] Adams, J. F., *Lectures on Lie Groups*, Benjamin, The University of Chicago Press, Chicago (1983).

[2] Arnold, V., *Méthodes mathematiques de la mécanique classique*, MIR, Moscow (1978).

[3] Duistermaat, J. J., *On global action angle coordinates*, Commun. Pure Appl. Math. **33**:6, 687–706 (1980).

[4] Golin, S., Knauf, A., Marmi, S., *The Hannay angles: Geometry, adiabaticity and an example*, Commun. Math. Phys. **123**, 95–122 (1989).

---

[21]As far as we know, this has not been done by the experts in the analytic issues of Hamiltonian systems.

[5] Hannay, J. H., *Angle variable holonomy in adiabatic excursion of an integrable hamiltonian*, J. Phys. A, **18**, 221–230 (1985).

[6] Klein, F., *Vorlesungen über höhere Geometrie*, Grund. Math. Wiss. 22, Springer-Verlag, Berlin (1926).

[7] Koiller, J., *Classical adiabatic angles for slowly varying mechanical systems*, Contemp. Math. **97**, 159–185 (1989).

[8] Lochak, P., Meunier, C., *Multiphase Averaging for Classical Systems*, Springer-Verlag, New York (1988).

[9] Montgomery, R., *The connection whose holonomy is the classical adiabatic angles of Hannay and Berry and its generalization to the nonintegrable case*, Commun. Math. Phys. **120**, 269–294 (1988).

[10] Neishtadt, A., *Averaging and passage through resonances*, in Proc. Intl. Congress Math., International Mathematical Union, Kyoto, 1271–1283 (1991).

[11] Tomita, A, Chiao, R. Y., *Observation of Berry's topological phase by use of an optical fiber*, Phys. Rev. Lett. **57**, 937–940 (1986).

[12] Weinstein, A., *Cohomology of symplectomorphism groups and critical values of hamiltonians*, Math. Z. **201**, 75–82 (1989).

[13] Weinstein, A., *Connections of Berry and Hannay type for moving lagrangian submanifolds*, Adv. Math. **82**, 133–159 (1990).

Part 3: Holonomy for Gyrostats

11   The Hamiltonian

Let X, Y, Z be the principal axis of inertia of the carrier. For simplicity, consider just one flywheel on the Z-axis of the inertia ellipsoid of the main body. Let λ be the double eigenvalue of the flywheel inertia matrix, λ_3 its third eigenvalue; $\mathcal{M}$ is the flywheel mass, $\dot{\theta}$ its angular velocity, and d the distance between the centers of mass of the carrier and flywheel.

Proposition 11.1 *The kinetic energy is a quadratic form in the Lie algebra $\mathbb{R} \times s0(3)$, given by*

$$E = \frac{1}{2}(I\mathbf{\Omega}, \mathbf{\Omega}) + \frac{1}{2}(\lambda + \mathcal{M}d^2)(\Omega_1^2 + \Omega_2^2) + \frac{1}{2}\lambda_3\Omega_3^2 + \lambda_3\dot{\theta}\Omega_3 + \frac{1}{2}\lambda_3\dot{\theta}^2. \quad (69)$$

Denoting by I_1, I_2, I_3 the eigenvalues of I_{mb}, the inertia matrix of the main body,

$$E = \frac{1}{2}(\dot{\theta}, \Omega_1, \Omega_2, \Omega_3)I_g(\dot{\theta}, \Omega_1, \Omega_2, \Omega_3)^t, \quad (70)$$

where I_g, the inertia operator of the *gyrostat*, is given by

$$I_g = \begin{pmatrix} \lambda_3 & 0 & 0 & \lambda_3 \\ 0 & \lambda + \mathcal{M}d^2 + I_1 & 0 & 0 \\ 0 & 0 & \lambda + \mathcal{M}d^2 + I_2 & 0 \\ \lambda_3 & 0 & 0 & \lambda_3 + I_3 \end{pmatrix}.$$

Using the coordinates

$$\mathbf{\Omega_g} = (\dot{\theta}, \Omega_1, \Omega_2, \Omega_3)^\dagger \quad (71)$$

for $\mathcal{G}$ (Lie algebra of $S^1 \times SO(3)$) and

$$\mathbf{M}_g = (I, M_1, M_2, M_3) \quad (72)$$

for its dual $\mathcal{G}^*$, it follows that the Legendre transformation is

$$\mathbf{M}_g = I_g \, \mathbf{\Omega_g}. \quad (73)$$

Consider the pairing $\mathcal{G}^* \times \mathcal{G}$, given by $2h = (\mathbf{M}_g, \mathbf{\Omega_g})$.

Proposition 11.2 *The energy of the gyrostat is expressed in terms of the momenta as*

$$h = (\mathbf{M}_g, I_g^{-1}\mathbf{M}_g). \tag{74}$$

Exercise 11.1 *Inverting I_g, obtain*

$$I_g^{-1} = \begin{pmatrix} \frac{1}{\lambda_3} + \frac{1}{I_3} & 0 & 0 & -\frac{1}{I_3} \\ 0 & \frac{1}{I_1 + \lambda + \mathcal{M}d^2} & 0 & 0 \\ 0 & 0 & \frac{1}{I_2 + \lambda + \mathcal{M}d^2} & 0 \\ -\frac{1}{I_3} & 0 & 0 & \frac{1}{I_3} \end{pmatrix}.$$

Moreover, since $\boldsymbol{\Omega_g} = I_g^{-1}\mathbf{M}_g$,

$$\Omega_1 = \frac{M_1}{I_1 + \lambda + \mathcal{M}d^2}, \quad \Omega_2 = \frac{M_2}{I_2 + \lambda + \mathcal{M}d^2}, \quad \Omega_3 = -\frac{1}{I_3} + \frac{M_3}{I_3}, \tag{75}$$

$$\dot{\theta} = \left(\frac{1}{I_3} + \frac{1}{\lambda_3}\right)I - \frac{1}{I_3}M_3. \tag{76}$$

Proposition 11.3

$$h = \frac{1}{2}\left(\frac{M_1^2}{I_1 + \lambda + \mathcal{M}d^2} + \frac{M_2^2}{I_2 + \lambda + \mathcal{M}d^2} + \frac{(M_3 - I)^2}{I_3} + \frac{I^2}{\lambda_3}\right). \tag{77}$$

Trajectories are intersections of a fixed momentum sphere $\|\mathbf{M}\| = J$ with the energy ellipsoids (77) with varying energies h, where I is taken as a parameter. The centers of the ellipsoids vary along the Z-axis. The possible phase portraits as a function of the parameter I are depicted in Figure 9.

Exercise 11.2 *Derive the Hamiltonian (69). See section 15 below.*

Is there a Poinsot description for gyrostats? The answer is "more or less," since the invariable plane moves up and down and the herpolhodes are not planar curves.

Exercise 11.3 *Check this formula, which can be used to define the inertia ellipsoid:*

$$\alpha\Omega_1^2 + \beta\Omega_2^2 + \gamma\Omega_3^2 = 2h - \frac{I^2}{\lambda_3}, \tag{78}$$

Figure 9. Reduced phase portrait for a rigid body with one flywheel. Top left: Euler system $I = 0$. For sufficiently big I, there are no unstable equilibria (bottom right). The satellite is then stabilized.

where

$$\alpha = I_1 + \lambda + \mathcal{M}d^2 \,, \ \beta = I_2 + \lambda + \mathcal{M}d^2 \,, \ \gamma = I_3. \tag{79}$$

Proposition 11.4 *(Modified Poinsot description). Replace the carrier by its inertia ellipsoid. After a polhode period T, the plane perpendicular to $\mathbf{m}$ and touching the inertia ellipsoid attains the same height. In particular, the holonomy angle $\Delta\phi$ still makes sense.*

Proof. A short calculation gives

$$(\omega, \mathbf{M}) = \text{const.} + \Omega(t)I, \tag{80}$$

showing that the polhodes are planar curves only when $I = 0$. ∎

12 Derivation of Formula (14)

We follow Montgomery's paper [15], pointing out the modifications where necessary.

12.1   Where and What to Integrate?

Configuration space is parametrized by three Euler angles

$$\varphi = \angle(x, \text{nodes}), \ \ \psi = \angle(\text{nodes}, X), \ \ \sigma = \angle(Z, z), \tag{81}$$

and one more angle θ, which gives the position of the flywheel.

Considering, on the Lie group $G = SO(3) \times S^1$, the left equivalence $T^*G \sim G \times \mathcal{G}^*$ gives the coordinates

$$(\varphi, \psi, \sigma, \theta, M_1, M_2, M_3, I).$$

We define an important closed curve $C = C_1 \cup C_2$, in a *three*-dimensional submanifold P of phase space $T^*(SO(3) \times S^1)$, fixing the constant values $\mathbf{m}, I, h$. Curve C_1 is described by the true dynamics in the interval $0 \leq t \leq T$, where T is the period of $\mathbf{M}(t)$ in the reduced dynamics; more precisely, C_1 is given by $(\varphi(t), \psi(t), \sigma(t), \theta(t), M(t), I)$.

For C_2 we fix $\mathbf{M}(t) = \mathbf{J}$ and $I(t) = I$, where without loss of generality, $\mathbf{J}$ is taken parallel to $(0, 0, 1)$. The coordinates φ and θ vary from 0 to $\Delta\varphi$ and from 0 to $\Delta\theta$, respectively. The remaining variables are fixed. C_1 and C_2 meet at the starting and ending points.

Exercise 12.1 *The action p dq (canonical form defined on phase space) can be written as*

$$pdq = (\mathbf{M}_g, d\mathbf{\Omega_g}) = I\, d\theta + \mathbf{M} \cdot d\mathbf{\Omega}. \tag{82}$$

Here $d\mathbf{\Omega}$ is a vector of 1-forms. Find them[22].

Hint. The calculations for $d\mathbf{\Omega}$ are in [15].

12.2   Applying Stokes's Theorem

Consider any two-dimensional surface Σ in P bounded by C. We compute the line integral of $p\, dq$ along C and equate it with the integral of $d(p\, dq)$ over Σ.

---

[22]To get the holonomy, one does not need the explicit expressions. The experts will tell you that the left and right "KAKS brackets" differ just in sign.

12.3   Line Integral over C_1

Since $(\mathbf{m}_g, \omega_{\mathbf{g}}) = 2h$, we get

$$\int_{C_1} (\mathbf{m}_g, \omega_{\mathbf{g}}) dt = 2hT. \tag{83}$$

12.4   Line Integral over C_2

Here we get one more term, besides the term $J\Delta\varphi$ in Montgomery's paper:

$$\int_{C_2} (I, 0, 0, J) \cdot (d\theta, 0, 0, d\varphi) = J\Delta\varphi + I\Delta\theta. \tag{84}$$

12.5   The Surface Integral over Σ

Actually, we compute the integral over the region $\Upsilon = \mathbf{M}(\Sigma)$, where $(\mathbf{M}, I) : T^*(SO(3) \times S^1) \to S^2 \times \mathbb{R}$. The first component $\mathbf{M}$ sends rotation matrices R to points $R^{-1}\mathbf{J}$ of the sphere:

$$\int\int_{\Sigma} d(pdq) = \int\int_{\mathbf{M}(\Sigma)-\Upsilon} J d\Upsilon = J\Upsilon, \tag{85}$$

where Υ is the solid angle enclosed by the trajectory of the reduced system.

13   Holonomy for the Gyrostat

Using Stokes theorem and equations (83), (84), (85), we obtain

$$\Delta\varphi = \frac{2hT - I\Delta\theta}{J} - \Upsilon. \tag{86}$$

Observe that $\Delta\theta$ can be obtained by a quadrature using equation (76):

$$\Delta\theta = \int_0^T \dot{\theta}(t) dt = \left(\frac{1}{I_3} + \frac{1}{\lambda_3}\right) IT - \frac{1}{I_3} \int_0^T M_3(t) dt. \tag{87}$$

Remark 13.1 *For the rigid body, Levi [10] made the beautiful observation that the polhode, viewed on the inertia ellipsoid, is transformed into the curve $M(t)$ in the momentum sphere by Gauss's mapping of elementary differential geometry. This is the starting point of his derivation of the* *holonomy formula*[23]. *We have also extended this approach for gyrostats* *[5].*

14   Final Remarks

Still another derivation of the holonomy for the rigid body follows from a general reconstruction formula, due to Marsden, Montgomery, and Ratiu [14], using connections on principal bundles.

In this example, the motion occurs simply by *inertia* in the configuration space of carrier *plus* flywheels. However, we can formulate an interesting *control problem*, where we allow "protocoled" flywheel motions and we desire a prescribed carrier holonomy. We anticipate that two flywheels suffice to achieve any reorientation.

Problem: Locate the flywheels optimally and find the optimal flywheel motions $\dot{\theta}_i(t)$ to achieve the desired reorientation. For instance, if we have one flywheel, located in the Z-axis, we get the following time-dependent Lagrangian on $TSO(3)$:

$$E = \frac{1}{2}(I\boldsymbol{\Omega}, \boldsymbol{\Omega}) + \frac{1}{2}(\lambda + \mathcal{M}d^2)(\Omega_1^2 + \Omega_2^2) + \frac{1}{2}\lambda_3\Omega_3^2 + \lambda_3\dot{\theta}\Omega_3. \qquad (88)$$

Notice the presence of the control variable $\dot{\theta}$ in the last term.

Another classic example is Kirchhoff's problem of the motion of a solid body through incompressible, inviscid, irrotational fluid. Here one has geodetic motions of a left-invariant metric on the group of rigid motions of three-dimensional space. The problem has six degrees of freedom; it is nonintegrable except at exceptional cases. See [2] and references therein. It would be instructive to compute the relevant holonomies for the integrable cases and to see what happens in the nearly integrable situations. A novel feature here is the possibility of *translational holonomy*, that is, how much the rigid body translates after one period of a periodic orbit of the reduced problem.

---

[23] Reasoning in reverse, Levi gave a "mechanical proof" for the Gauss–Bonnet theorem [11].

15 Derivation of the Hamiltonian

Following Arnold [3], we use the following conventions for reference frames and vectors: corresponding vectors in frames K and k will be denoted by the same letter, capitalized in the former. Thus:

$\mathbf{q} \in k$ represents a point of the body viewed in space.

$\mathbf{Q} \in K$ is the same point viewed in the body frame. $\mathbf{q}=R\mathbf{Q}$.

$\mathbf{v} = \dot{\mathbf{q}} \in k$ is the velocity vector.

$\mathbf{V} \in K$ is the same vector, but viewed in the body frame. $\mathbf{v}=R\mathbf{V}$.

$\omega \in k$ is the angular velocity viewed "from space."

$\Omega \in K$ is the angular velocity viewed "from the body." $\mathbf{w} = R\Omega$.

$\mathbf{m} \in k$ is the angular momentum viewed from space.

$\mathbf{M} \in K$ the angular momentum viewed from the body. $\mathbf{m}=R\mathbf{M}$.

Let f be the isomorphism between skew symmetric matrices and vectors in $\mathbb{R}^3$, $f : \mathcal{A} \to \mathbb{R}^3$, $f(A_{\mathbf{w}}) = \mathbf{w} = (w_1, w_2, w_3)$, $\mathbf{w} \in \mathbb{R}^3$, where

$$A_{\mathbf{w}} = \begin{pmatrix} 0 & -w_3 & w_2 \\ w_3 & 0 & -w_1 \\ -w_2 & w_1 & 0 \end{pmatrix}, \ A_{\mathbf{w}} \in \mathcal{A}.$$

We denote the vector product with the same symbol $[,]$ as commutators of matrices. We have a Lie algebra isomorphism, namely,

$$f([A, B]) = [f(A), f(B)].$$

Contribution of the Carrier

$$E_{carrier} = \frac{1}{2} \sum_i \mu \mathbf{v}_i^2$$
$$= \frac{1}{2} \sum_i \mu \mathbf{V}_i^2$$

$$= \frac{1}{2} \sum_i \mu [\mathbf{\Omega}, \mathbf{Q}_i]^2$$

$$= \frac{1}{2} \sum \mu([\mathbf{\Omega}, \mathbf{Q}_i], [\mathbf{\Omega}, \mathbf{Q}_i])$$

$$= \frac{1}{2} \sum \mu([\mathbf{Q}_i, [\mathbf{\Omega}, \mathbf{Q}_i], \mathbf{\Omega})$$

$$= \frac{1}{2} \sum_i (\mathbf{M}_i, \mathbf{\Omega})$$

$$= \frac{1}{2} (\mathbf{M}, \mathbf{\Omega})$$

$$= \frac{1}{2} (I\mathbf{\Omega}, \mathbf{\Omega}).$$

Here, as in the case of the usual free rigid body, we can assume that the positive definite symmetric matrix I is in diagonal form.

Contribution of the Flywheel

A point in the flywheel is written as

$$\mathbf{q} = RG\tilde{\mathbf{Q}} + R\mathbf{P}.$$

Differentiating, we get

$$\dot{\mathbf{q}} = \dot{R}(G\tilde{\mathbf{Q}}) + R(G\tilde{\mathbf{Q}})' + \dot{R}\mathbf{P}$$

$$= (\dot{R}R^{-1})RG\tilde{\mathbf{Q}} + R(\dot{G}\tilde{\mathbf{Q}} + G\dot{\tilde{\mathbf{Q}}}) + \dot{R}R^{-1}\mathbf{p}$$

$$= \dot{R}R^{-1}(\mathbf{q} - \mathbf{p}) + R(\dot{G}G^{-1})G\tilde{\mathbf{Q}} + \dot{R}R^{-1}\mathbf{p}.$$

Therefore,

$$\dot{\mathbf{q}} = [\omega, \mathbf{q} - \mathbf{p}] + R[\Theta, (\mathbf{Q} - \mathbf{P})] + [\omega, \mathbf{p}].$$

where $\Theta = f(\dot{S}S^{-1})$, $\tilde{\Theta} = f(S^{-1}\dot{S})$. Let E_{cat} be the kinetic energy of the flywheel. Then we have:

$$E_{cat} = \sum \frac{1}{2} \mu([\omega, \mathbf{q} - \mathbf{p}], [\omega, \mathbf{q} - \mathbf{p}])$$

$$+ \sum \mu([\omega, \mathbf{q} - \mathbf{p}], R[\Theta, \mathbf{Q} - \mathbf{P}])$$

$$+ \sum \frac{1}{2} \mu([\omega, \mathbf{p}], [\omega, \mathbf{p}])$$

$$+ \sum \frac{1}{2} \mu(R[\Theta, \mathbf{Q} - \mathbf{P}], R[\Theta, \mathbf{Q} - \mathbf{P}])$$

$$+ \sum \mu([\omega, \mathbf{q} - \mathbf{p}], [\omega, \mathbf{p}])$$
$$+ \sum \mu(R[\Theta, \mathbf{Q} - \mathbf{P}], [\omega, \mathbf{p}]).$$

Calculations in a Particular Case

For simplicity, we place the flywheel to the Z-axis of the inertia ellipsoid of the main body. Clearly, the inertia operator of the flywheel alone in the base $\tilde{K}$ (I_{cat}) is diagonal, with two equal eigenvalues (λ) and one possibly distinct (λ_3). The equality of the first two eigenvalues reflects the *material symmetry* of the flywheel implying that

$$G I_{cat} G^{-1} = I_{cat}.$$

This is true because G is a matrix of the type

$$\begin{pmatrix} \cos \dot{\theta} t & -\sin \dot{\theta} t & 0 \\ \sin \dot{\theta} t & \cos \dot{\theta} t & 0 \\ 0 & 0 & 1 \end{pmatrix}.$$

Thus $\Theta = f(G^{-1}\dot{G}) = \dot{\theta}(0, 0, 1)$.

Nonvanishing Terms

$$\sum \frac{1}{2} \mu([\omega, \mathbf{q} - \mathbf{p}], [\omega, \mathbf{q} - \mathbf{p}]) = \sum \frac{1}{2} \mu([\Omega, \mathbf{Q} - \mathbf{P}])^2$$
$$= \sum \frac{1}{2} \mu([\mathbf{Q} - \mathbf{P}, [\Omega, \mathbf{Q} - \mathbf{P}]], \Omega)$$
$$= \frac{1}{2} (G I_{cat} G^{-1} \Omega, \Omega)$$
$$= \frac{1}{2} \lambda(\Omega_1^2 + \Omega_2^2) + \frac{1}{2} \lambda_3 \Omega_3^2.$$

From (89), we get

$$\sum \mu([\omega, \mathbf{q} - \mathbf{p}], R[\Theta, \mathbf{Q} - \mathbf{P}]) = \sum \mu([\Omega, \mathbf{Q} - \mathbf{P}], [\Theta, \mathbf{Q} - \mathbf{P}])$$
$$= \sum \mu([\mathbf{Q} - \mathbf{P}, [\Theta, \mathbf{Q} - \mathbf{P}]], \Omega)$$

$$= (I_{cat}\mathbf{\Theta}, \mathbf{\Omega})$$
$$= (\mathbf{\Omega}, \lambda_3 \mathbf{\Theta})$$
$$= \lambda_3 \Omega_3 \dot{\theta}.$$

On the other hand, since $|\mathbf{P}| = d$, it follows that

$$\sum \mu([\omega, \mathbf{p}], [\omega, \mathbf{p}]) = \frac{1}{2}\mathcal{M}(\|\mathbf{\Omega}\|^2 \|\mathbf{P}\|^2 - (\mathbf{\Omega}, \mathbf{P})^2)$$
$$= \frac{1}{2}\mathcal{M}\{(\Omega_1^2 + \Omega_2^2 + \Omega_3^2)d^2 - \Omega_3^2 d^2\}$$
$$= \frac{1}{2}\mathcal{M}d^2(\Omega_1^2 + \Omega_2^2),$$

where $\mathcal{M}$ is the mass of the flywheel. Finally, we obtain an expected term. Developing (89) in detail, we get:

$$\sum \frac{1}{2}\mu(R[\mathbf{\Theta}, \mathbf{Q} - \mathbf{P}], R[\mathbf{\Theta}, \mathbf{Q} - \mathbf{P}]) = \sum \frac{1}{2}\mu([\mathbf{\Theta}, \mathbf{Q} - \mathbf{P}], [\mathbf{\Theta}, \mathbf{Q} - \mathbf{P}])$$
$$= \sum \frac{1}{2}\mu(\mathbf{Q} - \mathbf{P}, [\mathbf{\Theta}, \mathbf{Q} - \mathbf{P}], \mathbf{\Theta})$$
$$= \frac{1}{2}(I_{cat}\mathbf{\Theta}, \mathbf{\Theta})$$
$$= \frac{1}{2}(\lambda_3\mathbf{\Theta}, \mathbf{\Theta})$$
$$= \frac{1}{2}\lambda_3\dot{\theta}^2.$$

Vanishing Terms

Since the center of mass of the flywheel is located at $\mathbf{P} \in K$, we have $\tilde{\mathbf{P}} = 0$. Thus, from (89) we conclude that:

$$\sum \mu([\omega, \mathbf{q} - \mathbf{p}], [\omega, \mathbf{p}]) = \sum \mu([\mathbf{\Omega}, \mathbf{Q} - \mathbf{P}], [\mathbf{\Omega}, \mathbf{P}])$$
$$= \sum \mu([\mathbf{Q} - \mathbf{P}, \mathbf{\Omega}], [\mathbf{P}, \mathbf{\Omega}])$$
$$= \sum \mu(\mathbf{P}, [\mathbf{\Omega}, [\mathbf{Q} - \mathbf{P}, \mathbf{\Omega}]])$$
$$= \mathcal{M}(\mathbf{P}, [\mathbf{\Omega}, [G\tilde{\mathbf{P}}, \mathbf{\Omega}]])$$
$$= 0.$$

Similarly,

$$\sum \mu(R[\Theta, \mathbf{Q} - \mathbf{P}], [\omega, \mathbf{p}]) = \sum \mu([\Theta, \mathbf{Q} - \mathbf{P}], [\mathbf{\Omega}, \mathbf{P}])$$
$$= \mathcal{M}([\Theta, G\tilde{\mathbf{P}}], [\mathbf{\Omega}, \mathbf{P}])$$
$$= 0.$$

Contribution of the Flywheel

Summarizing our calculations,

$$E_{flywheel} = \frac{1}{2}\lambda_3 \dot{\theta}^2$$
$$+ \frac{1}{2}(\lambda + \mathcal{M}d^2)(\Omega_1^2 + \Omega_2^2)$$
$$+ \frac{1}{2}\lambda_3 \Omega_3^2$$
$$+ \lambda_3 \dot{\theta}\Omega_3.$$

References

[1] Abraham, R., Marsden, J. E., *Foundations of Mechanics*, Addison-Wesley, Boston (1978).

[2] Aref, H., Jones, S. W., *Chaotic motion of a solid through ideal fluid*, Phys. Fluids A **5**, 3026–3028 (1993).

[3] Arnold, V. I., *Mathematical Methods of Classical Mechanics*, Springer-Verlag, New York (1978).

[4] Bloch, A., Krishnaprasad, P. S., J. Marsden, J. E., Sanchez de Alvarez, G., *Stabilization of rigid body dynamics by internal and external torques*, Automatica **28**, 745–746 (1994).

[5] Almeida, M. L. B. P., *Holonomias para corpos rígidos e girostatos*, M.Sc. thesis, Instituto de Matemática, Universidade Federal do Rio de Janeiro (1993).

[6] Goldstein, H., *Classical Mechanics*, Addison-Wesley, Boston (1980).

[7] Hubert, C. H., *An attitude acquisition technique for dual-SPIN spacecraft*, Ph.D thesis, Cornell University, Ithaca, NY (1980).

[8] Landau, L., E. Lifchitz, E., *Mécanique*, MIR, Moscow (1966).

[9] Leimanis, E., *The General Problem of the Motion of Coupled Rigid Bodies about a Fixed Point*, Springer-Verlag, New York (1965) .

[10] Levi, M., *Geometric phases in the motion of rigid bodies*, Arch. Rational Mech. Anal. **122**, 213–229 (1993).

[11] Levi, M., *A "bicycle wheel" proof of the Gauss–Bonnet theorem, dual cones and some mechanical manifestations of the Berry phase*, Expos. Math. **12**, 145–164 (1993).

[12] Krishnaprasad, P. S., *Lie Poisson structures on dual spin spacecraft and asymptotic stability*, Nonlinear Anal. **9:10**, 1011–1035 (1985).

[13] Marsden, J. E., *Lecture Notes on Mechanics*, London Math. Soc. Lect. Notes Series vol. **174**, Cambridge University Press, Cambridge, UK (1992).

[14] Marsden, J. E., Montgomery, R., Ratiu, T., *Reduction, symmetry, and phases in mechanics*, Mem. Amer. Math. Soc. **88**, 436 (1990).

[15] Montgomery, R., *How much does the rigid body rotate? A Berry's phase from the 18th century*, Amer. J. Phys. **59**, 394–398 (1991).

[16] Wittenburg, J., *Dynamics of Systems of Rigid Bodies*, B.G. Teubner, Stuttgart (1977).

Part 4: Microswimming

The lives of protozoa and bacteria may be unfamiliar to us "higher" forms of life, and therefore dismissed as uninteresting. We should not be so arrogant: as Stephen Jay Gould says, to bacteria, we are mountains full of exploitable goodies. They are much better fit for survival!

E. coli[24], the most common intestinal bacteria, are approximately 2×10^{-4}cm in length and 10^{-4} cm wide. They have six flagellar filaments emerging from random points on the cell body. At such microscopic sizes, water is so viscous that inertia plays no role.

Reynolds number, given by

$$R_e = LV/\mu = \frac{\text{Inertial Forces}}{\text{Viscous Forces}}, \tag{89}$$

where L, V, μ are, respectively, a characteristic length, velocity, and kinematic viscosity, measures the relative importance of inertia to viscosity in fluid dynamical problems. If $R_e \ll 1$, viscosity effects dominate. Reynolds number for swimming microorganisms is typically between 10^{-2} for protozoa and 10^{-5} for motile bacteria. They propel themselves in an inertialess environment.

"*Earnest*" teaches many lessons to the biologist, to the mechanical or electrical engineer, and even to the psychologist. *E. coli* moves in a "biased" random way: runs alternating with tumbles. In a favorable direction, the runs are statistically longer. But the runs are not shorter in a bad direction, so *Earnest* is an optimist. For the basic biophysics of microswimming, see E. Purcell's beautiful talk, "Life at Low Reynolds number" [19] and Berg's book, *Random Walks in Biology* [1].

Here are some strategies observed in nature:

- *E. coli* swim using helical propellers that rotate.
- Small nematodes and spermatozoan tails swim using planar undulations.
- *Spirochetes* swim using internal flagella.
- *Synechoccocus* swim using traveling compression waves.

---

[24] *E.* stands for Escherichia, but we prefer rather *Earnest*.

• Many protozoa swim by waving a layer of densely packed cilia (Figure 10).

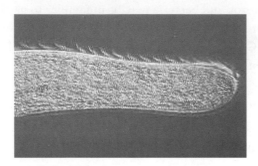

Figure 10. Metacrony in *Spirostomum*. Cilia work in coordination through a space-temporal wave along the cell surface (Dusenbery [6]).

16   Historical Remarks

The implications of low Reynolds number for microswimming were realized only in 1930 (Ludwig [17]). Perturbation methods were used by Taylor ([21], 1951) in his treatment of the infinite swimming sheet. This problem was also discussed in Blake ([4], 1971), and Childress ([5], 1981). Lighthill ([16], 1952) calculated the swimming velocity of a squirming spherical cell. Blake [2] corrected some of Lighthill's formulas and adapted Lighthill's model to explain ciliary propulsion by replacing the loci of the cilia tips by a continuous envelope. For certain densely ciliated organisms, such as *Opalina*, this model provides good results.

Blake ([4], 1971) discussed the problem of swimming in two dimensions. The swimming velocity of a cell with circular geometry, due to small amplitude, symmetric traveling waves was approximated using perturbation techniques. Shapere and Wilczek ([20], 1989) interpreted the problem in differential geometric terms, namely, as a connection on a principal bundle. This description places the problem in a broader class of kinematical problems that includes satellite reorientation and the ability of a falling cat to land on her feet (Montgomery 1990, [18]).

17   The Configuration Space

The configuration space, which we will denote by $\mathcal{Q}$, describes the organism in its environment and is the space of all parametrized embeddings q : $S^2 \to \mathbb{R}^3$. The image $\Sigma = q(S^2)$ represents the outer membrane of the organism or the ciliary "envelope," [25] and will be referred to as a "located shape." It is important to keep in mind that q is a *parametrized* embedding: reparametrizations of the *same* geometric shape represent *different* states of the organism[26].

By composition on the image, there is a left action on $\mathcal{Q}$ by the group $SE(3)$ of Euclidean motions

$$SE(3) \hookrightarrow \mathcal{Q} \to \mathcal{S}. \tag{90}$$

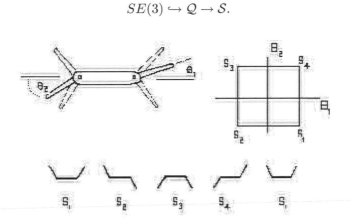

Figure 11. In what direction will Purcell's *animat* go? See [19].

The base space of the bundle is $\mathcal{S} = \mathcal{Q}/SE(3)$, the space of "unlocated shapes." The tangent space $T_q\mathcal{Q}$ at an embedding $\Sigma = q(S^2)$ consists of all vector fields along Σ,

$$\vec{v} : \Sigma \to \mathbb{R}^3, \tag{91}$$

and represents an infinitesimal boundary motion. Purcell's "cult paper" [19] has a delightful example with a challenge to the reader; see Figure 11.

---

[25] Flagellary motion can also be studied, using techniques from slender body theory.

[26] Certain organisms (e.g., *synechoccocus*) are thought to swim using what can be mathematically described as time-dependent self-reparametrizations: traveling compression and expansion waves along the outer membrane [8].

The two most important features are the following:

- $\mathcal{Q}$ possesses a very natural Riemannian metric, given by the hydrodynamical power expenditure. Remarkably, the physical constraints for self-locomotion, namely, no net force or torque being exerted on the fluid, coincide with the horizontal spaces of the associated mechanical connection A.

- Swimming velocity is computed in terms of the *curvature* coefficients via the small amplitude approximation (104) given below.

To compute the curvature for a specific shape, we need the following ingredients:

- Solutions to Stokes equations with boundary conditions specified on the base shape.

- An expression for the Lie bracket of these vector fields in the exterior of the shape.

- An expression for the connection form A.

- A splitting of the basis vectors into horizontal and vertical.

This program was started by our group in the papers [11, 12, 13, 14]. Collaborations for future work are welcome. The horizontal distributions for the indicated geometries can be inferred from the work by the following authors:

	Geometry	Restrictions
Taylor (1951) [21]	Planar	Axially symmetric
Lighthill (1952) [16]	Spherical	Axially symmetric
Blake (1970)	Spherical, circular, and cylindrical	Axially symmetric
Shapere–Wilczek (1989) [20]	Spherical and circular	None
Ehlers (1995)	Elliptical	None

18 Hydrodynamics

Navier–Stokes equations (without body forces) are given by

$$\nabla \cdot \hat{v} = 0, \qquad \rho \frac{D\hat{v}}{Dt} = -\nabla p + \mu \nabla^2 \hat{v}, \qquad (92)$$

where ρ is the density and $\hat{v}$, p the velocity and pressures of the fluid external to the organism described by the shape $\Sigma \in \mathcal{Q}$.

In the Stokesian realm where inertial effects are neglected, we use the Stokes approximation. These linear partial differential equations of elliptic type are

$$\nabla \cdot \hat{v} = 0, \qquad -\nabla p + \mu \nabla^2 \hat{v} = 0. \qquad (93)$$

The nonslip assumption gives the boundary condition $\vec{v} \in T_\Sigma \mathcal{Q}$, defining a unique solution, which we denote $(\Sigma, \hat{v}, p)$.

Exercise 18.1 *Recall that the group of Euclidean motions acts on $\mathcal{Q}$. Let R be a rotation matrix, then whenever $\hat{v}(\vec{r})$, $p(\vec{r})$ is a solution with prescribed boundary values at Σ, $R^T \hat{v}(R\vec{r} + \vec{b})$, $p(R\vec{r} + \vec{b})$ is also a solution with prescribed values on $R\Sigma + \vec{b}$.*

Remark 18.2

(i) *At each instant the neutrally buoyant, low Reynolds number swimmer does not exert zero net forces and torques on the fluid [5]. The inertial terms in the equations of motion that would account for such an imbalance are not present (the nonlinear terms in the Navier–Stokes equations). It is this condition that will define a geometrical connection on the space of located shapes.*

(ii) *Because acceleration does not play a role, the equations of motion are time independent. The boundary condition depends on time, but the fluid velocities depend only on the instantaneous boundary velocity field. A given change in shape leads to an instantaneous motion through the whole fluid no matter how fast it is carried out (so long as R_e remains $\ll 1$).*

(iii) *The "scallop theorem" [19]: because of time independence, reciprocal motions lead to no net translation. Stokes flows are reversible: one* *can stir then unstir fluids if $R_e \ll 1$!!! There are no low Reynolds number "scallops"; low Reynolds number swimmers must have at least two-degrees of freedom.*

The stress tensor σ is given by

$$\sigma_{ij} = -p\delta_{ij} + \mu \left(\frac{\partial u_i}{\partial x_j} + \frac{\partial u_i}{\partial x_j} \right). \tag{94}$$

Exercise 18.3 *Show, from Stokes equations, that the stress tensor σ is divergence-free.*

Given any mathematical surface S outside Σ with normal $\vec{n}$, one computes the field of surface forces (stresses) along S:

$$\vec{f} = \sigma(\hat{v}) \cdot \vec{n}.$$

The classic "Lorentz reciprocity theorem" [9] says that the operator

$$\vec{v} \to \vec{f}$$

on the space of vector fields along Σ is self-adjoint and positive. Hence we have the following definition.

Definition 18.4 *The hydrodynamical power dissipation*

$$P = \int_\Sigma \vec{f} \cdot \vec{v} \, dS$$

defines a Riemannian metric on $\mathcal{Q}$, which we call the power metric.

19   The Momentum Mapping

Proposition 19.1 *The momentum map $\mu : TQ \to se(3)^*$ is given by*

$$\mu(\vec{v}_\Sigma) \cdot (\vec{w}, \vec{b}) = \vec{w} \cdot \int_\Sigma \vec{r} \times \sigma(\hat{v}) \cdot \vec{n} \, dS + \vec{b} \cdot \int_\Sigma \sigma(\hat{v}) \cdot \vec{n} \, dS \tag{95}$$

$$= \vec{w} \cdot \int_\Sigma \vec{r} \times \vec{f} \, dS + \vec{b} \cdot \int_\Sigma \vec{f} \, dS = \vec{w} \cdot \vec{T} + \vec{b} \cdot \vec{F},$$

where we have represented an element of $se(3)$ as a pair of vectors $(\vec{w}, \vec{b})$. Here $\vec{F} = \int_\Sigma \vec{f} \, dS$ is the total force, and $\vec{T} = \int_\Sigma \vec{r} \times \vec{f} \, dS$ is the total torque.

Exercise 19.1 *Prove this formula, using the abstract nonsense definition (16). The identification $TQ \equiv T^*Q$ is given by the power metric*

$$\langle \vec{v}, \vec{w} \rangle = \int_\Sigma \vec{v} \cdot \sigma(\hat{w}) \vec{n} \, dS.$$

Hint. For any element $\xi = (\vec{w}, \vec{3}) \in se(d)$ of the Lie algebra, ξ_P is an infinitesimal rigid motion of the shape Σ and is described by the *rigid motion vector field*, $\xi_P(\Sigma) = (\vec{w} \times \vec{r} + \vec{b})|_\Sigma$. Therefore,

$$
\begin{aligned}
\mu(\vec{v}_\Sigma) \cdot (\vec{w}, \vec{b}) &= \int_\Sigma (\vec{w} \times \vec{r} + \vec{b}) \cdot \sigma(\hat{v}) \vec{n} \, dS \\
&= \int_\Sigma (\vec{w} \times \vec{r}) \cdot \sigma(\hat{v}) \vec{n} \, dS + \int_\Sigma \vec{b} \cdot \sigma(\hat{v}) \vec{n} \, dS \\
&= \vec{w} \cdot \int_\Sigma \vec{r} \times \sigma(\hat{v}) \vec{n} \, dS + \vec{b} \cdot \int_\Sigma \sigma(\hat{v}) \vec{n} \, dS.
\end{aligned}
$$

■

Exercise 19.2 *Since σ is divergence-free, the total force and the total torque can be computed by integrating along any other mathematical surface surrounding Σ.*

Remark 19.3 *The power metric is degenerate in dimension $d = 2$. In fact, the Stokes paradox says that there is no Stokesian flow vanishing at infinity for a uniformly translating cylinder, or, what is the same, there is no Stokesian flow past a cylinder that is uniform at infinity. One can remedy this by admitting flows with logarithmical singularities at infinity, but we do not pursue this approach. Rather, in two dimensions we admit that for translations of a shape $\vec{v} = \vec{b}$, its Stokes extension is $\hat{v} = \vec{b}$, in which the fluid moves rigidly as a whole with constant pressure $p \equiv 0$. It follows that $\langle \vec{v}, \vec{b} \rangle = 0$ for all $\vec{v}$. In other words, the Legendre transform that associates forces to velocities has a nontrivial kernel generated by the rigid translations.*

Definition 19.4 *Vertical vector fields are those generated by infinitesimal rigid motions of the shape, and horizontal vector fields form the orthogonal complement to the vertical space, with respect to the power metric.*

Proposition 19.2 *For $d = 3$, a vector $\vec{v}_\Sigma$ is horizontal if and only if $\mu(\vec{v}_\Sigma) = 0$. For $d = 2$, $\vec{v}_\Sigma$ is horizontal if and only if $\vec{v}$ is not a rigid translation and $\mu(\vec{v}_\Sigma) = 0$.*

The following result is very useful for the calculations.

Proposition 19.3 *In order to compute the momentum map, it is enough to find the Stokes extension of a finite number of boundary vector fields, namely, those given by infinitesimal rigid motions $\vec{w} \times \vec{r} + \vec{b}$.*

Proof.   Let $\hat{b}$ and $\hat{w}$ denote the Stokes extensions of translations and rotations, respectively. First we consider translations

$$\mu(\vec{v})(\vec{0}, \vec{b}) = \vec{b} \cdot \int_\Sigma \sigma(\hat{v}) \cdot \vec{n} \, dS = \int_\Sigma \vec{b} \cdot \sigma(\hat{v}) \cdot \vec{n} \, dS = \int_\Sigma \vec{v} \cdot \sigma(\hat{b}) \cdot \vec{n} \, dS,$$

and similarly for the rotations,

$$\mu(\vec{v})(\vec{w}, \vec{0}) = \vec{w} \cdot \int_\Sigma \vec{r} \times \sigma(\hat{v}) \cdot \vec{n} \, dS = \int_\Sigma \vec{w} \cdot (\vec{r} \times \sigma(\hat{v}) \cdot \vec{n}) \, dS$$
$$= \int_\Sigma (\vec{w} \times \vec{r}) \cdot \sigma(\hat{v}) \cdot \vec{n} \, dS = \int_\Sigma \vec{v} \cdot \sigma(\hat{w}) \cdot \vec{n} \, dS.$$

∎

Definition 19.5 *Take a basis $\vec{v}_k$, $k = 1, \ldots, 6$ for the vertical vector fields, that is, the three unit translations and the three unit infinitesimal rotations. The 6×6 matrix I with entries $\langle \vec{v}_i, \vec{v}_j \rangle$ is called in fluid dynamics the resistance matrix. In our language it corresponds to the locked inertia tensor.*

Decompose $\vec{v}$ into its horizontal and vertical parts, $\vec{v} = \vec{v}^h + \vec{v}^v$. Write $\vec{v}^v = \sum c_k \vec{v}_k$ so that

$$0 = \langle \vec{v}^h, \vec{v}_l \rangle = \langle \vec{v}, \vec{v}_l \rangle - \langle \vec{v}^v, \vec{v}_l \rangle = \langle \vec{v}, \vec{v}_l \rangle - \sum c_k \langle \vec{v}_k, \vec{v}_l \rangle.$$

Therefore,

$$I(\vec{v}^v) \cdot \vec{v}_l = \sum c_k I(\vec{v}_k) \cdot \vec{v}_l$$
$$= \langle \vec{v}, \vec{v}_l \rangle,$$

which says that

$$I_\Sigma(\vec{v}^v) = \langle \vec{v}, \cdot \rangle = \mu(\vec{v}).$$

In the case $d = 3$, I is invertible and so

$$(\vec{b}, \vec{w}) = A(\vec{v}) = I^{-1}(\mu(\vec{v})). \tag{96}$$

Definition 19.6 *The Stokes extension* $\vec{v}_{(\vec{b}, \vec{w})}$ *associated to the rigid motion* $(\vec{b}, \vec{w})$ *$(= A(\vec{v}))$ is called the counterflow of the velocity field* $\vec{v}$.

In practice, we search for a counterflow with boundary condition $(\vec{b}, \vec{w})$ whose Stokes resistance $I((\vec{b}, \vec{w}))^\dagger$ equals $\mu(\vec{v})$. Then we subtract:

$$\vec{v} - \vec{v}_{(\vec{b}, \vec{w})} \text{ is horizontal.}$$

How about the curvature of the connection? There is a beautiful "master formula" by Shapere and Wilczek [20], derived more rigorously in Ehlers' thesis [7]:

$$\mathcal{F}_\Sigma(\vec{u}, \vec{v}) = A([\vec{v}^h, \vec{u}^h]), \tag{97}$$

where $[\,,\,]$ is the Lie bracket of vector fields.

Exercise 19.7 *Prove the "master formula" (97).*

Hint. Start with Cartan's formula $d\omega(u_o, v_o) = v \cdot \omega(u) - u \cdot \omega(v) - \omega[u, v]$ for the exterior derivative of a 1-form on a manifold Q. Recall that in the right-hand side, u and v are arbitrary extensions of vectors $u_o, v_o \in T_q Q$. Using the solution of the Stokes equations, extend the tangent vector $\vec{v}_m^h \in T_\Sigma Q$ to a neighborhood of Σ. Since the stress tensor is divergence-free, the Stokes extension of the horizontal projection $\vec{v}^h$ at the shape Σ remains horizontal for deformed shapes $\Sigma(t)$. Hence the first two terms in Cartan's formula vanish. It remains to show that the Lie bracket of vector fields in the infinite-dimensional manifold Q can be computed using the familiar Lie bracket of vector fields[27]. ∎

Exercise 19.8 *What is needed from fluid mechanics in order to study motions due to self-reparametrizations of* Σ *(vector fields tangent to* Σ*)?*

---

[27]This is so plausible that it could remain unnoticed, but proving this fact requires some abstract thinking.

Solution.   In (97) the Lie bracket of tangential vector fields does not require the Stokes extensions. So fluid mechanics enters (i) to obtain the resistance matrix I, and (ii) to find total force and torque for a given tangential vector field. Figure 12 and 13 explain qualitatively the counterintuitive fact that the motion of the organism is in the same direction as the waves of contraction/expansion.

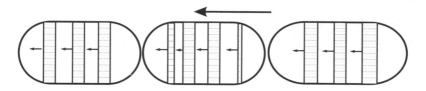

Figure 12. The tangential traveling wave mechanism. The black and white areas represent regions of contraction and expansion on the cell's outer membrane. The cell travels in the same direction of the wave.

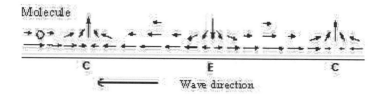

Figure 13. What happens to the test molecule? The wave train moves to the left. The molecule tends to move more to the right than to the left. Why? Explain qualitatively.

Exercise 19.9 *(Lighthill's Fugo [12]). Consider a spherical cell of radius r_1 that draws an angle of $\delta\phi$ of its outer membrane towards one pole, shrinks its radius to r_2, extends its membrane towards the other pole for $\delta\phi$, then expands again. The cell will translate further through the fluid during the first leg than during the third, thereby producing a net translation after a complete cycle. Quantify this.*

Solution.   Parametrize the sphere S by (ϕ, θ), where ϕ is the azimuthal coordinate. Let $u_T(\phi)$ be the tangential deformation field in the direction of the meridians. We need the following information from hydrodynamics: For a sphere of radius a, the stress vector corresponding to a rigid translation $\hat{\mathbf{U}}$ is constant in the same direction of $\hat{\mathbf{U}}$ at every point of the sphere (this property seems to hold uniquely for the sphere). Its value, incidentally, is $3\mu\hat{U}/2a$.

Exercise 19.10 *The corresponding total force is $\hat{\mathbf{F}} = 6\pi\mu a\hat{\mathbf{U}}$, the famous Stokes drag for the sphere.*

Exercise 19.11 *Show that the countervelocity associated with the deformation field u^T is simply the average over the sphere surface*

$$\vec{U} = -\frac{1}{4\pi a^2} \int_S u_T \sin\phi \, dS. \tag{98}$$

Continuing There is no translation associated with expansion and contraction legs of the swimming stroke, so we need only calculate the translation associated with the first and third legs of the stroke. Suppose that the boundary vector fields during these legs are

$$u_T = \pm c\frac{\partial}{\partial\phi}.$$

On each of the first and third legs the membrane shrinks/expands by $r_i\delta\phi$. Note that the durations in time are different:

$$\Delta t_i = \delta\phi r_i/c.$$

The countervelocities, in the direction of the z-axis, are in absolute value

$$\frac{1}{4\pi a^2} \int_S c \sin\phi \, dS = \frac{c\pi}{4}.$$

Therefore, the distance traveled on each leg i is in absolute value

$$\frac{c\pi}{4}\frac{\delta\phi r_i}{c} = \frac{\pi\delta\phi r_i}{4},$$

so the distance traveled by the organism per stroke is

$$\frac{\pi\delta\phi}{4}(r_2 - r_1).$$

If the cell swims at a rate of f strokes per second, then the velocity is

$$\frac{\pi\delta\phi f}{4}(r_2 - r_1). \tag{99}$$

■

20 Swimming = Holonomy

A time periodic swimming stroke is given by a loop in shape space with $s(0) = s(T)$. The swimmer can make progress because the horizontal distribution is *nonholonomic*[28]. After a complete swimming stroke, the swimmer assumes its original shape but its position in the fluid has changed by a rigid motion which is an element of the Euclidean group:

$$q(T) = g(T) \cdot q(0). \tag{100}$$

The quantity $g(T)$, which represents the net rotation and translation due to the swimming stroke, is called the *holonomy* of the connection.

In terms of the Stokes connection form, $g(t)$ satisfies the differential equation

$$g^{-1}(t)_* g'(t) = -A_{q(t)}(q'(t)). \tag{101}$$

Exercise 20.1 *Prove this formula.*

Hint. Here $q(t)$ is an arbitrary curve of located shapes with $\pi q(t) = s(t)$. There is a unique curve $g(t)$ in G such that $g(t)q(t)$ is the horizontal lift of $s(t)$. This means that

$$A\left(\frac{d}{dt} g(t)q(t)\right) = 0.$$

Use the equivariance property of the connection to finish the proof. ∎

The solution to this differential equation is written *formally* as a *path ordered* integral [20]

$$g = \text{Path}\exp\left(-\int_0^T A_{q(t)}(q'(t))\, dt\right). \tag{102}$$

Note that the path ordering is necessary because the group of Euclidean motions is not Abelian.

---

[28]From the viewpoint of control theory, one would like the horizontal distributions to be as far from integrable (in the Frobenius sense) as possible (otherwise the organism would be constrained to a submanifold of its environment!)

Proposition 20.1 *If a swimming stroke is given as*

$$\Sigma(t) = \Sigma + \sum_j a_j(t)\,\vec{v}_j, \ 0 \le t \le T, \tag{103}$$

where Σ is a base shape, $\{\vec{v}_j\}$ is a basis for the vector fields along Σ, and a_j the associated amplitude functions, then after one stroke the net translation and rotation are given by

$$g = I + \sum_{m<n} \mathcal{F}_{mn} \int_0^T a_m(t)\dot{a}_n(t)\,dt + O(|a|^3). \tag{104}$$

Here $\mathcal{F}$ is the curvature of the connection:

$$\mathcal{F}_{mn} = A([\vec{v}_n^h,\ \vec{v}_m^h]), \tag{105}$$

where $[\,,\,]$ is the Lie bracket of vector fields.

Remark 20.2 (i) *The term $\sum_{m<n}\mathcal{F}_{mn}\int_0^T a_m(t)\dot{a}_n(t)\,dt$ (which is quadratic in the amplitude) is the first term in the expansion for the path-ordered integral (102).* (ii) *In most cases of interest, symmetry of the wave patterns implies that the motions of the cell are along a fixed axis. In this case the motion is independent of path in the sense that the trajectory can be broken into infinitesimal pieces rearranged and reassembled into the same trajectory. If, for example, the motion involves a rotation around one axis and a translation along another, time averaging may give erroneous results. To see where the problem occurs, consider the infinitesimal motion consisting of a rotation about the z-axis and a translation along the x-axis; the net infinitesimal motion then depends on the order in which these are taken. When reconstructing the finite motion from infinitesimal ones, this "path ordering" must be taken into account.*

21   Nonspherical Self-Reparametrizing Cells

The difficulty in computing the swimming velocities for an *arbitrary deformation* of a shape is that solutions to the Stokes equations must be developed with boundary conditions prescribed on that complicated shape. It is possible, in principle, to carry out this program for cells whose average shape is geometrically simple (prolate spheroidal, for example), and even for the general ellipsoid, but the extremely complicated analysis limits the usefulness of that method.

Calculations by Kurt Ehlers (not published before) will be presented for the following restricted situation:

The physical shape Σ will be constant, but its parametrization $q(t)$ will be time dependent. For simplicity we will write $\Sigma(t)$ for $q(t)(S^2)$; One would like to determine the importance of the particular shape, but with the same "size" (volume or surface area) on the swimming performance.

From Exercise 19.8, if we restrict to *tangential* wave forms (those thought to be responsible for the motions of *Synechococcus*), then the only solutions to the Stokes equations are those necessary to compute the stress tensor for *streaming* flow past the shape. This is a very significant simplification, since the solutions for streaming flow past objects of various geometries have been computed [9].

We present approximate formulae for the propulsive velocity of a *spheroidal cell* that swims using travelling surface waves. We show that a prolate spheroid swims faster along its axis of symmetry than a sphere or oblate spheroid of the same "size" (volume or surface area). We also derive a formula for the swimming velocity of an oblate spheroid that swims in a direction perpendicular to its axis of symmetry.

21.1   Prolate and Oblate Spheroids

In our analysis we will use the solution, due to Happel–Brenner [9], of Stokes equations with boundary conditions on a nearly spherical cell given by (here θ is the azimuthal coordinate)

$$r(\phi, \theta) = a(1 + \epsilon f(\phi, \theta)). \tag{106}$$

The Stokes extension is given as a power series in the nondimensional parameter ϵ.

Consider an ellipsoidal cell whose shape is defined by the equation

$$\frac{x^2 + y^2}{a^2} + \frac{z^2}{a^2(1 - \epsilon)^2} = 1. \tag{107}$$

If $\epsilon > 0$, then the spheroid is oblate; if $\epsilon < 0$, the spheroid is prolate. To first order in ϵ the cell's shape has polar coordinate (P_0 and P_2 are Legendre polynomials)

$$r(\theta) = a(1 - \epsilon \cos^2 \theta) = a \left(1 - \epsilon \left(\frac{1}{3} P_0(\cos \theta) + \frac{2}{3} P_2(\cos \theta) \right) \right). \quad (108)$$

Now consider travelling waves on the outer membrane of the cell of the form

$$\theta \to \theta + \eta \sin(n\theta - \omega t). \quad (109)$$

These can be envisioned as trains of waves of contractions and expansions travelling down the cell body. The main result of our analysis is as follows.

Proposition 21.1 *The swimming velocity of an oblate/prolate spheroid with major and minor axis a and $a - \epsilon$ is*

$$U = \frac{\pi}{8} \eta^2 n\omega a \left(1 - \frac{1}{2} \epsilon \right) + O(\epsilon^2, \eta^4). \quad (110)$$

Here η is the amplitude of the wave, n is the wave number, and ω is the frequency.

By normalizing the volume of the cell and amplitude of the oscillation, we show that to first order in ϵ, a prolate spheroid swims faster than an oblate spheroid by a factor of nearly 2ϵ.

21.2 Outline of the Calculation

Let (r, ϕ, θ) be spherical coordinates where θ is the azimuthal coordinate. Consider a spheroidal cell described by the coordinate

$$r = a \left(1 + \epsilon \sum_{k=0}^{\infty} f_k(\phi, \theta) \right), \quad (111)$$

where the $f(\phi, \theta)$'s are spherical harmonic functions that are $O(1)$ with respect to the dimensionless parameter ϵ. In the special case of axial symmetry about the z-axis, r is independent of the coordinate ϕ.

In view of Exercise 19.8, in order to find the counterflow, we need (i) the total force associated to a tangential boundary condition $\vec{U}$ on the spheroid, and (ii) the Stokes drag corresponding to a rigid translation $\vec{t}$ of the spheroid.

Since the stress tensor is divergence free, instead of integrating over the spheroidal surface (111) with $\epsilon \neq 0$, we integrate over the sphere corresponding to $\epsilon = 0$ (Exercise 18.3). However[29] *we need to find the corresponding "virtual" velocities $\tilde{U}$ and $\tilde{t}$ at the sphere.*

We use a Taylor expansion to determine an approximation (to arbitrary order in ϵ) to the velocity field on the sphere leading to the original velocity field when its Stokes extension is restricted to the spheroid:

$$\mathbf{v} = \sum_{i=0}^{\infty} \epsilon^i \mathbf{v}^{(i)}, \quad p = \sum_{i=0}^{\infty} \epsilon^i p^{(i)}. \tag{112}$$

Substituting these into the Stokes equations and equating terms of like power, one finds that for each i

$$\nabla^2 \mathbf{v}^{(i)} = \frac{1}{\mu} \nabla p^{(i)}, \quad \nabla \cdot \mathbf{v}^{(i)} = 0. \tag{113}$$

The boundary conditions take the form $\mathbf{v}^{(i)} = 0$ at ∞, and

$$\sum_{i=0}^{\infty} \epsilon^i \mathbf{v}^{(i)} = \vec{U} \tag{114}$$

on the deformed spheroid.

Exercise 21.1 *Use the implicit function theorem to obtain a recursive formula for the boundary condition on the sphere,*

$$\mathbf{v}^{(0)} = \hat{\mathbf{U}}, \tag{115}$$

and

$$\mathbf{v}^{(i)} = -\sum_{j=1}^{i} \frac{1}{j!} a^j f^j(\phi, \theta) \left(\frac{\partial^j \mathbf{v}^{(i-j)}}{\partial r^j} \right), \tag{116}$$

and setting $r = a$.

It is possible to solve these boundary value problems on the sphere.

---

[29] There is no free lunch.

This is the crux of the calculation[30]. For the reader's benefit, and to give an idea of the algebraic complexity, we give Lamb's general solution of the Stokes equations in terms of solid spherical harmonics [15].

Proposition 21.2

$$\mathbf{v} = \sum_{m \leq -1} \nabla \phi_m + \sum_{n \leq -2} \nabla \times (\vec{r} \xi_n)$$

$$+ \frac{1}{\mu} \sum_{n \leq -2} \left[\frac{r^2}{2(2n+1)} \nabla p_n + \frac{n}{(n+1)(2n+1)(2n+3)} r^{2n+3} \nabla \left(\frac{p_n}{r^{2n+1}} \right) \right]. \tag{117}$$

Exercise 21.2

 (i) *Show that p_{-2} and ξ_{-2} give rise to the rigid translations and rotations of the sphere;*

 (ii) *Show all the other terms yield no net force or torque on any shape;*

 (iii) *Show that only ϕ_{-1} produces a change of volume.*

Remark 21.3 *We checked the code by computing the force, F_z, required to rigidly push the cell in the z direction with velocity $\hat{U}$. We obtained*

$$F_z = 6\pi\mu\hat{U}a \left(1 - \frac{1}{5}\epsilon + O(\epsilon^2) \right), \tag{118}$$

which is in agreement with [9][31].

21.3   Results for Prolate and Oblate Spheroids

Recall that our ellipsoidal cell is defined by the equation

$$\frac{x^2 + y^2}{a^2} + \frac{z^2}{a^2(1-\epsilon)^2} = 1. \tag{119}$$

---

[30]Computer algebra is of great help. We can provide the details by e-mail to the interested reader.

[31]It is interesting to note that this linear approximation for the Stokes drag is never farther than 6% in error from the exact solutions as computed by Payne and Pell. This is true even in the extreme cases of the rod ($\epsilon = -1$) and the disk ($\epsilon = 1$) [9].

If $\epsilon > 0$ then the spheroid is oblate, if $\epsilon < 0$ the spheroid is prolate. To first order in ϵ the cell shape has polar coordinates

$$r(\theta) = a[1 - \epsilon\cos^2\theta] = a\left[1 - \epsilon\left(\frac{1}{3}P_0(\cos\theta) + \frac{2}{3}P_2(\cos\theta)\right)\right] \qquad (120)$$

Consider travelling waves on the outer membrane of the cell,

$$\theta \to \theta + \eta\sin(n\theta - \omega t). \qquad (121)$$

These can be envisioned as trains of waves of contractions and expansions travelling down the cell body.

Proposition 21.3 *To second order, the propulsive velocity is*

$$U = \frac{\pi}{8}\eta^2 n\omega a\left(1 - \frac{1}{2}\epsilon\right) + O(\epsilon^2, \eta^4). \qquad (122)$$

Letting $\epsilon = 0$ in (122), we recover the result for a sphere of radius a [5].

This method is not limited to problems with axial symmetry such as that considered above. This is significant since many microorganisms are not axially symmetric in shape. As an example, consider a spheroid whose axis of symmetry is the z-axis that swims along the x-axis[32]. The cell is described by the equation

$$\frac{x^2 + y^2}{a^2} + \frac{z^2}{a^2(1 - \epsilon)^2} = 1. \qquad (123)$$

The drag is

$$F_x = 6\pi\mu\hat{U}a\left(1 - \frac{2}{5}\epsilon + O(\epsilon^2)\right). \qquad (124)$$

Proposition 21.4 *If the travelling wave is now described by*

$$\phi \to \phi + \eta\sin(n\phi - \omega t), \qquad (125)$$

then the velocity of propulsion is

$$U = \frac{\pi}{8}\eta^2 n\omega a\left(1 - \frac{3}{32}\epsilon\right) + O(\epsilon^2, \eta^4). \qquad (126)$$

---

[32]Some microorganisms are disk-like and swim edgewise.

21.4   Comparisons of Swimming Velocities

It is interesting to compare the swimming velocities of prolate spheroids, spheres, and oblate spheroids of the same size. Equal volume and equal surface area are equivalent conditions when calculating to first order, so we will normalize all three spheroids so that their volumes and surface areas are $\frac{4}{3}\pi a^3$ and $4\pi a^2$, respectively. The sphere, prolate spheroid, and the oblate spheroid have equatorial radii a, $a(1-\frac{1}{3}d)$, and $a(1+\frac{1}{3}d)$, respectively, where $d = |\epsilon|$ in (119). We have also normalized[33] the amplitude of oscillation so that the amplitude of oscillation at the equator for all three spheroids is $a\eta$. Other physiological constraints on the size and elastic properties of the membrane may be appropriate when applying the theory to a specific organism. The normalized velocities are:

$$U_{oblate} = \frac{\pi}{8}\eta^2 n\omega a\left(1 - \frac{20}{24}\epsilon\right),$$

$$U_{sphere} = \frac{\pi}{8}\eta^2 n\omega a,$$

$$U_{prolate} = \frac{\pi}{8}\eta^2 n\omega a\left(1 + \frac{20}{24}\epsilon\right).$$

To first order, prolate spheroid swims by a factor of nearly ϵ faster than a sphere and nearly twice that compared to an oblate spheroid.

The cyanobacterium *Synechococcus* is thought to swim using a tangential mechanism as described by (121). These organisms are $2\mu m$ long and $1\mu m$ in diameter. Observed swimming velocities are on the order of $25m\mu m/sec$. Reasonable parameters for such an organism are

$$n = 30, \eta = .02\mu m, \omega = 800s^{-1}.$$

If $\epsilon = .2$, then the prolate spheroid swims nearly $5\mu m/sec$ faster than the sphere, and $10\mu m/sec$ faster than the oblate spheroid of the equivalent sizes.

---

[33]Velocities for other choices of normalizations are easily approximated using (122).

22   Final Remarks

The linear correction to Stokes law for prolate and oblate spheroids is amazingly accurate. It would be interesting to compute the swimming velocity of a prolate spheroid using prolate spheroidal harmonics to compare the velocity with those predicted by our model. The Stokes function for streaming flow has been computed by Sampson [9] and the stress tensor can be calculated.

Keller and Wu [10] presented an alternative method for investigating the effect of shape on swimming performance at low Reynolds number. This method assumes a constant vector field on a "porous" prolate spheroidal shell surrounding the cell. A form for the velocity field was chosen such that solutions to the Stokes equations could be easily obtained and such that the resulting flow agreed quantitatively with observations of actual swimming microorganisms (*paramecia*). A general porous cell model could be made by using the technique for asymmetrical cells together with this approach. It is possible that more complicated shapes and/or more complicated flows could be treated in this way. Because the flow in this approach is steady, it may be possible to compute higher order terms as well.

References

[1] Berg, H. C., *Random Walks in Biology*, expanded edition, Princeton University Press, Princeton, NJ (1993).

[2] Blake, J. R., *A spherical envelope approach to ciliary propulsion*, J. Fluid Mech. **46**, 199–208 (1971).

[3] Blake, J. R., *Infinite models for ciliary propulsion*, J. Fluid Mech. **49**, 209–222 (1971).

[4] Blake, J. R., *Self propulsion due to oscillations on the surface of a cylinder at low Reynolds number*, Bull. Austral. Math. Soc. **3**, 255–264 (1971).

[5] Childress, S., *Mechanics of Swimming and Flying*, Cambridge University Press, Cambridge, UK (1981).

[6] Dusenbery, D. B, *Life at Small Scale*, Scientific American Library (1996).

[7] Ehlers, K. M., *The geometry of swimming and pumping at low Reynolds Number*, Ph.D. Thesis, University of California, Santa Cruz (1995).

[8] Ehlers, K. M., Samuel, A., Berg, H., Montgomery, R., *Do cyanobacteria swim using traveling surface waves?*, Proc. Nat. Acad. Sci., **93** 8340–8343 (1996).

[9] Happel, J., Brenner, H., *Low Reynolds Number Hydrodynamics*, Prentice-Hall, Englewood Cliffs, NJ (1965).

[10] Keller, S., Wu, T., *A porous prolate-spheroidal model for ciliated microorganisms*, J. Fluid Mech. **80:2**, 259–278 (1977).

[11] Koiller, J., Montgomery, R., Ehlers, K., *Problems and progress in Microswimming*, J. Nonlinear Science **6**, 507–541 (1996).

[12] Koiller, J., Raupp. M. A., Fernandez, J. D., Ehlers, K., Montgomery, R., *Spectral methods for Stokes flows*, Comput. Appl. Math., **17:3**, 343–371 (1998).

[13] Koiller, J., Fernandez, J. D., *Efficiencies of nonholonomic locomotion problems*, Rep. Math. Phys., **42:1/2**, 165–183 (1998).

[14] Koiller, J., Ehlers, K., Cherman, A., Delgado. J., Montgomery, R., Duda, F., *Low Reynolds number swimming in two dimensions*, in Proc. HAMSYS98, Ernesto Lacomba, J. Llibre, eds. World Scientific, River Edge, NJ, to appear.

[15] Lamb, H. *Hydrodynamics*. Cambridge University Press, Cambridge, UK (1895).

[16] Lighthill, M. J., *On the squirming motion of nearly spherical deformable bodies through liquids at very small Reynolds number*, Commun. Pure Appl. Math. **5**, 109–118 (1952).

[17] Ludwig, W., *Zur theorie der flimmerbewegung (dynamik, nutzeffekt, energiebilanz)*, Z. vergl. Physiol. **13**, 397–504 (1930).

[18] Montgomery, R., *Nonholonomic control and gauge theory*, in Z. Li, J. F. Canny, eds., *Nonholonomic motion planning*, Kluwer, Norwell, MA (1993), pp.343–377.

[19] Purcell, E. M., *Life at low Reynolds number*. Amer. J. Phys. **45**, 3–11 (1977).

[20] Shapere, A., Wilczek, F., *Geometry of self-propulsion at low Reynolds number*. J. Fluid Mech. **198**, 557–585 (1989).

[21] Taylor, G. I., *Analysis of the swimming of microscopic organisms*, Proc. Roy. Soc. Lond. Ser. A **209**, 447–461 (1951).

Photograph: Carlos Oliveira

Bifurcation from Families
of Periodic Solutions

Jack K. Hale
CENTER FOR DYNAMICAL SYSTEMS AND NONLINEAR STUDIES
GEORGIA INSTITUTE OF TECHNOLOGY
ATLANTA, GEORGIA 30332, USA
e-mail: hale@math.gatech.edu

and

Plácido Táboas
I.C.M.C./UNIVERSIDADE DE SÃO PAULO
C. POSTAL 668, SÃO CARLOS, SP 13560-970, BRASIL
e-mail: pztaboas@icmsc.sc.usp.br

1   Introduction

Problems in nonlinear analysis and their applications frequently reduce to a nonlinear equation in some Banach space. There is no general procedure to solve these equations, but the so-called alternative method is a reduction method successfully applied in numerous instances. In this note we intend to reinforce this contention by applying the alternative method to describe a bifurcation phenomenon that occurs in some examples that come up from nonlinear oscillations theory. It is an unusual application of the method, since the solutions of a two-parameter problem emanate from a one-parameter family of solutions existing for a critical point in the space of parameters. When the parameters approach this critical point along special curves, branches of solutions approach the one-parameter family of solutions but do not tend to any particular member of the family.

The ideas behind the method that is introduced to discuss the oscillation problem have been extended in many directions. The zeros of functions that enjoy some symmetry properties often occur as families lying on a smooth manifold. It is then natural to consider perturbations of the functions that depend upon a parameter λ and then determine which points on the manifold lie on curves (parametrized by λ) of zeros of the perturbed function. Such problems arise in many applications and the theory below has been extended to treat such problems. See the article by Chillingworth [Ch 98] for recent developments and references.

Another important problem in applications occurs when an ordinary differential equation (ODE) (or partial differential equation, PDE) has a homoclinic orbit Γ, that is, an orbit that limits to a saddle equilibrium point as time approaches $+\infty$ as well as $-\infty$. The first problem is to determine those global-in-time solutions that remain in a neighborhood of Γ when the vector field is subjected to perturbations that are 2π-periodic in time. Chow, Hale, and Mallet-Paret solved this problem completely for a second-order equation in [CHM-P 80] using a natural extension of the method below. The existence of such solutions implies complicated dynamics often referred to as chaos.

Many further extensions have occurred, and we refer to the paper of Lin [Li 88] for interesting results and references.

We begin with a brief description of the alternative method that follows closely the presentation of Chow and Hale in [CH 82]. See also the book of Hale [Ha 80].

2   The Alternative Method

Let X be a Banach space. From now on the notation X_P means that there exists a continuous projection $P : X \to X$, whose range $\mathcal{R}(P)$ is the closed subspace X_P of X. If A is a linear map, we denote by $\mathcal{N}(A)$ its null space. The identity operator in any Banach space X will be denoted by I, as this will not cause any confusion.

Let X and Z be Banach spaces, $A : X \to Z$ a continuous linear map and $N : X \to Z$ a continuous map. N is not necessarily a linear map. By a nonlinear equation in some Banach space, as we mentioned at the beginning of the former section, we mean precisely an equation in the unknown $x \in X$:

$$Ax - Nx = 0. \tag{1}$$

We need the following fundamental lemma.

Lemma 2.1 *Let X and Z be Banach spaces, $A : X \to Z$ a continuous linear map, $N : X \to Z$ a continuous map, and suppose*

$$\mathcal{N}(A) = X_U, \quad \mathcal{R}(A) = Z_E.$$

Then there exists a bounded linear map $K : Z_E \to X_{I-U}$, called the right inverse of A, satisfying $AK = I$ in Z_E, $KA = I - U$ in X. If we split $x = y + z$, where $y \in X_U$ and $z \in X_{I-U}$, then (1) is equivalent to the system

$$\begin{cases} z - KEN(y + z) = 0, \\ (I - E)N(y + z) = 0. \end{cases} \tag{2}$$

Proof. The map A is one-to-one from X_{I-U} onto Z_E. Therefore, the existence of a right inverse $K : Z_E \to X_{I-U}$, satisfying $AK = I$ in Z_E, $KA = I - U$ in X, is obvious. The boundedness of K is a straightforward consequence of the continuity of the projections U and E and the closed graph theorem (see [Ka 70, III. §5.4], for instance). Taking $x = y + z$, with

$y \in X_U$ and $z \in X_{I-U}$, we have

$$KAx = KAy + KAz = z. \qquad (3)$$

Decomposing (1) into complementary subspaces by the projections E and $I - E$, one reaches the equivalent system:

$$\begin{cases} E(A - N)(y + z) = 0, \\ (I - E)(A - N)(y + z) = 0. \end{cases} \qquad (4)$$

We notice that the right-hand side of the first equation of (4) lies in Z_E, which is the domain of the one-to-one map K. The one-to-oneness of K ensures that, if we apply K to both sides of the first equation in (4), we obtain an equivalent equation. Doing this, recalling that $EA = A$, and using (3), we simplify the resulting equations to arrive to the equivalent (2). ∎

In the circumstances of Lemma 2.1, one can treat (1) by dealing with (2) in the unknowns $y \in X_U$ and $z \in Z_{I-U}$. If it is possible to solve the first equation of (2) for z as a function of y, $z = z^*(y)$, then (1) reduces to the equivalent one:

$$(I - E)N\big(y + z^*(y)\big) = 0. \qquad (5)$$

In this case, (5) is called *the bifurcation equation* and its left-hand side, meant as a function of y, is called *the bifurcation function*. This procedure in dealing with (1) is called *the alternative method in the null space of A*. The alternative method is a method that reduces the determination of the solutions of (1) to the solution of another equation in the unknown y, which lies in the subspace X_U of X. In applications, the solution $z^*(y)$ of the first equation of (2) is usually provided by the implicit function theorem. The method is especially useful in cases where the map A is a Fredholm operator, that is, where the null space $\mathcal{N}(A) = X_U$ has finite dimension n and the range $\mathcal{R}(A)$ has finite codimension m. In these cases, (5) becomes a finite-dimensional problem, a nonlinear system of m equations in n unknowns.

Let us consider now an important specialization of Equation (1). Let X, Z, and Λ be Banach spaces, $A : X \to Z$ a bounded linear operator, and $N : X \times \Lambda$ a generally nonlinear operator, continuous together with its partial Fréchet derivative in X. Suppose that

$$N(0,0) = 0, \qquad \partial N(0,0)/\partial x = 0 \qquad (6)$$

and consider a version of (1) depending on a parameter:

$$Ax - N(x, \lambda) = 0. \qquad (7)$$

We consider (7) for (x, λ) varying in a neighborhood of $(0,0)$ and suppose, as in Lemma 2.1, that $\mathcal{N}(A) = X_U$ and $\mathcal{R}(A) = Z_E$. So, if $x = y + z$, with $x \in X_U$, $z \in X_{I-U}$, Lemma 2.1 implies that (7) is equivalent to the system

$$\begin{cases} z - KEN(y + z, \lambda) = 0, \\ (I - E)N(y + z, \lambda) = 0, \end{cases} \qquad (8)$$

for some right inverse $K : Z_E \to X_{I-U}$, satisfying $AK = I$ in Z_E, $KA = I - U$ in X.

The implicit function theorem applied to the first equation of (8) ensures the existence of a neighborhood V of $(0,0)$ in X_U and a function $z^* : V \to X_U$, $z^*(y, \lambda)$, which is continuous together with its Fréchet derivative with respect to y, such that $z^*(0,0) = 0$ and $z^*(y, \lambda)$ satisfies the first equation of (8). Moreover, $z^*(y, \lambda)$ is the unique solution of this equation in a neighborhood of $z = 0$ in X_{I-U}, for $(y, \lambda) \in V$. Indeed, the function $z^*(y, \lambda)$ has the same regularity as the operator $N(x, \lambda)$ near $(x, \lambda) = (0,0)$; that is, if the function $N(x, \lambda)$ has continuous partial derivatives up to order k (or is analytic), then $z^*(y, \lambda)$ has continuous partial derivatives up to order k (or is analytic).

Substituting $z^*(y, \lambda)$ in the second equation of (8), one can finally state that (7) reduces to the bifurcation equation

$$(I - E)N (y + z^*(y, \lambda), \lambda) = 0 \qquad (9)$$

in the unknowns $(y, \lambda) \in V$.

Following Chow and Hale [CH 82], the alternative method in the null space of A for equations in the form of (7), in a neighborhood of $(x, \lambda, x) = (0,0)$, will be called the method of Liapunov–Schmidt. The reader is referred to [CH 82] and [Ha 80] for extensions of this method as well as numerous applications and other references.

As a simple illustration of the method of Liapunov–Schmidt, we consider the determination of all solutions in $\mathbb{R}^n$ near the origin of the equation

$$0 = f(x, y, \lambda),$$
$$By = g(x, y, \lambda),$$

$x \in \mathbb{R}$, $y \in \mathbb{R}^{n-1}$, $\lambda \in \mathbb{R}$, where $\det B \neq 0$ and f, g together with their first derivatives vanish at $(x, y, \lambda) = (0, 0, 0)$. It is clear that the second equation has a unique solution $y^*(x, \lambda)$ for x near zero, $y^*(0, 0) = 0$, $\partial y^*(0, 0)/\partial x = 0$, and this function is as smooth as g. Therefore, the equation has a solution near the origin if and only if x satisfies the scalar equation

$$F(x) \equiv f(x, y^*(x, \lambda), \lambda) = 0.$$

The function $F(x)$ is the bifurcation function. It is instructive to take the example $f(x, y, \lambda) = \lambda + x^2 + \text{h.o.t.}[1]$ $g(x, y, \lambda)$ arbitrary, to see that the bifurcation function is $F(x, \lambda) = \lambda + x^2 + \text{h.o.t}$ and there are no solutions for $\lambda > 0$ and two solutions for $\lambda < 0$. For $f(x, y, \lambda) = \lambda + y \cdot Ay$ and $g(x, y, \lambda) = cx^2 + \text{h.o.t.}$, the bifurcation function is $F(x, \lambda) = \lambda + B^{-1}c \cdot ABcx^2 + \text{h.o.t.}$ and there are no solutions if $\lambda B^{-1}c \cdot ABc > 0$ and two solutions if this quantity is negative.

3 Bifurcation near Families of Solutions

In the applications given below, the equations corresponding to (1) involve an operator A that is provided by a linear variational equation associated with an ODE along a fixed periodic orbit. The domain of A is the Banach space $Y = \mathcal{P}(T)$ of the continuous periodic functions in $(-\infty, \infty)$, with period $T > 0$, endowed with the supremum norm. A crucial point is to define a continuous projection P from Y to the range $\mathcal{R}(A)$ of A, that is, to write $\mathcal{R}(A) = Y_P$. In order to face this question we recall some basic facts of the general theory of ordinary differential equations. The main role is played by the Fredholm alternative. We state it below and give a proof that was extracted from [Ha 80, Chapter IV].

---

[1]higher order terms

Consider a linear homogeneous equation,

$$\dot{x} = B(t)x, \tag{10}$$

where B is a continuous periodic $n \times n$ matrix function of $t \in (-\infty, \infty)$, with period $T > 0$ (from now on we will say T-periodic). We denote by $X(t; \tau)$ the fundamental matrix of (10) such that $X(\tau; \tau) = I$. The adjoint equation to (10) is

$$\dot{y} = -yB(t). \tag{11}$$

We recall that if $X(t)$ is a fundamental matrix of (10), its inverse $X^{-1}(t)$ is a fundamental matrix of the adjoint (11). Let us fix $X(t)$ as the principal matrix of (10), that is, $X(t) = X(t; 0)$.

Since B is T-periodic, a necessary and sufficient condition for a solution y of (11) to be T-periodic is that $y(0) = y(T)$ and this condition is equivalent to $y(0)[X^{-1}(T) - I] = 0$. Therefore, (11) has a nontrivial T-periodic solution if and only if $[X^{-1}(T) - I]$ is a singular matrix. It is also easily seen that x is a T-periodic solution of (10) if and only if $[X(T) - I]x(0) = 0$. Therefore, since the null spaces of $X(T) - I$ and $X^{-1}(T) - I$ have the same dimension, the foregoing remarks imply that the spaces of T-periodic solutions of (10) and its adjoint (11) have the same dimension.

The Fredholm Alternative. *Let B be an $n \times n$ continuous T-periodic matrix function defined on $\mathbb{R}$ and consider a function $f \in \mathcal{P}(T)$. Then the nonhomogeneous linear equation*

$$\dot{x} = B(t)x + f(t) \tag{12}$$

has a T-periodic solution if and only if

$$\int_0^T y(t)f(t)\, dt = 0 \tag{13}$$

for every solution y of the adjoint equation (11) such that $y^* \in \mathcal{P}(T)$ (y^* denotes the transpose of y). If the orthogonality condition (13) holds, the nonhomogeneous equation (12) has an r-parameter family of T-periodic solutions, where r is the dimension of the space of T-periodic solutions of the homogeneous equation (10).

Proof. A solution x of (12) is a T-periodic solution if and only if $x(0) = x(T)$. For $x(0) = x_0$, the solution $x(t)$ is given by the variation of constants formula

$$x(t) = X(t; 0)x_0 + \int_0^T X(t; s)f(s)\, ds. \tag{14}$$

Recalling that $X(t; \tau)X(\tau; s) = X(t; s)$ for any t, τ, s in $\mathbb{R}$, one sees that condition $X(T) = x_0$ is equivalent to

$$\left[X^{-1}(T; 0) - I\right]x_0 = \int_0^T X^{-1}(s; 0)f(s)\, ds. \tag{15}$$

Let us write (15) in its shorthand form,

$$Dx_0 = b, \tag{16}$$

where D is the matrix $X^{-1}(T; 0) - I$ and b is the vector $\int_0^T X^{-1}(s; 0)f(s)\, ds$. There exists a solution x_0 of (16) if and only if b lies in the range $\mathcal{R}(D)$ of the linear map defined by D; that is, b is in the space $[b_1, b_2, \ldots, b_n]$, spanned by the columns b_1, $b_2, \ldots$, b_n, of D. But this, by its turn, is equivalent to $ab = 0$ for every vector a such that $aB = 0$ (that is, $a^* \perp b$ for every a such that $a^* \in [b_1, b_2, \ldots, b_n]^\perp$). Since $X^{-1}(t; 0)$ is the principal matrix of the adjoint equation (11), the set of the vectors a such that $aB = 0$ coincides with the set of initial values of all T-periodic solutions of (11). Thus, the condition (15) is equivalent to $ab = 0$ for every a such that $aX^{-1}(t; 0)$ is a T-periodic solution (11). Recalling the definition of b, this can be paraphrased as the following: the existence of an $x_0 \neq 0$ satisfying the condition (15) is equivalent to

$$\int_0^T aX^{-1}(s; 0)f(s)\, ds = 0$$

for every a such that $aX^{-1}(t; 0)$ is a T-periodic solution of (11). This completes the proof of the first part of the Fredholm alternative.

To prove the remaining assertion, suppose that (13) holds and that ϕ is a particular T-periodic solution of (10). Then the general form of T-periodic solutions of (12) is given by $x = z + \phi$, where z varies in the space of T-periodic solutions of (10).   ∎

Let $h : \mathbb{R}^n \to \mathbb{R}^n$ be a C^1 function and recall that a *first integral* of the ordinary differential equation

$$\dot{x} = h(x) \tag{17}$$

is a C^1 function $u : \mathbb{R}^n \to \mathbb{R}$, nonconstant on an open set, such that

$$\nabla u(x) \cdot h(x) = 0, \qquad x \in \mathbb{R}^n,$$

where ∇u denotes the gradient of u. This means that the derivative of u along the solutions $x(t)$ of (17) is zero $\big(du(x(t))/dt = 0\big)$; that is, each orbit of (17) remains in some level surface of u.

Denoting by $x(t; x_0)$ the solution of (17) such that $x(0; x_0) = x_0$, recall that $x(t; x_0)$ is a C^1 function of both variables. Substituting $x = x(t; x_0)$ in (17) and differentiating with respect to x_0, one sees that $V(t; x_0) := \partial x(t; x_0)/\partial x_0$ is the principal matrix of the linear variational equation of (17) along $x(t; x_0)$:

$$\dot{x} = Hx, \qquad \big(H := dh\big(x(t; x_0)\big)/dx\big). \tag{18}$$

Applications of the Fredholm alternative in the line of problems treated below may depend on the following fundamental lemma, which is Lemma 2.5, chap. VIII, of [Ha 80].

Lemma 3.1 *If u is a first integral of class C^2 of (17), then ∇u is a solution of the adjoint equation to (18):*

$$\dot{w} = -wH. \tag{19}$$

Proof. Since $u\big(x(t; x_0)\big) = x_0$, for every t and x_0 for which $x(t; x_0)$ exists, it follows that $u_x\big(t, x(t; x_0)\big)V(t; x_0) = u_x(0; x_0)$ for all t. Differentiating this relation with respect to t,

$$0 = \dot{\nabla} uV(t; x_0) + \nabla \dot{V}(t; x_0) = \big[\dot{\nabla} + \nabla uH\big]V(t; x_0)$$

and, since $V(t; x_0)$ is a nonsingular matrix, the proof is complete. ∎

In the sequel, we give two applications of the method of Liapunov–Schmidt to the study of existence and bifurcation of periodic solutions of ODEs. These equations can be seen as perturbations of an autonomous equation near a center. We are interested in studying periodic solutions of the perturbed equations near a fixed T-periodic orbit of this center. The perturbations depend on t and are T-periodic. As was pointed out in section 2, the method reduces an equation in a Banach space (in this case $\mathcal{P}(T)$) to a system of equations where the unknowns are the coordinates of the sought periodic solution in the null space of a linear operator. The problems we deal with here have an interesting peculiarity: the role of one of the mentioned coordinates in the null space is played by the phase shifts of the periodic solution corresponding to the fixed orbit in the center.

3.1   Interaction of Forcing and Damping in a Nonlinear Oscillator

In this section we consider a problem in nonlinear oscillations that appeared in [HT 78]. Given a 2π-periodic solution of a nonlinear oscillator and a neighborhood of this solution, we wish to know if there are 2π-periodic solutions under the effect of a small damping and a small 2π-periodic forcing. We begin by formulating the problem in a suitable Banach space, where the operators A and N of the equation corresponding to (7) are defined in terms of the vector field in the equation, the perturbation, and a representation of the fixed 2π-periodic orbit. We make a nonstandard choice of the right inverse of A in order to use the phase as a relevant parameter.

Let us consider the equation of a nonlinear oscillator:

$$\ddot{x} + g(x) = 0, \tag{20}$$

where g is a sufficiently smooth function to ensure the continuity of all derivatives that we will consider below and satisfies $xg(x) > 0$, if $x \neq 0$. By considering the two-dimensional first-order system equivalent to (20), it is known that the origin of the $x\dot{x}$-plane is a center; that is, every nontrivial orbit is periodic and encircles the origin $(0,0)$. Each orbit has an amplitude $\rho > 0$ and a period $2\pi/\omega(\rho)$ (we define ρ by the property that the corresponding orbit passes through the point $(\rho, 0)$, $\rho > 0$). The function $\omega : (0, \infty) \to (0, \infty)$ is smooth and gives the frequency as a function of the amplitude.

There is no loss of generality in assuming the existence of a 2π-periodic solution p of (20), since we can make a time rescaling. We denote by $\Gamma \subset \mathbb{R}^2$

the orbit of this solution in the $x\dot{x}$-plane; that is, $\Gamma = \{ (p(t), \dot{p}(t)) \mid 0 \le t < 2\pi \}$. Let f be a 2π-periodic continuous function and consider the following perturbation of (20):

$$\ddot{x} + g(x) = -\lambda\dot{x} + \mu f(t), \tag{21}$$

where λ and μ are real parameters. The term $\lambda\dot{x}$ might represent a damping (for $\lambda > 0$) or an excitation effect (for $\lambda < 0$). In any case it destroys periodicity, but the periodic forcing $\mu f(t)$ may compensate for this effect and give rise to 2π-periodic solutions of (21). This fact motivates the following questions:

For a sufficiently small neighborhood W of the orbit Γ in the $x\dot{x}$-plane, is there a region V in the parameter space for which there are 2π-periodic solutions of (21) whose orbits lie in W for each $(\lambda, \mu) \in V$? How does the number of such solutions vary when (λ, μ) varies in V? How do they behave when the parameter (λ, μ) approaches $(0,0)$?

In what follows, we solve completely the first two questions and give a good idea of what happens with respect to the third one.

There exist two functions $\omega : (0, \infty) \to (0, \infty)$, $\phi : \mathbb{R} \times (0, \infty) \to \mathbb{R}^2$, as smooth as the function g, such that $\phi(\theta + 2\pi, \rho) = \phi(\theta, \rho)$ for any $(\theta, \rho) \in \mathbb{R} \times (0, \infty)$, such that the general nontrivial solution x of (20) is represented as

$$x(t) = \phi\big(\omega(\rho)(t + \alpha), \rho\big), \tag{22}$$

where the real parameter α is the phase shift and $\rho > 0$ is the amplitude of x. In fact, let us denote by $u(\cdot; r)$ the solution of (20) having amplitude $r > 0$, such that $u(0; r) = r$ and $\dot{u}(0; r) = 0$. If $T(r)$ denotes the period of $u(\cdot; r)$, define the frequency $\omega(\rho) := 2\pi/T(\rho)$ and set $\phi(\theta, \rho) := u\big(\theta T(\rho)/2\pi; \rho\big)$. These functions ω and ϕ have the required properties.

Let $\rho_0 > 0$ be the amplitude of the 2π-periodic solution p of (10) considered above and assume that p has phase shift $\alpha = 0$; that is, $\omega(\rho_0) = 1$ and $p(t) = \phi(t, \rho_0)$ for every $t \in \mathbb{R}$.

A standing hypothesis is that the frequency $\omega(\rho)$ varies with ρ; that is,

(H1) $\omega'(\rho_0) \neq 0.$

Notice that hypothesis (H1) excludes the linear oscillator but, despite its generality, this is not the only excluded case. There are examples of nonlinear restoring forces g giving rise to solutions of (20) with constant period. This subject goes back to 1919, with Appel [Ap 19], and has been considered by many authors since that time.

We consider the two-dimensional system equivalent to (21):

$$\begin{cases} \dot{x} = y, \\ \dot{y} = -g(x) - \lambda y + \mu f(t), \end{cases} \tag{23}$$

as a perturbation of the system equivalent to (20):

$$\begin{cases} \dot{y} = x, \\ \dot{y} = -g(x). \end{cases} \tag{24}$$

Let us denote by $P^{\perp}(t)$ the normal vector to the orbit Γ at the point $P(t) = (p(t), \dot{p}(t))$ given by $P^{\perp}(t) = (\ddot{p}(t), -\dot{p}(t))$ for all t, $0 \leq t < 2\pi$. We can choose the neighborhood W of Γ as a tubular neighborhood; that is, every point x of W can be represented by coordinates $(\tau, \sigma) \in [0, 2\pi) \times \mathbb{R}$ defined by

$$x = P(\tau) + \sigma P^{\perp}(\tau).$$

Each 2π-periodic solution $u(t) = (x(t), \dot{x}(t))$ of (23) that remains in W is associated to a unique number α, $0 \leq \alpha < 2\pi$, in such a way that $u(\alpha) = P(0) + \sigma P^{\perp}(0)$. That is, α is the first moment in the interval $[0, 2\pi)$ when the point $u(t)$ reaches the fiber $\mathcal{F}(0) : P(0) + \xi P^{\perp}(0)$, $\xi \in \mathbb{R}$, of the tubular neighborhood. The number α will be referred to as the phase of the solution u of (23) and of the corresponding solution x of (21). This solution can be represented as

$$x(t) = p(t - \alpha) + z(t - \alpha),$$

where the increment $z(t) = x(t + \alpha) - p(t)$, $t \in \mathbb{R}$, satisfies

$$\big(z(0), \dot{z}(0)\big) \cdot \dot{P}(0) = 0.$$

Therefore, $\dot{z}(0) = 0$. From now on we denote $Z(t) := \big(z(t), \dot{z}(t)\big)$.

Since there is a one-to-one correspondence between the solutions $x(t)$ of (21) and the solutions $x(t + \alpha)$ of

$$\ddot{x} + g(x) = -\lambda \dot{p}(t) + \mu f(t + \alpha), \tag{25}$$

our problem reduces to the study of 2π-periodic solutions $x(t)$ of (25) of the form

$$x(t) = p(t) + z(t), \quad Z(0) \cdot \dot{P}(0) = 0, \tag{26}$$

with small $|Z(t)|$. By inserting (26) into (25), we finally arrive at the study of small 2π-periodic solutions $z(t)$ of

$$\begin{cases} \ddot{z} + g'\big(p(t)\big)z = -\lambda \dot{p}(t) - \lambda \dot{z} + \mu f(t + \alpha) + G(t, z), \\ Z(0) \cdot \dot{P}(0) = 0. \end{cases} \tag{27}$$

with small $|Z(t)|$, where the remainder $G(t, z)$ is a continuous function, 2π-periodic in t for each fixed z, such that $G(t, z) = O\big(|z|^2\big)$, uniformly in t, as $z \to 0$.

We remark that, as a consequence of (H1), the 2π periodic function $\dot{p}$ spans the space of 2π-periodic solutions of the linear variational equation of (20) along the solution $p(t)$:

$$\ddot{v} + g'\big(p(t)\big)v = 0. \tag{28}$$

In fact, substituting $p(t)$ in (20) and differentiating the equation with respect to t we get

$$\ddot{p}(t) + g'\big(p(t)\big)\dot{p}(t) = 0,$$

which shows that $\dot{p}$ is, indeed, a 2π-periodic solution of (28). Now we insert in (20) the general solution represented by $\phi(\theta, \rho)$ in (22) and differentiate the equation with respect to ρ at $\rho = \rho_0$ (recall that $\omega(\rho_0) = 1$ and $p(t) = \phi(t, \rho_0)$). This shows that $q(t) := \big[\partial \phi(\omega(\rho)t, \rho)/\partial \rho\big]_{\rho=\rho_0}$ is another solution of (28). But

$$q(t) = \phi_\theta(t, \rho_0)\omega'(\rho_0)t + \phi_\rho(t, \rho_0)$$

and, according to (H1), $q(t)$ is an unbounded solution of (28), being linearly independent of $\dot{p}$. Therefore, the solution $\dot{p}$ spans the space of 2π-periodic solutions of (28).

The Fredholm alternative states that if φ is a continuous 2π-periodic function, then the nonhomogeneous equation associated with (28);

$$\ddot{v} + g'\big(p(t)\big)v = \varphi(t), \tag{29}$$

has a 2π-periodic solution if and only if

$$\int_0^{2\pi} \varphi(t)u(t)\, dt = 0$$

for every 2π-periodic solution of the adjoint equation to (28). Since (28) is self-adjoint, this condition is equivalent to

$$\int_0^{2\pi} \varphi(t)\dot{p}(t)\, dt = 0. \tag{30}$$

Consider the Banach space $Z = \mathcal{P}_{2\pi}$ of the continuous 2π-periodic functions with the supremum norm $||\cdot||$ and let $X = \mathcal{P}_{2\pi}^{(2)}$ be its algebraic subspace of all C^2 functions $\varphi \in Z$, endowed with a C^2-norm. Now we use the differential equation in (27) as a motivation to define the bounded linear operator $A : X \to Z$ and the nonlinear continuous operator $N : X \times \mathbb{R}^3 \to Z$ by

$$Az = \ddot{z} + g'\big(p(\cdot)\big)z,$$
$$N(z, \lambda, \mu, \alpha) = -\lambda\dot{p}(\cdot) - \lambda\dot{z} + \mu f(\cdot + \alpha) + G(\cdot, z) \tag{31}$$

for every $(z, \lambda, \mu, \alpha) \in X \times \mathbb{R}^3$. Since $\dot{p}$ spans the space of all 2π-periodic solutions of (28), the null space of A is given by $\mathcal{N}(A) = [\dot{p}] \subset X$ and its range $\mathcal{R}(A)$ satisfies $Z = \mathcal{R}(A) \oplus [\dot{p}]$. That is, A is a Fredholm operator.

Setting

$$\nu^{-1} := \int_0^{2\pi} \big(\dot{p}(t)\big)^2 dt,$$

we define the projections $Q : Z \to Z$ and $P : X \to X$ by

$$Q\varphi = \nu \left[\int_0^{2\pi} \dot{p}(t)\varphi(t)\, dt \right] \dot{p}, \quad \varphi \in Z,$$
$$P\varphi = |\ddot{p}(0)|^{-2}\ddot{p}(0)\ddot{\varphi}(0)\dot{p}, \quad \varphi \in X. \tag{32}$$

Now we can formulate our problem (27) in its shorthand form:

$$Az = N(z, \lambda, \mu, \alpha),$$
$$Pz = 0. \tag{33}$$

Recall that $X_Q = X_P = \mathcal{N}(A)$ and that (29) has a 2π-periodic solution if and only if (30) is satisfied. Therefore, $\mathcal{R}(A) = Z_{I-Q}$, and we are ready to apply the method of Liapunov–Schmidt to the first equation of (33), with $I - Q$ playing the role of the projection E of the foregoing section and Q in the role of the projection U. Thus, decomposing the first equation of (33) by the projections Q and $I - Q$, we reach the following system:

$$(I - Q)Az = (I - Q)N(z, \lambda, \mu, \alpha),$$
$$QAz = QN(z, \lambda, \mu, \alpha), \tag{34}$$
$$Pz = 0.$$

Now we make a slight modification in the procedure of the method of Liapunov–Schmidt. We define a right inverse of A by using the projection P instead of Q. That is, noticing that A is one-to-one from X_{I-P} onto Z_{I-Q}, the reasoning of the proof of Lemma 2.1 can immediately be adapted to show that there exists a bounded linear operator $K : Z_{I-Q} \to X_{I-P}$ such that $AK = I$ in Z_{I-Q} and $KA = I - P$ in X. Since K is one-to-one in Z_{I-Q}, we can apply it to both sides of the first equation of (34) and arrive at the following equivalent system, after some straightforward simplifications:

$$z - Pz = K(I - Q)N(z, \lambda, \mu, \alpha),$$
$$0 = QN(z, \lambda, \mu, \alpha), \tag{35}$$
$$Pz = 0,$$

which, in turn, is equivalent to

$$z = K(I - Q)N(z, \lambda, \mu, \alpha),$$
$$0 = QN(z, \lambda, \mu, \alpha), \tag{36}$$

because any solution z^* of (36) lies in the range of K and, therefore, must satisfy the condition $Pz^* = 0$.

In what follows, we consider $[0, 2\pi)$ as a compact manifold by identifying it to S^1; that is, we take $[0, 2\pi)$ as the quotient topological space $\mathbb{R}/2\pi\mathbb{Z}$. Besides, we notice that

$$N(0, 0, 0, \alpha) = 0, \quad \partial N(0, 0, 0, \alpha)/\partial z = 0$$

for every $\alpha \in [0, 2\pi)$. Therefore, according to the implicit function theorem, for each $\alpha_0 \in [0, 2\pi)$, there exist neighborhoods V_{α_0} of $\alpha_0 \in [0, 2\pi)$, V of $(\lambda, \mu) \in \mathbb{R}^2$, and U of $0 \in X$, and a function $z^* : V_{\alpha_0} \times V \to U$ such that $z^*(\alpha, \lambda, \mu)$ satisfies the first equation of (36). Considering the compactness of S^1 we may construct a finite number of neighborhoods like $V_{\alpha_i} \times V$, $\alpha_i \in [0, 2\pi)$, $i = 0, \ldots, k$, where the implicit function theorem can be used to extend z^* to $[0, 2\pi) \times V$. Notice that the uniqueness of the implicit function plays a fundamental role in this process of extension of z^*. Now we extend z^*, by periodicity in α, to $\mathbb{R} \times V$. Substituting z^* in the second equation of (36), this system is reduced to the scalar equation:

$$QN\big(z^*(\alpha, \lambda, \mu), \lambda, \mu, \alpha\big) = 0, \quad \alpha \in \mathbb{R}.$$

This is the bifurcation equation. Therefore, if $|(\lambda, \mu)|$ is small, we can say that (21) has a 2π-periodic solution near the solution $p(t - \alpha)$ of (20) if and only if (α, λ, μ) satisfies this equation which, according to definitions of N and of the projection Q, can be explicitly written as

$$\nu \int_0^{2\pi} \dot{p}\left[-\lambda \dot{z}^*(\alpha, \lambda, \mu) - \lambda \dot{p} + \mu f(\cdot + \alpha) + G(\cdot, z^*(\alpha, \lambda, \mu))\right] = 0. \quad (37)$$

The right-hand side of (37) defines the bifurcation function $F(\alpha, \lambda, \mu)$ in such a way that $F(\alpha, 0, 0) = 0$. Expanding F, the bifurcation equation takes finally the form

$$F(\alpha, \lambda, \mu) = -\lambda + h(\alpha)\mu + \text{h.o.t.} = 0, \quad (38)$$

where "h.o.t." indicates higher order terms in (λ, μ) and the coefficient $h(\alpha)$ of μ is given by

$$h(\alpha) = \int_0^{2\pi} \dot{p}(t) f(t + \alpha)\, dt. \quad (39)$$

Recall that we are assuming the functions involved in (21) are sufficiently regular to ensure the smoothness of h needed below. Before we proceed, we need some simplifying generic hypotheses on the function h:

(H2) $h'(\alpha) \neq 0$, except at the points $\alpha_1 < \alpha_2 < \cdots < \alpha_N$ in $[0, 2\pi)$.

Notice that the integer N must be even:

(H3) $$h''(\alpha_j) \neq 0, \quad j = 1, \ldots, N,$$

(H4) $$h(\alpha_j) \neq h(\alpha_k), \quad \text{if} \quad j \neq k, \quad 1 \leq j, k \leq N.$$

At this point, by a speculative analysis it is already possible to give in advance an idea of the nature of the bifurcation diagram in the plane (λ, μ), of the 2π-periodic solutions of (21) whose orbits lie in the neighborhood W of the orbit Γ. In fact, the dominant part of (38), for small $|(\lambda, \mu)|$, can be written as

$$(\lambda, \mu) \cdot \big(-1, h(\alpha)\big) = 0, \tag{40}$$

where the dot indicates the usual scalar product.

Let $\alpha_m, \alpha_M \in \{\alpha_1, \alpha_2, \ldots, \alpha_N\}$ be the points defined by

$$h(\alpha_m) = \inf_{0 \leq \alpha < 2\pi} h(\alpha), \qquad h(\alpha_M) = \sup_{0 \leq \alpha < 2\pi} h(\alpha). \tag{41}$$

Therefore, the range of the point $(-1, h(\alpha))$, when α varies in the interval $[0.2\pi)$, is the line segment $\big[(-1, h(\alpha_m)), (-1, h(\alpha_M))\big]$. According to Figure 1, there are solutions (α, λ, μ) of (40) for $\alpha \in [0, 2\pi)$ if and only if the vector (λ, μ) lies in one of the sectors where (λ, μ) is orthogonal to some $(-1, h(\alpha))$. In this case, the number of 2π-periodic solutions of (21) with orbit in W and phase $\alpha \in [0, 2\pi)$ is the number of α's in $[0, 2\pi)$ for which the orthogonality condition (40) holds. The right part of Figure 1 shows the bifurcation diagram when the graph of the function h has the shape indicated in the left part of the figure. The numbers in parenthesis indicate the number of solutions existing when (λ, μ) lies in the respective sector. As is seen in Figure 1, when $(\lambda, \mu) \neq (0, 0)$ crosses the boundary of any of the sectors, the number of solutions changes by two.

The next two theorems provide more precise conclusions on the bifurcation diagram. The first one states that the local bifurcation diagram near $(\lambda, \mu) = (0, 0)$, corresponding to the true bifurcation equation (37), is

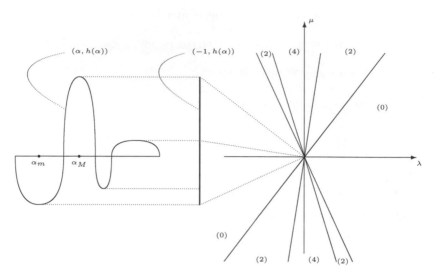

Figure 1. Sketch of a bifurcation diagram.

a slight perturbation of a diagram like the one given in Figure 1; that is, there are smooth curves through the origin, defining sectorial regions of the $\lambda\mu$-plane where the corresponding equation (21) has a constant number of 2π-periodic solutions with orbits in W. When $(\lambda, \mu) \neq (0,0)$ crosses one of these curves, the number of such 2π-periodic solutions changes by two. The second theorem describes the behavior of these 2π-periodic solutions when (λ, μ) approaches $(0,0)$. It states that, depending on how $(\lambda, \mu) \to (0,0)$, the solutions approach the family $p(\cdot + \alpha)$, $\alpha \in [0, 2\pi)$, of solutions of the unperturbed equation (20), but it is possible that they do not approach a particular solution of this family. We keep the notation introduced in the considerations above. For each $\alpha \in [0, 2\pi)$, we denote by r_α the line of the $\lambda\mu$-plane, orthogonal to $(-1, h(\alpha))$, given by

$$r_\alpha : \quad (\lambda, \mu) \cdot (-1, h(\alpha)) = 0.$$

Since all of the functions of α in consideration are 2π-periodic, we always consider α varying in $[0, 2\pi)$ with the topology defined by $\mathbb{R}/2\pi\mathbb{Z}$, as before.

Theorem 3.1 *If the hypotheses (H1)–(H4) are satisfied, then there are neighborhoods W of Γ, V of $(0,0)$ in the $\lambda\mu$-plane, and curves $\mathcal{C}_j \subset V$,* *tangent at $(0,0)$ to the lines r_{α_j}, $j = 1, 2, \ldots, N$, $C_j \cap C_k = (0,0)$, if $j \neq k$, such that the number of 2π-periodic solutions of (21) with orbit in W changes by two when (λ, μ) crosses each of these curves. Moreover, if C_m and C_M are the curves tangent to the lines r_{α_m} and r_{α_M}, respectively, they partition V in two "conic" regions, S and $S^c = V \setminus S$, S being closed in V. If $(\lambda, \mu) \in S^c$, there is no 2π-periodic solution of (21) with orbit in W. If $(\lambda, \mu) \in S \setminus (C_m \cup C_M)$, there are at least two such solutions of (21). These solutions are pairwise distinct if $(\lambda, \mu) \notin C_j$, $j = 1, \ldots, N$.*

Remark 3.1 *The set S is precisely characterized by the following: $S \subset V$ is closed in V, its boundary is $\partial S = C_m \cup C_M$, and it has the property that, if $h(\alpha_m) < h(\alpha) < h(\alpha_M)$, then the connected component of $r_\alpha \cap S$ containing $(0,0)$ is a closed line segment with $(0,0)$ in its interior.*

Proof. The proof is based in the bifurcation equation (38) that is written in the form

$$(\lambda, \mu) \cdot (-1, h(\alpha)) + R(\alpha, \lambda, \mu) = 0, \qquad (42)$$

with $(\lambda, \mu) \in \mathbb{R}^2$ and $\alpha \in [0, 2\pi)$. For $(\lambda, \mu) \neq (0,0)$, let $(\lambda, \mu) = ru$, where r is a positive number and u a unit vector. Defining $v(\alpha) := (-1, h(\alpha))$, we can rewrite (42) as

$$F(r, u, \alpha) := v(\alpha) \cdot u + V(r, u, \alpha) = 0, \qquad (43)$$

where $V(r, u, \alpha) = R(\alpha, ru)/r$. Notice that, since $V(0, u, \alpha) = 0$, the only solutions of (43) for $r = 0$ are those $(0, u_0, \alpha_0)$ such that u_0 and $v(\alpha_0)$ are orthogonal.

For each $\alpha_0 \in [0, 2\pi)$, let us choose a unit vector u_0 orthogonal to $v(\alpha_0)$ and seek the solutions $(r, u, \alpha) \in \mathbb{R} \times S^1 \times \mathbb{R}$ of (43) near $(0, u_0, \alpha_0)$. We extend the function F for $r < 0$, setting $V(r, u, \alpha) := -R(-ru, \alpha)/r$ for $r < 0$, in such a way that $F(r, u, \alpha)$ becomes meaningful in a complete neighborhood of $(0, u_0, \alpha_0)$ in $\mathbb{R} \times S^1 \times \mathbb{R}$.

There are two cases to consider.

First, suppose $h'(\alpha_0) \neq 0$. Therefore, $\partial F(0, u_0, \alpha_0)/\partial \alpha = v'(\alpha_0) \cdot u_0 \neq 0$ and an application of the implicit function theorem leads to the existence of a positive number $\bar{\delta}$ such that to each $(r, u) \in \mathbb{R} \times S^1$, with $0 \leq r, |u - u_0| < \bar{\delta}$, is associated a unique $\alpha^*(r, u)$ in a neighborhood of

$\alpha_0 \in \mathbb{R}$, with α^* being a C^1 function such that $F(r, u, \alpha^*(r, u)) = 0$. So, $(r, u, \alpha^*(r, u))$ is the unique solution of (43) in a suitable neighborhood of $(0, u_0, \alpha_0)$. Therefore, in this case there is no bifurcation of solutions of (43) near $(0, u_0, \alpha_0)$.

Second, suppose $h'(\alpha_0) = 0$. In this case $v'(\alpha_0) = 0$ and the implicit function theorem cannot be applied in the same way as in the former case. However, according to (H3), we have $h''(\alpha_0) \neq 0$ and, therefore, $\partial^2 F(0, u_0, \alpha_0)/\partial \alpha^2 = v''(\alpha_0) \cdot u_0 \neq 0$. Again the implicit function theorem can be applied to ensure the existence of a number $\bar{\delta} > 0$ such that to each $(r, u) \in \mathbb{R} \times /S^1$, with $0 \leq r$, $|u - u_0| < \bar{\delta}$, is associated a unique α^* in a neighborhood of $\alpha_0 \in \mathbb{R}$, being α^* a C^1 function such that $\partial F(r, u, \alpha^*(r, u))/\partial \alpha = 0$, with $\alpha^*(0, u_0) = \alpha_0$. Therefore, $M(r, u) := F(r, u, \alpha^*(r, u))$ is a minimum value or a maximum value of the function $F(r, u, \cdot)$, depending upon whether $v''(\alpha_0) \cdot u_0$ is positive or negative, respectively. The partial derivative $\partial M(0, u_0)/\partial u$ is the linear functional from TS^1, the tangent space of S^1 in u_0, that associates to each vector ν, orthogonal to u_0, the number $v(\alpha_0) \cdot \nu$. Since $(0, u_0, \alpha_0)$ satisfies (43), it follows that the vectors $v(\alpha_0)$ and ν are collinear and, therefore, $v(\alpha_0) \cdot \nu \neq 0$. Thus, by the implicit function theorem, there exists a unique, continuously differentiable function, $u = u^*(r)$, $0 \leq r < \hat{\delta}$, $u^*(0) = 0$, for some constant $\hat{\delta} > 0$, such that $M(r, u^*(r)) = 0$. The vector u_0 is tangent to the curve $\mathcal{C} : (\lambda, \mu) = ru^*(r)$, $0 \leq r < \hat{\delta}$ at $(0, 0)$; therefore, $\mathcal{C}$ is tangent to the line $r_{\alpha_0} : (\lambda, \mu) \cdot v(\alpha_0)$.

To show that $\mathcal{C}$ has the properties required for curves $\mathcal{C}_j$, $j = 1, 2, \ldots, N$, in the statement of the theorem, let us suppose, to fix a case, that $M(r, u)$ is a minimum value of $F(r, u, \cdot)$, that is, that $v''(\alpha_0) \cdot u_0 > 0$. The case $v''(\alpha_0) \cdot u_0 < 0$ is analogous. Consider the path $u(t) = (\cos t, \sin t)$, $0 \leq t < 2\pi$, with $u(t_0) = u_0$. Then,

$$[\partial M(r, u(t))/\partial t]_{(r,t)=(0,t_0)} = v(\alpha_0) \cdot (-\sin t_0, \cos t_0) \neq 0,$$

since $(-\sin t_0, \cos t_0)$ and $v(\alpha_0)$ are collinear (because $(-\sin t_0, \cos t_0)$ and u_0 are orthogonal). Thus, if $(\bar{r}, \bar{t})$ lies in some neighborhood of $(0, t_0)$, the number $[\partial M(r, u(t))/\partial t]_{(\bar{r},\bar{t})}$ does not change sign; that is, $M(\bar{r}, \cdot)$ is a strictly monotone function of the variable t. This means that there are two simple solutions of (43) when u lies on one side of $\bar{u} = u(\bar{t})$, and there is no solution when u is on the other side.

Using a compactness argument in $[0, 2\pi)$ (recall that we are considering an identification topology in $[0, 2\pi)$), after a finite number of applications of the implicit function theorem we arrive at the complete description of the bifurcation diagram in a neighborhood V of $(\lambda, \mu) = (0, 0)$. Doing that, it is necessary to notice that there is no solution of (43) when (λ, μ) lies in the sectors of V that contain the λ axis. This property characterizes the region S^c. $\blacksquare$

Theorem 3.1 states that, for $(\lambda, \mu) \neq (0, 0)$ in V, there exists a finite number of 2π-periodic solutions of (21) and gives a description of how these solutions bifurcate when (λ, μ) varies in the neighborhood V. When $(\lambda, \mu) = (0, 0)$, equation (21) is the autonomous equation of the free oscillator (20), which has a *continuum* of 2π-periodic solutions, $\{p(\cdot + \alpha)\}_{\alpha \in [0, 2\pi)}$. The next theorem explains the behavior of the 2π-periodic solutions of (21) as (λ, μ) approaches $(0, 0)$. We keep the notation above.

Theorem 3.2 *Suppose the hypotheses (H1) (H4) are satisfied. Let γ be a continuous curve given by $\lambda = \lambda(\beta)$, $\mu = \mu(\beta)$, $0 \leq \beta \leq 1$, contained in the set S, satisfying $(\lambda(\beta), \mu(\beta)) = (0, 0)$ if and only if $\beta = 0$. Let $x(\cdot; \beta) \in \mathcal{P}_{2\pi}^{(2)}$ be a solution of (21), corresponding to $(\lambda, \mu) = (\lambda(\beta), \mu(\beta))$, whose orbit $\bigcup_{t \in \mathbb{R}} \{(x(t; \beta), x(t; \beta))\}$ is contained in W. Choose $x(\cdot; \beta)$ in such a way that $(x(\cdot; \beta), \dot{x}(\cdot; \beta)) \in \mathcal{P}_{2\pi} \times \mathcal{P}_{2\pi}$ depends continuously on β. If*

$$S(\gamma) := \{(x(\cdot; \beta), \dot{x}(\cdot; \beta)) \in \mathcal{P}_{2\pi} \times \mathcal{P}_{2\pi} \; ; \; 0 < \beta \leq 1\},$$

then $S(\gamma)$ is relatively compact in $\mathcal{P}_{2\pi} \times \mathcal{P}_{2\pi}$ and every limit point of $S(\gamma)$, as $\beta \to 0$, is a 2π-periodic solution of (24), $p(\cdot + \alpha_0, \dot{p}(\cdot + \alpha_0))$, for some $\alpha_0 \in [0, 2\pi)$. Moreover, if

$$h_m(\gamma) = \liminf_{\beta \to 0} \frac{\lambda(\beta)}{\mu(\beta)}, \quad h_M(\gamma) = \limsup_{\beta \to 0} \frac{\lambda(\beta)}{\mu(\beta)},$$

then there exists an interval $I(\gamma) = [\alpha_1, \alpha_2] \subset [0, 2\pi)$ such that $h(I(\gamma)) = [h_m(\gamma), h_M(\gamma)]$ and

$$\overline{S(\gamma)} \setminus S(\gamma) = \{p(\cdot + \alpha, \dot{p}(\cdot + \alpha)) \in \mathcal{P}_{2\pi} \times \mathcal{P}_{2\pi} \; ; \; \alpha \in I(\gamma)\}, \qquad (44)$$

where $\overline{S(\gamma)}$ is the closure of $S(\gamma)$ in $\mathcal{P}_{2\pi} \times \mathcal{P}_{2\pi}$.

Proof. $S(\gamma)$ is relatively compact as a consequence of the Ascoli–Arzelá theorem, since the orbits of $(x(\cdot;\beta), \dot{x}(\cdot;\beta))$, $\beta \in [0,1]$, lie in W and, therefore (23) implies that $\{(\dot{x}(\cdot;\beta), \ddot{x}(\cdot;\beta)), \beta \in [0,1]\}$, is a uniformly bounded family.

Let $\beta_n \in [0,1)$, $n = 1, 2, \ldots$, be a sequence with $\beta_n \to 0$, as $n \to \infty$, in such a way that $(x(\cdot;\beta_n), \dot{x}(\cdot;\beta_n))$ is convergent in $\mathcal{P}_{2\pi} \times \mathcal{P}_{2\pi}$. The integral form of (23) implies

$$x(t;\beta_n) = x(0;\beta_n) + \int_0^t \dot{x}(s;\beta_n)\,ds$$

$$\dot{x}(t;\beta_n) = \dot{x}(0;\beta_n) - \int_0^t g(x(s;\beta_n))\,ds \qquad (45)$$

$$-\lambda(\beta_n)\int_0^t \dot{x}(s;\beta_n)\,ds + \mu(\beta_n)\int_0^t f(s)\,ds.$$

Recalling that $(\lambda(\beta_n), \mu(\beta_n)) \to (0,0)$ and taking limits in (45), as $n \to \infty$, we must have $(x(\cdot;\beta_n), \dot{x}(\cdot;\beta_n)) \to (p(\cdot + \alpha_0), \dot{p}(\cdot + \alpha_0))$ for some $\alpha_0 \in [0, 2\pi)$.

To see that α_0 is such that $h(\alpha_0) \in [h_m(\gamma), h_M(\gamma)]$, it suffices to notice that, for each $n = 1, 2, \ldots$, there exists $\alpha(\beta_n) \in [0, 2\pi)$ such that $(\alpha(\beta_n), \lambda(\beta_n), \mu(\beta_n))$ satisfies the bifurcation equation (42) and, therefore,

$$h(\alpha(\beta_n)) - \frac{\lambda(\beta_n)}{\mu(\beta_n)} + \frac{1}{\mu(\beta_n)} R(\alpha(\beta_n), \lambda(\beta_n), \mu(\beta_n)). \qquad (46)$$

The continuous choice of $(x(\cdot;\beta), \dot{x}(\cdot;\beta))$ forces $\alpha(\beta_n) \to \alpha_0$, as $n \to \infty$. Since $R(\alpha, \lambda, \mu) = O(|(\lambda, \mu)|^2)$, as $(\lambda, \mu) \to (0,0)$, taking the limit of (46), as $n \to \infty$, we get

$$h(\alpha_0) = \lim_{n \to \infty} \frac{\lambda(\beta_n)}{\mu(\beta_n)};$$

therefore, $\overline{S(\gamma)} \setminus S(\gamma) \subset \{p(\cdot + \alpha, \dot{p}(\cdot + \alpha)) \in \mathcal{P}_{2\pi} \times \mathcal{P}_{2\pi} ; \ \alpha \in I(\gamma)\}$.

Conversely, suppose that $\alpha_0 \in I(\gamma)$ is given. We can choose a sequence of solutions of (46), $(\alpha(\beta_n), \lambda(\beta_n), \mu(\beta_n))$, with $\beta_n \in (0,1]$ and $\beta_n \to 0$ as $n \to \infty$. Thus, $(\alpha(\beta_n), \lambda(\beta_n), \mu(\beta_n)) \to (\alpha_0, 0, 0)$ as $n \to \infty$, and therefore $\{p(\cdot + \alpha, \dot{p}(\cdot + \alpha)) \in \mathcal{P}_{2\pi} \times \mathcal{P}_{2\pi} ; \ \alpha \in I(\gamma)\} \subset \overline{S(\gamma)} \setminus S(\gamma)$. ∎

Choosing the curve $\gamma \subset S$ in such a way that $h_m(\gamma) < h_M(\gamma)$, the trajectories of its limit points, $(x(\cdot;\beta), \dot{x}(\cdot;\beta))$, in $\mathbb{R}^3$ cover a spiral-shaped strip on a cylindrical surface, as is shown in Figure 2. The limit points of the initial conditions $(x(0,\beta), \dot{x}(0,\beta))$, as $\beta \to 0$, form an arc Γ in the phase plane $(x, \dot{x})$.

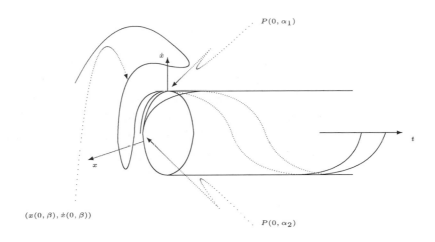

Figure 2. Limit of $\big(x(\cdot;\beta), \dot{x}(\cdot;\beta)\big)$, as $\beta \to 0$.

We have the following immediate corollary of Theorem 3.2.

Corollary 3.1 *Let* $\gamma : \lambda = \lambda(\beta)$, $\mu = \mu(\beta)$ *be a continuous curve contained in* S *and* $x(\cdot;\beta) \in \mathcal{P}^2_{2\pi}$ *a solution of (21), such that all the hypotheses of Theorem 3.2 are satisfied. A necessary and sufficient condition for*

$$(x(\cdot;\beta), \dot{x}(\cdot;\beta)) \to (p(\cdot + \alpha_0), \dot{p}(\cdot + \alpha_0)), \quad \beta \to 0,$$

for some fixed $\alpha_0 \in [0, 2\pi)$, *is that there exists the limit* $\lim_{\beta \to 0}\big(\lambda(\beta)/\mu(\beta)\big) = h_0$. *In this case, the phase* α_0 *is given by the equation* $h(\alpha_0) = h_0$.

3.2   The Predator–Prey Model of Lotka–Volterra with Periodic Forcing

The ideas of the previous section can be applied to more general abstract problems concerning perturbations of a center where the period as a function of the amplitude has no singular points. Our objective now is to show another example that exhibits distinctive aspects with respect to the previous one.

The classical predator–prey Lotka–Volterra model is described by the system

$$\dot{x}_1 = ax_1 - bx_1x_2,$$
$$\dot{x}_2 = cx_1x_2 - dx_2, \tag{47}$$

where $a, b, c, d > 0$ are constants, $x_1, x_2 \geq 0$ are the populations of prey and predators, a and d being their growth and decline rates, respectively. The constants b and d represent predation effects. The system (47) has two equilibria: a saddle, $(0,0)$, and a center, $(c/d, a/b)$. The significant quadrant $Q_1 := \{(x_1, x_2) : x_1, x_2 > 0\}$ is invariant and the orbits in Q_1 are periodic and encircle the equilibrium $(c/d, a/b)$. Suppose that $e(t) = (e_1(t), e_2(t))$ is a T-periodic solution of (47) with orbit $\Gamma \subset Q_1$. If f_1 and f_2 are two smooth T-periodic functions and ε_1, ε_2 are real parameters, we formulate the same questions from the previous section for the perturbed system:

$$\dot{x}_1 = ax_1 - bx_1x_2 + \varepsilon_1 f_1(t),$$
$$\dot{x}_2 = cx_1x_2 - dx_2 + \varepsilon_2 f_2(t); \tag{48}$$

that is:

For a sufficiently small neighborhood W of the orbit Γ in the x_1x_2-plane, is there a region V in the parameter space for which there are T-periodic solutions of (48) whose orbits lie in W for each $(\varepsilon_1, \varepsilon_2) \in V$? How does the number of such solutions vary when $(\varepsilon_1, \varepsilon_2)$ varies in V? How do they behave when the parameter $(\varepsilon_1, \varepsilon_2)$ approaches $(0,0)$?

These questions were investigated in [Ta 87] following the same procedure of the previous section. The answer to these questions, however, differs from the results obtained there in some fundamental points. Since none of the parameters ε_1, ε_2 represents a dissipation or a gain of energy, it is possible that (48) possesses T-periodic solutions with orbit in W, for all $(\varepsilon_1, \varepsilon_2)$ in a neighborhood of $(0,0)$; that is, in this problem the region

V might be a full neighborhood of $(0,0)$ in the parameter space. There is also the possibility of no bifurcation of T-periodic solutions of (48); that is, the number of such solutions with orbit in W might be constant when $(\varepsilon_1, \varepsilon_2) \neq (0,0)$ varies in a neighborhood of $(0,0)$.

We consider the shorthand form of (47):

$$\dot{x} = g(x), \qquad (49)$$

where

$$x = \begin{pmatrix} x_1 \\ x_2 \end{pmatrix}, \qquad g(x) = \begin{pmatrix} ax_1 - bx_1x_2 \\ cx_1x_2 - dx_2 \end{pmatrix},$$

and the forced system, (48), is denoted by

$$\dot{x} = g(x) + \hat{\varepsilon}f(t), \qquad (50)$$

where

$$f(t) = \begin{pmatrix} f_1(t) \\ f_2(t) \end{pmatrix}, \qquad \hat{\varepsilon} = \begin{pmatrix} \varepsilon_1 & 0 \\ 0 & \varepsilon_2 \end{pmatrix}.$$

Here again we have a representation of the nonconstant solutions of (49) in the quadrant Q_1 in terms of the amplitude and the phase. Such a solution x is given by $x(t) = \phi(\omega(\rho)(t + \alpha), \rho)$, where $\phi : \mathbb{R} \times (0, \infty) \to Q_1$ and $\omega : (0, \infty) \to (0, \infty)$ are smooth functions satisfying $\phi(\theta+1, \rho) = \phi(\theta, \rho)$ for all $(\theta, \rho) \in \mathbb{R} \times (0, \infty)$. The number $\rho > 0$ is the amplitude, $\omega(\rho)$ is the frequency, and $\alpha \subset [0, 1/\omega(\rho))$ is the phase of the solution x.

We assume the T-periodic solution e has amplitude ρ_0 and phase $\alpha = 0$; that is, the period of e is $T = [\omega(\rho_0)]^{-1}$ and $e(t) = \phi(\omega(\rho_0)t, \rho_0)$. We can ensure that

$$\omega'(\rho_0) > 0; \qquad (51)$$

see F. Rothe [Ro 93], for instance. Thus, arguing in the same way as in the previous section, we can show that the function $\dot{e}$ spans the space of T-periodic solutions of the linear variational equation of (49) along the periodic solution e:

$$\dot{x} = g'(e(t))x, \qquad (52)$$

where

$$g'(e(t)) = \begin{pmatrix} a - be_2(t) & -be_1(t) \\ ce_2(t) & ce_1(t) - d \end{pmatrix}.$$

Now we define an appropriate setting to formulate our problem in such a way as to use the method of Liapunov–Schmidt. We take W as a tubular neighborhood of Γ, so that each point $x \in W$ has coordinates (τ, σ), $0 \leq \tau < T$, $|\sigma| < \sigma_0$ ($\sigma_0 > 0$ is a constant), such that

$$x = e(\tau) + \sigma e^\perp(\tau), \tag{53}$$

where $e^\perp(\tau) := (\dot{e}_2(\tau), -\dot{e}_1(\tau))$. Any T-periodic solution x of (50) that remains in W has a phase α, $0 \leq \alpha < T$, defined as the first moment in the interval $[0, T)$ for which the solution reaches the fiber corresponding to $\tau = 0$. That is, the coordinates of $x(\alpha)$ in W are of the form $(0, \sigma)$. Therefore, $x(\alpha) = e(0) + \sigma e^\perp(\alpha)$, and such a solution is represented as

$$x(t) = e(t - \alpha) + z(t - \alpha),$$
$$z(0) \qquad \cdot \dot{e}(0) = 0. \tag{54}$$

Notice that the shift $x(t) \to x(t + \alpha)$ defines a one-to-one correspondence between solutions of (50) and solutions of

$$\dot{x}(t) = g(x) + \hat{\varepsilon} f(t + \alpha). \tag{55}$$

By using this fact we can transfer the parameter α from the condition (54) to the differential equation. In other words, we can reformulate our problem so as to investigate the T-periodic solutions of (55) in W, with the condition (54) replaced by

$$x(t) = e(t) + z(t),$$
$$z(0) \quad \cdot e(0) = 0, \tag{56}$$

with small $|z(t)|$. Substituting (56) into (55), we finally arrive at the problem of investigating small periodic solutions z of

$$\dot{z} = g'(e(t))z + \hat{\varepsilon} f(t + \alpha) + R(z),$$
$$z(0) \cdot \dot{e}(0) = 0, \tag{57}$$

where $R(z) = O(|z|^2)$ as $z \to 0$.

The use of the Fredholm alternative differs slightly from the way that it was applied in the previous section because the linear variational equation (52) is not self-adjoint in this case. We need the following well-known first integral of (49):

$$u(x) = a \ln x_2 + d \ln x_1 - cx_1 - bx_2.$$

It is a consequence of Lemma 3.1 that

$$\eta(t) := \big(d/e_1(t) - c,\, a/e_2(t) - b\big) \tag{58}$$

is a T-periodic solution of the adjoint equation to (52):

$$\dot{z} = -zg'(e(t)), \tag{59}$$

and therefore spans the space of T-periodic solutions of (59).

Let $Z = \mathcal{P}_T$ be the Banach space of the T-periodic continuous functions, $\phi : \mathbb{R} \to \mathbb{R}^2$, with the norm $|\phi| := \sup |\phi(t)|$, $t \in \mathbb{R}$, and consider its algebraic subspace $X = \mathcal{P}_T^{(1)}$ of the C^1 functions $\phi \in Z$, endowed with a C^1 norm. Let us define the continuous projections $P : X \to X$ and $Q : Z \to Z$ by

$$P\phi = |\dot{e}(0)|^{-2}(\phi(0) \cdot \dot{e}(0))\dot{e}, \qquad \phi \subset X,$$
$$Q\psi = \left(\nu \int_0^T \eta(t) \cdot \phi(t)\, dt\right) \eta, \qquad \phi \in Z,$$

where

$$\nu = \left(\int_0^T \eta^2(t)\, dt\right)^{-1}.$$

Let us define the linear map $L : X \to Z$ and the map $N : X \times \mathbb{R}^3 \to Z$ by

$$Lz = \dot{z} - g'(e)z, \quad z \in X,$$
$$N(z, \varepsilon, \alpha) = \hat{\varepsilon}f(\cdot + \alpha) + R(z), \quad (z, \varepsilon, \alpha) \in X \times \mathbb{R}^3.$$

Now the problem (57) can be rewritten as the system

$$Lz = N(z, \varepsilon, \alpha),$$
$$Pz = 0, \tag{60}$$

and, according to the definitions of P and Q, we have $\mathcal{R}(L) = Z_{I-Q}$ and $\mathcal{N}(L) = X_P$. There exists a right inverse $K : Z_{I-Q} \to X$ of L satisfying $LK = I$ in Z_{I-Q} and $KL = I - P$ in X.

Decomposing the first equation of (60) in the complementary subspaces QZ and $(I-Q)Z$ and taking into account the properties of the right inverse K, one sees that this equation is equivalent to the system

$$z = Pz + K(I - Q)N(z, \varepsilon, \alpha),$$
$$0 = QN(z, \varepsilon, \alpha).$$

Substituting the second equation of (60) into this system, we arrive at (60) being equivalent to

$$z = K(Q - I)N(z, \varepsilon, \alpha),$$
$$0 = QN(z, \varepsilon, \alpha). \tag{61}$$

We consider the interval $[0,T)$ as a compact manifold identifying it with S^1; in other words, we take $[0, T)$ as the quotient topological space $\mathbb{R}/T\mathbb{Z}$. Using the compactness of this space, by a finite number of applications of the implicit function theorem, we can ensure the existence of a neighborhood V of $\varepsilon = 0$ in $\mathbb{R}^2$ and of a unique smooth function $z = z^*(\varepsilon, \alpha)$ from $V \times [0, T)$ into X such that $z(0, \alpha) = 0$ for all $0 \le \alpha < T$, and $z^*(\varepsilon, \alpha)$ is the unique solution of the first equation of the system (61) in a neighborhood of $z = 0$. Substituting $z = z^*(\varepsilon, \alpha)$ into the second equation of (61) we obtain the bifurcation equation

$$\nu \int_0^T \eta(t) \left[\hat\varepsilon f(t + \alpha) + R(z^*(\varepsilon, \alpha)(t)) \right] dt = 0. \tag{62}$$

According to (58) this equation is

$$\left[d \int_0^T \frac{f_1(t + \alpha)}{e_1(t)} \, dt - cM_1 \right] \varepsilon_1$$

$$+ \left[a \int_0^T \frac{f_2(t + \alpha)}{e_2(t)} \, dt - bM_2 \right] \varepsilon_2 + S(\varepsilon, \alpha) = 0, \tag{63}$$

where the numbers M_1, M_2 are given by

$$M_i = \int_0^T f_i(t) \, dt, \quad i = 1, 2,$$

and $S(\varepsilon, \alpha)$ is a T-periodic function of α for each fixed ε, with $S(\varepsilon, \alpha) = O(|\varepsilon|^2)$, as $\varepsilon \to 0$.

Considering the vector $v(\alpha) = (v_1(\alpha), v_2(\alpha))$ for $\alpha \in [0, T]$, given by

$$v_1(\alpha) = d \int_0^T \frac{f_1(t + \alpha)}{e_1(t)} \, dt - cM_1,$$

$$v_2(\alpha) = a \int_0^T \frac{f_2(t + \alpha)}{e_2(t)} \, dt - bM_2,$$

the bifurcation equation takes the form

$$v(\alpha) \cdot \varepsilon + S(\varepsilon, \alpha) = 0. \tag{64}$$

This equation is like the bifurcation equation for the oscillator in its form (42). There the role of the closed curve $v(\alpha)$, $0 \leq \alpha < T$, is played by the curve $(-1, h(\alpha))$, $0 \leq \alpha < 2\pi$.

If we neglect the higher order terms $S(\varepsilon, \alpha)$, by an analysis of the approximate equation

$$v(\alpha) \cdot \varepsilon = 0, \tag{65}$$

we can anticipate that the bifurcation diagram has the same nature as the diagram obtained for the oscillator. For each $\varepsilon \neq 0$ in $\mathbb{R}^2$, the solutions of (65) are the numbers $\overline{\alpha}$, $0 \leq \overline{\alpha} < 0$, for which $v(\overline{\alpha})$ is orthogonal to ε. We assume the following hypotheses on the closed curve $v(\alpha)$ of the ε-plane (we use the abbreviation l.i. for "linearly independent"):

(H5) $\qquad\qquad\qquad v(\alpha) \neq 0, \quad 0 \leq \alpha < 0.$

(H6) $\quad v(\alpha), v'(\alpha)$ are l.i. except possibly at points $\alpha_1 < \alpha_2 < \cdots < \alpha_n$.

(H7) $\qquad\qquad v(\alpha_i), v(\alpha_j)$ are l.i. if $i \neq j$, $1 \leq i, j \leq n$.

(H8) $\quad v''(\alpha)$ isn't aligned simultaneously with $v(\alpha)$ and $v'(\alpha)$, $0 \leq \alpha < T$.

A precise description of the bifurcation diagram of T-periodic solutions of (48) with orbits in a neighborhood of Γ is provided by the next theorem. The proof of this theorem can be carried out following the steps of the proof of Theorem 3.1 with obvious adaptations.

Theorem 3.3 *If the hypotheses (H5)–(H8) are satisfied, then there are neighborhoods W of Γ, V of $(0,0)$ in the ε-plane such that:*

(i) *If there exist numbers $\alpha_i \in [0,T)$, $i = 1,\dots,n$, satisfying hypothesis (H6), then there are curves γ_i tangent to the lines $\ell_i = \{\varepsilon : \varepsilon \cdot v(\alpha_i) = 0\}$ at $\varepsilon = 0$, $i = 1,\dots,n$, respectively. For each such curve γ, $V \setminus \gamma$ has exactly two connected components and $\gamma_i \cap \gamma_j = \{0\}$ if $i \neq j$. If $\varepsilon \neq 0$ crosses such a curve, the number of T-periodic solutions of (48) with orbit in W changes by two.*

(ii) *There is a nonempty subset S of V such that the equation (48) has at least two T-periodic solutions with orbit in W. Either $S = V$ or S is the union of two open sectorial subsets of V whose boundary ∂S in V is given by $\partial S = \gamma_i \cup \gamma_j$ for some i, j, $1 \leq i, j \leq n$.*

Corollary 3.2 *If the vector $v(\alpha)$, $\alpha \in [0,T)$, describes a closed curve with no double points such that $v(\alpha)$ and $v'(\alpha)$ are l.i. for every $\alpha \in [0,T)$, then there exist neighborhoods W of Γ and V of $\varepsilon = 0$ such that (48) has exactly two T-periodic solutions with orbit in W when ε varies in V.*

The central picture shown in Figure 3 depicts a situation considered in Corollary 3.2. If the shape and location of the curves $v(\alpha)$ are as in the other two pictures, the corresponding diagrams are included in Theorem 3.3.

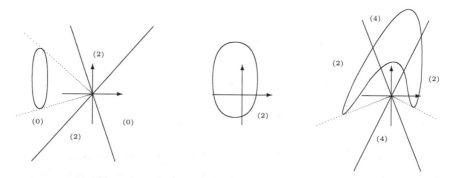

Figure 3. Examples of bifurcation diagrams.

Remark 3.2 *Theorem 3.3 suggests the following interesting inverse problem: Given an even number of lines through the origin of the ε-plane, are there functions f for which the bifurcation diagram of T-periodic solutions of (48) near Γ is determined by bifurcation curves tangent to these lines in the origin? The reference [Ta 87] provides a solution for this problem when there is no bifurcation curves. In [GT 95], more general cases are investigated.*

Remark 3.3 *A theorem analogous to Theorem 3.2 can be stated in the case of the predator–prey model to describe the limiting behavior of T-periodic solutions of (48) as ε → 0.*

References

[Ap 19] P. Appel, *Traité de Mécanique Rationnelle*, Vol. 1, Gauttier-Villars, Paris (1919).

[Ch 98] D. R. J. Chillingworth, *Generic multiparameter bifurcation from a manifold.* Preprint, March 19, 1999.

[CH 82] S. N. Chow and J. Hale, *Methods of Bifurcation Theory*, Springer-Verlag, New York (1982).

[CHM-P 80] S. N. Chow, J. K. Hale, and J. Mallet-Paret, *An example of bifurcation to homoclinic orbits*, J. Differential Equations, 37 (1980), 351–373.

[GT 95] M. A. Giongo and P. Táboas, *Inverse problems on bifurcation from families of periodic solutions*, Appl. Anal., 59 (1995), 185–199.

[Ha 80] J. Hale, *Ordinary Differential Equations*, Krieger, New York (1980).

[HT 78] J. Hale and P. Táboas, *Interaction of damping and forcing in a second order equation*, Nonlinear Anal., 2 (1978), 77–84.

[Ka 70] T. Kato, *Perturbation Theory for Linear Operators*, Springer-Verlag, New York (1970).

[Li 88] X. B. Lin, *Using Melnikov's method to solve Silnikov's problem.* Proc. Roy. Soc. Edinburgh, 116A (1988), 319–382.

[Ro 93] F. Rothe, *Remarks on periods of planar Hamiltonian systems,* SIAM J. Math. Anal., 24:1 (1993), 129–154.

[Ta 87] P. Táboas, *Periodic solutions of a forced Lotka-Volterra equation,* J. Math. Anal. Appl., 124 (1987), 82–97.

Index